T0192253

Lecture Notes in Computer Science 14297

Founding Editors

Gerhard Goos
Juris Hartmanis

Editorial Board Members

Elisa Bertino, *Purdue University, West Lafayette, IN, USA*
Wen Gao, *Peking University, Beijing, China*
Bernhard Steffen ⓘ, *TU Dortmund University, Dortmund, Germany*
Moti Yung ⓘ, *Columbia University, New York, NY, USA*

The series Lecture Notes in Computer Science (LNCS), including its subseries Lecture Notes in Artificial Intelligence (LNAI) and Lecture Notes in Bioinformatics (LNBI), has established itself as a medium for the publication of new developments in computer science and information technology research, teaching, and education.

LNCS enjoys close cooperation with the computer science R & D community, the series counts many renowned academics among its volume editors and paper authors, and collaborates with prestigious societies. Its mission is to serve this international community by providing an invaluable service, mainly focused on the publication of conference and workshop proceedings and postproceedings. LNCS commenced publication in 1973.

Jarosław Byrka · Andreas Wiese

Editors

Approximation and Online Algorithms

21st International Workshop, WAOA 2023
Amsterdam, The Netherlands, September 7–8, 2023
Proceedings

 Springer

Editors
Jarosław Byrka
University of Wrocław
Wrocław, Poland

Andreas Wiese
Technical University of Munich
Munich, Germany

ISSN 0302-9743 ISSN 1611-3349 (electronic)
Lecture Notes in Computer Science
ISBN 978-3-031-49814-5 ISBN 978-3-031-49815-2 (eBook)
https://doi.org/10.1007/978-3-031-49815-2

© The Editor(s) (if applicable) and The Author(s), under exclusive license
to Springer Nature Switzerland AG 2023

This work is subject to copyright. All rights are reserved by the Publisher, whether the whole or part of the material is concerned, specifically the rights of translation, reprinting, reuse of illustrations, recitation, broadcasting, reproduction on microfilms or in any other physical way, and transmission or information storage and retrieval, electronic adaptation, computer software, or by similar or dissimilar methodology now known or hereafter developed.
The use of general descriptive names, registered names, trademarks, service marks, etc. in this publication does not imply, even in the absence of a specific statement, that such names are exempt from the relevant protective laws and regulations and therefore free for general use.
The publisher, the authors, and the editors are safe to assume that the advice and information in this book are believed to be true and accurate at the date of publication. Neither the publisher nor the authors or the editors give a warranty, expressed or implied, with respect to the material contained herein or for any errors or omissions that may have been made. The publisher remains neutral with regard to jurisdictional claims in published maps and institutional affiliations.

This Springer imprint is published by the registered company Springer Nature Switzerland AG
The registered company address is: Gewerbestrasse 11, 6330 Cham, Switzerland

Paper in this product is recyclable.

Preface

The 21st Workshop on Approximation and Online Algorithms (WAOA 2023) focused on the design and analysis of algorithms for online and computationally hard problems. Both kinds of problems have a large number of applications in a variety of fields. The workshop took place in Amsterdam, the Netherlands, during 7–8 September 2023, and it was a success: it featured many interesting presentations and provided opportunities for stimulating interactions. WAOA 2023 was part of ALGO 2023, which also hosted ESA, ALGOCLOUD, ALGOWIN, ATMOS, and IPEC.

The topics of WAOA 2023 were: algorithmic game theory, algorithmic trading, coloring and partitioning, competitive analysis, computational advertising, computational finance, cuts and connectivity, FPT-approximation algorithms, geometric problems, graph algorithms, inapproximability results, mechanism design, network design, packing and covering, paradigms for the design and analysis of approximation and online algorithms, resource augmentation, and scheduling problems.

In response to the call for papers we received 43 submissions. Each of the submissions was reviewed by at least three referees, and some submissions were reviewed by more than three referees. The submissions were mainly judged on originality, technical quality, and relevance to the topics of the conference. Based on the reviews, the program committee selected 16 papers. This volume contains the final revised versions of these papers as well as an invited contribution by our plenary speaker Nicole Megow. The EasyChair conference system was used to manage the electronic submissions, the reviewing process, and the electronic program committee discussions. It made our task much easier.

We would like to thank all the authors who submitted papers to WAOA 2023 and all attendees of WAOA 2023, including the presenters of the accepted papers. A special thank you goes to the plenary speaker Nicole Megow for accepting our invitation and giving a very nice talk. We would also like to thank the PC members and the external reviewers for their diligent work in evaluating the submissions and their contributions to the electronic discussions. Furthermore, we are grateful to all the local organizers of ALGO 2023, especially Solon Pissis who served as the local chair of the organizing committee.

October 2023

Jarosław Byrka
Andreas Wiese

Organization

General Chair

Solon Pissis — CWI and Vrije Universiteit Amsterdam, The Netherlands

Program Committee Chairs

Jarosław Byrka — University of Wrocław, Poland
Andreas Wiese — Technical University of Munich, Germany

Steering Committee

Evripidis Bampis — Sorbonne Université, France
Thomas Erlebach — Durham University, UK
Christos Kaklamanis — University of Patras, Greece
Nicole Megow — University of Bremen, Germany
Laura Sanita — Eindhoven University of Technology, The Netherlands
Martin Skutella — Technical University of Berlin, Germany
Roberto Solis-Oba — University of Western Ontario, USA

Program Committee

Marek Adamczyk — University of Wrocław, Poland
Karl Bringmann — Saarland University, Germany
Jarosław Byrka (Co-chair) — University of Wrocław, Poland
Sami Davies — Northwestern University, USA
Guy Even — Tel-Aviv University, Israel
Andreas Emil Feldmann — University of Sheffield, UK
Zachary Friggstad — University of Alberta, Canada
Arindam Khan — Indian Institute of Science, Bangalore, India
Kamyar Khodamoradi — University of British Columbia, Canada
Max Klimm — Technical University of Berlin, Germany

Alexandra Lassota	École Polytechnique Fédérale de Lausanne, Switzerland
Ben Moseley	Carnegie Mellon University, USA
Tim Oosterwijk	Vrije Universiteit Amsterdam, The Netherlands
Kirk Pruhs	University of Pittsburgh, USA
Erik Jan van Leeuwen	Utrecht University, The Netherlands
Laura Vargas Koch	ETH Zürich, Switzerland
Andreas Wiese (Co-chair)	Technical University of Munich, Germany

Additional Reviewers

Steven Miltenburg
Václav Blažej
Malte Tutas
Atrayee Majumder
Frank Staals
Daniel Vaz
Saladi Rahul
Pieter Kleer
Karnati Venkata Naga Sreenivasulu
Lars Rohwedder
Ramin Mousavi
Malte Tutas
Dylan Hyatt-Denesik
Siddharth Gupta
Ariel Kulik
Svenja M. Griesbach
Hsiang-Hsuan Liu
Syamantak Das
Steven Miltenburg
Daniel Schmand
Leon Sering
Ramin Mousavi
Mahya Jamshidian
Foivos Fioravantes
Alexandros Hollender
Evangelos Kipouridis
Franziska Eberle
Debajyoti Kar
Alejandro Cassis
Alexander Lindermayr

Adrian Vetta
Nikhil Kumar
Sven Jäger
Arash Rafiey
Diptarka Chakraborty
Lennart Kauther
Jesper Nederlof
Jens Schlöter
Rakesh Mohanty
Kunal Dutta
Tung Anh Vu
André Nusser
Georg Anegg
Vasileios Nakos
Thomas Erlebach
Mathieu Mari
Yuri Faenza
Justin Ward
Théophile Thiery
Sebastian Berndt
Saswata Jana
Arka Ray
Christiane Schmidt
Leonidas Palios
Malin Rau
Arturo Merino
Martin Herold
Manolis Vasilakis
K. Subramani
Akbar Rafiey

Optimization Under Explorable Uncertainty: Adversarial and Stochastic Models (Invited Talk)

Nicole Megow

University of Bremen, Germany

Abstract. In the traditional frameworks for optimization under uncertainty, an algorithm has to accept the incompleteness of input data. Clearly, more information or even knowing the exact data would allow for significantly better solutions. How much more information suffices for obtaining a certain solution quality? Which information shall be retrieved? Explorable uncertainty is a framework in which parts of the input data are initially unknown, but can be obtained at a certain cost using queries. An algorithm can make queries until it has obtained sufficient information to solve a given problem. The challenge lies in balancing the cost for querying and the impact on the solution quality.

In this talk, I will give an overview on the field of explorable uncertainty with a focus on combinatorial optimization problems. I will include problems such as finding a minimum spanning tree in a graph of uncertain edge cost, finding the minimal elements in intersecting sets, and finding a set of minimum total value. The latter can be seen as a subproblem of a more complex problem such as solving a knapsack or matching problem under explorable uncertainty. I will discuss an adversarial online model and recent advances on a stochastic variant.

Contents

Approximation Ineffectiveness
of a Tour-Untangling Heuristic

Bodo Manthey and Jesse van Rhijn[(✉)]

University of Twente, Enschede, The Netherlands
{b.manthey,j.vanrhijn}@utwente.nl

Abstract. We analyze a tour-uncrossing heuristic for the Euclidean Travelling Salesperson Problem, showing that its worst-case approximation ratio is $\Omega(n)$ and its average-case approximation ratio is $\Omega(\sqrt{n})$ in expectation. We furthermore evaluate the approximation performance of this heuristic numerically on average-case instances, and find that it performs far better than the average-case lower bound suggests. This indicates a shortcoming in the approach we use for our analysis, which is a rather common method in the analysis of local search heuristics.

Keywords: Travelling salesperson problem · Local search · Probabilistic analysis

1 Introduction

The Travelling Salesperson Problem (TSP) is a classic example of an NP-hard combinatorial optimization problem [8]. Different variants of the problem exist, with one of the most studied variants being the Euclidean TSP. In this version, the weight of an edge is given by the Euclidean distance between its endpoints. Even this restricted version is NP-hard [9].

Due to this hardness, practitioners often turn to approximation algorithms and heuristics for the TSP. One simple heuristic is 2-opt [1]. In each iteration of this heuristic, one searches for a pair of edges in the tour that can be replaced by a different pair, such that the total length of the tour decreases. Although this heuristic performs quite well in practice [1, Chapter 8], it may require an exponential number of iterations to converge even in the plane [6].

Interestingly, Van Leeuwen & Schoone showed that a restricted variant of 2-opt in which one only removes intersecting edges terminates in $O(n^3)$ iterations in the worst case [11]. For convenience, we refer to this variant as X-opt. More recently, da Fonseca et al. [5] analyzed this heuristic once more, extending the results to matching problems and showing a bound of $O(tn^2)$ for instances where all but t points are in convex position. Their work builds on previous related work on computing uncrossing matchings [2].

Supported by NWO grant OCENW.KLEIN.176.

© The Author(s), under exclusive license to Springer Nature Switzerland AG 2023
J. Byrka and A. Wiese (Eds.): WAOA 2023, LNCS 14297, pp. 1–13, 2023.
https://doi.org/10.1007/978-3-031-49815-2_1

The insight that removing intersecting edges improves the tour is a key intuition behind 2-opt. However, not all 2-opt iterations remove intersections. Indeed, 2-opt has proved extremely effective also for non-metric TSP instances, where there is no notion of intersecting edges at all.

This raises the question of approximation performance: can one get away with using X-opt instead of 2-opt at minimal cost to the approximation guarantee, thereby ensuring an efficient heuristic for TSP instances in the plane? The approximation ratio of 2-opt has long been known to sit between $\Omega\left(\frac{\log n}{\log \log n}\right)$ and $O(\log n)$ for d-dimensional Euclidean instances [4], and has recently been settled to $\Theta\left(\frac{\log n}{\log \log n}\right)$ for 2-dimensional instances [3]. However, no previous work seems to have discussed X-opt.

We analyze this simpler case here, showing an approximation ratio of $\Omega(n)$ in the worst case and $\Omega(\sqrt{n})$ in the average case. This answers our previously raised question in the negative; in order to obtain a good approximation ratio, one must allow for iterations that improve the tour without removing intersections. Especially the average-case result stands in stark contrast to the average-case approximation ratio of 2-opt, which is known to be $O(1)$ [4].

We also perform a numerical experiment, which presents a different picture from our formal results. To within the precision we are able to achieve, our experiments indicate an average-case approximation ratio of X-opt of $O(1)$. We consider this evidence that the techniques we use to obtain the average-case bound of $\Omega(\sqrt{n})$, which are standard techniques used to perform probabilistic analyses of local search heuristics, fall short of explaining the true practical performance of X-opt.

1.1   Definitions and Notation

Given two points $x, y \in \mathbb{R}^2$, we define $d(x, y)$ as the Euclidean distance between x and y. We define $L(x, y)$ as the line segment between x and y. By an abuse of notation, if $e = \{x, y\}$ with $x \neq y$, we write $L(e) = L(x, y)$. We write $\ell(e) = d(x, y)$ for the length of $L(e)$.

Let $X \subseteq \mathbb{R}^2$. For a set $E \subseteq \{e \in 2^X \mid |e| = 2\}$, we write $\ell(E) = \sum_{e \in E} \ell(e)$. If $L(e) \cap L(f) \neq \emptyset$ for some $e, f \in E$, i.e., some of the line segments represented by the edges in E intersect, then we say E is crossing. Conversely, if E is not crossing, then it is noncrossing. In particular, a local optimum for X-opt is exactly a noncrossing tour.

Given $A \subset \mathbb{R}^2$ and $x \in \mathbb{R}^2$, we define the distance between A and x as $d(x, A) = \min_{y \in A} d(x, y)$. Note that $d(x, A)$ might not exist, for instance, if A is an open set. However, we will only consider sets for which $d(x, A)$ is well-defined.

Let A be a rectangular region in $\mathbb{R}^2$. Let X be a set of n points in A. We call the four line segments that make up the boundary of A the edges of A. For an edge e of A, let $x_e = \arg\min_{x \in X} d(x, e)$ be the point closest to e. We call X nice for A if for each pair of edges e, f of A, it holds that $x_e \neq x_f$.

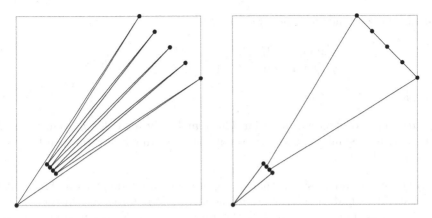

Fig. 1. The construction used in Theorem 1. Left: a noncrossing tour of length $\Omega(n)$. Right: a tour of length $O(1)$.

2   Worst Case

We construct a worst-case instance in which there exists a noncrossing tour with length $\Omega(n)$, as well as a tour of constant length. The construction we use is depicted in Fig. 1.

Theorem 1. *Let $n \in \mathbb{N}$ be even. For $\epsilon > 0$ sufficiently small, there exists an instance of the Euclidean TSP in the plane where X-opt has approximation ratio at least $\frac{n}{2} \cdot (1-\epsilon)$. In particular, the approximation ratio can be brought arbitrarily close to $\frac{n}{2}$.*

Proof. We place one point s at $(0,0)$. Next, we place $k = n/2$ points equally spaced along the line segment extending from $(1 - \epsilon/2, 1)$ to $(1, 1 - \epsilon/2)$. We label these points $\{y_i\}_{i=1}^{k}$, ordering them by increasing x-coordinate.

Consider the cone K with vertex at the origin, defined by all conic combinations of $\{y_1, y_k\}$. Define the height along the axis of K of a point a by the distance of a from the origin along the axis of K. We place $k - 1$ points along the line segment perpendicular to the axis of K at a height $\epsilon/\sqrt{8}$, excluding its endpoints. We label these points $\{x_i\}_{i=1}^{k-1}$, sorting them again by increasing x-coordinate. Note that it does not matter where exactly we place these points, as long as no two points are placed in the same location. Observe that we have now placed exactly $2k = n$ points inside $[0,1]^2$.

To draw a noncrossing tour, we start at s, and draw the edge $\{s, y_1\}$. We then draw the edges $\{y_i, x_i\}_{i=1}^{k-1}$. Lastly, we add the edges $\{x_{k-1}, y_k\}$ and $\{y_k, s\}$, which closes up the tour.

By construction, this tour contains no intersecting edges. To bound its length from below, observe that all edges have a length of at least $\sqrt{2} - \epsilon/\sqrt{2} = \sqrt{2}(1 - \epsilon/2)$. Thus, this tour has a length of at least $\sqrt{2}n(1 - \epsilon/2)$.

We now bound the length of the optimal tour from above, by

$$d(s, x_1) + d(x_1, y_1) + d(s, x_{k-1}) + d(x_{k-1}, y_k) + d(x_1, x_k) + d(y_1, y_k) \le 2\sqrt{2}(1 + \epsilon/2).$$

Putting these bounds together, we find a ratio of

$$\frac{\sqrt{2}n(1-\epsilon/2)}{2\sqrt{2}(1+\epsilon/2)} = \frac{n}{2} \cdot \frac{1-\epsilon/2}{1+\epsilon/2} \geq \frac{n}{2} \cdot (1-\epsilon),$$

as claimed.   □

Remark. The construction used for Theorem 1 only holds for n even. For odd n, we can use a similar construction, but the approximation ratio then becomes $(n-1)/2 \cdot (1-\epsilon)$.

A simple argument shows that the approximation ratio given in Theorem 1 is essentially as bad as one can get in the metric TSP. Given any instance, let x and y be those points separated by the greatest distance. Any tour must travel from x to y and back to x again, so any tour is of length at least $2 \cdot d(x,y)$. Moreover, every tour contains exactly n edges, so any tour has length at most $n \cdot d(x,y)$. Hence, the approximation ratio of any algorithm for the metric TSP is at most $n/2$.

3 Average Case

Although the worst-case construction of Sect. 2 shows that the uncrossing heuristic may yield almost as bad of an approximation as is possible for TSP, it is possible that the heuristic still shows good behavior on average. To exclude this possibility, we consider a standard average-case model wherein n points are placed uniformly and independently in the plane. We then construct a tour of length $\Omega(n)$ in expectation. We present our results in Theorem 9.

To simplify our arguments, it would be convenient if we could consider only a subset of all points, so that we can look at a linear-size sub-instance with nicer properties. Constructing a long noncrossing tour through this subset would be much easier.

Since an optimal tour through a set of points X is always at least as long as an optimal tour through a subset $Y \subset X$, it is tempting to conjecture that something similar holds for all noncrossing tours. Perhaps, given a noncrossing tour through $Y \subseteq X$, we can extend the tour to all of X without decreasing its length?

Unfortunately, this turns out to be false. We provide a counterexample in Fig. 2.

Theorem 2. *There exists a set of points in the plane, together with a noncrossing tour T through all but one of these points, such that all noncrossing tours through all points have length strictly less than $\ell(T)$.*

Proof. Consider the instance shown in Fig. 2, along with the tour $T = avcubwa$. Note that T passes through all points, except for the central point x. We construct this instance such that $d(b,c) = d(a,b) = d(a,c) = 1$, so that

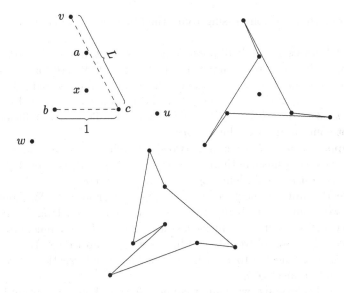

Fig. 2. Top left: instance described in Theorem 2. Bottom: the tour T from Theorem 2. Top right: the longest possible noncrossing tour through all points in the instance.

a, b and c form the vertices of an equilateral triangle. We also set the distances $d(v,c) = d(b,u) = d(a,w) = L$, where $L \gg 1$. We set the angle $\angle cub = \angle avc = \angle awb$ small, so that $d(c,u) = d(v,a) = d(b,w) = L - 1 + o(1)$; here, $o(1)$ means terms decreasing in L. For instance, setting this angle to $1/L$ would suffice. Then we have $d(w,c) = L - \frac{1}{2} + o(1)$. By construction, $d(x,a) = d(x,b) = d(x,c) = 1/\sqrt{3}$. We can furthermore compute $d(u,v) = d(v,w) = d(u,w) = \sqrt{3}L + o(1)$, $d(a,u) = d(b,v) = d(c,w) = L - \frac{1}{2} + o(1)$, and $d(x,u) = d(x,v) = d(x,w) = L - \frac{1}{4\sqrt{3}} + o(1)$. Since we will compare sums of these various distances, all inequalities in the remainder of the proof are understood to hold for sufficiently large L.

We classify the edges into types:

Type	1	2	3	4	5	6	7
Length (up to $o(1)$ terms)	$\frac{1}{\sqrt{3}}$	1	$L-1$	$L-\frac{1}{2}$	$L-\frac{1}{4\sqrt{3}}$	L	$\sqrt{3}L$

We furthermore call the edges of types 1 and 2 the short edges, of types 3 through 6 the long edges, and of type 7 the very long edges. Finally, we classify the points $\{a, b, c, u, v, w\}$ into the near points $N = \{a, b, c\}$ and the far points $F = \{u, v, w\}$.

We start by computing the length of T:

$$\ell(T) = 3L + 3(L - 1) + o(1) = 6L - 3 + o(1).$$

We now show that all noncrossing tours that include x are shorter than $6L - 3 + o(1)$.

As it is laborious to check all possible noncrossing tours, we begin by excluding some possibilities. First, suppose the tour contains k very long edges. Then the tour can contain at most $6 - 2k$ long edges, since each long edge has exactly one endpoint in F, and $k + 1$ of these have been used. Such a tour has a length of at most $k\sqrt{3} + (6 - 2k)L + O(1) < 6L - 3 + o(1)$. Thus, any sufficiently long tour cannot contain any very long edges.

Now suppose that a tour contains two short edges. Since very long edges are excluded, such a tour has length at most $5L + 2 + o(1) < 6L - 3 + o(1)$. Therefore, the tour must consist of six long edges and one short edge.

Suppose the unique short edge in our tour is of type 2. Without loss of generality, we assume it is $\{a, b\}$. Since all other edges are long, both a and b must connect to points in F. Suppose a connects to u. Then x must connect to u and w, since otherwise we would introduce a second short edge. But this makes it impossible to connect c to v without creating an intersection or a subtour. Thus, a cannot connect to u.

By identical reasoning, we cannot connect b to u. The only option is thus to connect a to v and b to w. As very long edges are excluded, v must connect to c (connecting to x would yield an intersection with $L(a, b)$) and x must connect to w. But this makes it impossible to connect u to the tour without creating an intersection. This shows that the short edge can only be of type 1.

Suppose the tour contains an edge of type 6. Without loss of generality, we assume this to be $\{b, u\}$. Then observe that w can only be connected to the rest of the instance by connecting it to b or c, since very long edges are excluded. Hence, connecting w would yield a subtour, and so no edges of type 6 are possible.

We now note that the tour cannot contain all three edges of type 5, since these all connect to x. Suppose the tour contains two such edges, say, $\{x, v\}$ and $\{x, w\}$. Then it is not possible to connect b to the rest of the instance without creating an intersection or a subtour. Hence, at most one edge of type 5 is permissible.

The longest tour now contains one edge of type 5, three edges of type 4, two edges of type 3 and one edge of type 1, which yields a length of

$$L - \frac{1}{4\sqrt{3}} + 3\left(L - \frac{1}{2}\right) + 2(L - 1) + \frac{1}{\sqrt{3}} + o(1) = 6L - 3.5 + \frac{3}{4\sqrt{3}} + o(1) < 6L - 3 + o(1).$$

Therefore, all noncrossing tours through all points of the instance have length strictly less than $\ell(T)$.   □

Theorem 2 shows that in constructing a tour through a random instance, we must carefully make sure to take all points into account. Any points we leave behind could reduce the length of the tour if we attempt to add them after constructing a long subtour.

As the worst-case tour consists of many long almost-parallel edges, we seek to construct a similar tour in random instances. Our strategy will consist of dividing part of the unit square into many long parallel strips, and forming noncrossing Hamiltonian paths within these strips. We then connect paths of adjacent strips without creating intersections, forming a long Hamiltonian path through all strips together. The endpoints can then be connected if we leave out some space for points along which to form a connecting path. See Fig. 3 for a schematic depiction of our construction.

Before we proceed to the proof of Theorem 9, we need some simple lemmas. We start with a lemma that bounds the probability that any region in $[0,1]^2$ contains too few points for our construction to work.

Lemma 3. *Let $A \subseteq [0,1]^2$, and let X be a finite set of n points placed independently uniformly at random in $[0,1]^2$. Let $N = |A \cap X|$. Then*

$$\mathbb{P}(N \leq k) \leq e^k e^{-\frac{k^2}{2n\,\mathrm{area}(A)}} e^{-n\cdot\mathrm{area}(A)/2}$$

for $k \leq n \cdot \mathrm{area}(A)$.

Proof. For $x \in X$, let $S(x)$ be an indicator variable taking a value of 1 iff $x \in A$, and 0 otherwise. Then $N = \sum_{x \in X} S(x)$. Let $\mu = \mathbb{E}(N) = n \cdot \mathrm{area}(A)$. By Chernoff's bound, for $\delta \in (0,1)$,

$$\mathbb{P}(N \leq (1-\delta)\mu) \leq e^{-\delta^2 \cdot n \cdot \mathrm{area}(A)/2}.$$

The result now follows from setting $\delta = 1 - k/\mu$, which implies $\delta^2 = 1 + k^2/\mu^2 - 2k/\mu$, and inserting this into the above bound. □

The following observation and lemma are required to form suitable Hamiltonian paths through subsets of our random instance.

Observation 4. *Let a,b,c,d be four distinct points in the plane, no three of which are collinear. Suppose $L(a,b)$ intersects $L(c,d)$. Then $L(a,d)$ cannot intersect $L(b,c)$. Moreover, $d(a,b) + d(c,d) > d(a,d) + d(b,c)$.*

Lemma 5. *Let X be a set of distinct points in the plane, and assume no three points in X are collinear. Fix distinct points $s,t \in X$. Then there exists a noncrossing Hamiltonian path through X with endpoints s and t.*

Proof. Fix an arbitrary, possibly crossing Hamiltonian path with endpoints s and t. Suppose the edges $e = \{a,b\}$ and $f = \{c,d\}$ intersect. By Observation 4, $e' = \{a,d\}$ and $f' = \{b,c\}$ do not intersect. We assume that the path remains connected if we exchange e and f for e' and f' (if this fails, then we swap a for b).

Observation 4 also shows that the length of the resulting path is strictly smaller than the length of the original path. We repeat this process, removing intersections until we obtain a noncrossing path. This process must terminate in a finite number of steps, since the number of Hamiltonian paths is finite and each step strictly decreases the path length. Hence, no path is seen twice.

It remains to show that s and t remain the endpoints throughout this process. Observe that a point is an endpoint of the path if and only if it has degree 1. Since the exchange operation preserves the degree of all vertices in the path, the endpoints of the path do not change. □

The next lemma is useful to connect Hamiltonian paths in neighboring rectangular regions.

Lemma 6. *Let A, B and C be distinct rectangular regions in the plane. Assume B shares an edge with A and with C, but A and C are disjoint except possibly in a single point. Let S_A, S_B and S_C be finite sets of points in A, B and C respectively. Let x_A and x_C be the points in B closest to A and C, and assume $x_A \neq x_C$. Let P be a noncrossing Hamiltonian path through S_B with endpoints x_A and x_C. Let P' be a path obtained by connecting x_A to any point in S_A and x_C to any point in S_C. Then P' is noncrossing.*

Proof. Without loss of generality, we assume the regions A, B and C are aligned at the horizontal axis. Moreover, we assume (again without loss of generality) that C borders B to the right. This leaves three cases to examine for A: it may border B to the left, to the bottom, or to the top. The latter two cases are identical, so we examine only the first two.

Suppose A borders B to the left. Let $y_A \in S_A$, and suppose we extend P by connecting y_A to x_A. Because P is noncrossing, we need only check whether the edge $e = \{y_A, x_A\}$ intersects any edge of P. Suppose such an edge, say e^*, exists. Then this implies that one endpoint of e^* must lie to the left of x_A, which is a contradiction.

Now consider the border of B with C, and extend the path by adding the edge $f = \{y_C, x_C\}$ for some $y_C \in S_C$. By the same reasoning, f cannot intersect any edge of P. It remains to check whether e can intersect f. Observe that $L(e)$ lies entirely to the left of x_A, which lies to the left of x_C, which in turn lies to the left of the entirety of $L(f)$. Thus, e and f cannot intersect.

Now suppose A borders B to the top. The argument from the previous case now fails, since we can no longer order the extending edges from left to right. Suppose that e intersects f in the point $q \in B$. We add a direction to e and f, taking x_A and x_C as their origin. Since x_C lies below x_A, we know that e must lie below f after passing through q, and stay above f until it reaches its endpoint. Since the endpoints of e and f lie outside of B, this implies that either the point where e exits B lies below the point where f exits B, or both edges exit B in the same border of B. This is a contradiction, and so we are done. $\square$

The proof of Lemma 6 fails when the points x_A and x_C are identical, or equivalently, when the point set contained in B is not nice for B. The following lemma shows that, provided B contains enough randomly placed points, this is unlikely to occur.

Lemma 7. *Let A be a rectangular region in the plane. Let X be a set of $n \geq 2$ points placed uniformly at random in A. The probability that X is not nice for A is at most $6/n$.*

Proof. Let e_i, $i \in [4]$, be any edge of A. Let E_{ij} denote the event that $x_{e_i} = x_{e_j}$ for $i \neq j$. Then we seek to bound $\mathbb{P}(\cup_{i=1}^{4} \cup_{j=i+1}^{4} E_{ij})$ from above.

Without loss of generality, we assume that e_i lies along the x-axis, and that $A \setminus e_i$ lies above e_i, i.e., e_i is the bottom edge of A.

Let e_j be a different edge of A. If e_j is the top edge of A, then the event E_{ij} occurs with probability 0, since it requires all points to lie on the same horizontal line.

Suppose now that e_j is the right edge of A, and let $y \in X \backslash \{x_{e_i}\}$. Observe that the horizontal coordinates of all points in X are independent uniform random variables. Thus, the probability that x_{e_i} is the point with the largest horizontal coordinate is $1/n$ by symmetry. Therefore, $\mathbb{P}(E_{ij}) = 1/n$.

To conclude, we apply a union bound to obtain

$$\mathbb{P}\left(\cup_{1 \leq i < j \leq 4} E_{ij}\right) \leq \sum_{1 \leq i < j \leq 4} \mathbb{P}\left(E_{ij}\right) = \frac{6}{n}$$

as claimed. $\square$

The final lemma we require follows from elementary calculus and probability theory.

Lemma 8. *Let X and Y be uniformly distributed over $[a, b]$. Then*

$$\mathbb{E}(|X - Y|) = \frac{b - a}{3}.$$

Proof. As X and Y are independent, we can compute the required quantity directly using their joint distribution. Let f_X and f_Y denote their respective probability density functions. We have

$$\mathbb{E}(|X - Y|) = \int_a^b \int_a^b |x - y| f_X(x) f_Y(y) dx dy$$

$$= \frac{1}{(b-a)^2} \int_a^b \int_a^b |x - y| dx dy.$$

Substituting $x = a + (b - a)\bar{x}$ and $y = a + (b - a)\bar{y}$, the integral reduces to $(b-a)^3 \int_0^1 \int_0^1 |\bar{x} - \bar{y}| d\bar{x} d\bar{y} = (b - a)^3/3$, completing the proof. $\square$

We are now in a position to prove Theorem 9. See Fig. 3 for a sketch of the construction we use in the proof.

Theorem 9. *Suppose a TSP instance is formed by placing n points uniformly at random in the unit square. Then the expected value of the ratio of the worst local optimum of X-opt and the optimal tour on this instance is $\Omega(\sqrt{n})$.*

Proof. We begin by partitioning the unit square into six rectangular regions. Let $c \in (0, 1)$ be a constant, to be fixed later. Let C_1 denote the square region with opposite points $(0, 0)$ and (c, c), and let C_2 similarly denote the square region with corner points $(1 - c, 0)$ and $(1, c)$. Next, let C_3 be the rectangular region with opposite corners $(c, 0)$ and $(1 - c, c)$. The region C_4 denotes the rectangular region with corner points $(0, c)$ and $(c, 1)$, while C_5 denotes the region with

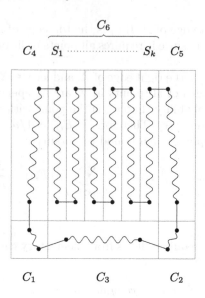

Fig. 3. The construction used in Theorem 9. Wavy lines represent Hamiltonian paths within the rectangular region they lie in, while straight lines represent edges.

corners $(1 - c, c)$ and $(1, 1)$. Finally, the region C_6 denotes the rectangle with corners (c, c) and $(1 - c, 1)$.

We divide C_6 into vertical strips of width $\alpha \cdot \frac{1-2c}{n}$ for some $\alpha > 0$ to be fixed later, so that there are $k = \lfloor n/\alpha \rfloor$ strips in total. We label the strips from left to right as $\{S_i\}_{i=1}^k$.

Let X be a set of n points placed uniformly at random in $[0, 1]^2$. Note that, by Lemma 3, the probability that any of $X_{C_i} := C_i \cap X$ for $i \in [5]$ contains fewer than 31 points is at most $5e^{-\Omega(n)}$. Hence, we assume for the remainder of the proof that these regions contain at least 31 points. The possibility that this is not true reduces the expected tour length by a factor of at most $1 - e^{-\Omega(n)}$, which does not affect the result.

Observe that each $x \in X_{C_i}$ is uniformly distributed over C_i. Since we assume that each X_{C_i} contains at least 31 points, we see by Lemma 7 that the probability that X_{C_i} is not nice for X is at most $6/31$. By a union bound, the probability that X_{C_i} is not nice for C_i for any $i \in [5]$ is at most $5 \cdot \frac{6}{31} = \frac{30}{31}$. Hence, we assume for the remainder that each X_{C_i} is nice for C_i, $i \in [5]$. The possibility that this is not true reduces the expected tour length by at most a factor of $\frac{1}{31}$, which does not affect the result.

Consider strip S_i. Let l_i be the point in $S_i \cap X$ with the smallest x-coordinate, the *leftmost* point, provided it exists. Similarly, r_i denotes the point in $S_i \cap X$ with the largest x-coordinate, or the *rightmost* point. Since the probability that any three points in X are collinear is 0, we can use Lemma 5 to establish the existence of a noncrossing Hamiltonian path through S_i with endpoints l_i and r_i. Let P_{S_i} be such a path.

After forming the paths P_{S_i} for all $i \in [k]$, we connect P_{S_i} to $P_{S_{i+1}}$ for $i \in [k-1]$. To do this, we simply connect r_i to l_{i+1}. The result is a Hamiltonian path P_6 through C_6, which contains the paths P_i as subpaths. By Lemma 6, we know that P_6 is noncrossing.

Next, we form noncrossing Hamiltonian paths through the remaining regions. For the region C_4, we let the endpoints of the path be the rightmost and bottom-most points in $X \cap C_4$. For C_5, the endpoints are the leftmost and bottom-most points. For C_1, we take the top-most and rightmost points, for C_2 the leftmost and top-most, and for C_3 the leftmost and rightmost. We label the path through region C_i by P_i, $i \in [6]$. Observe that these endpoints are distinct in all cases, since we assume that X_{C_i} is nice for C_i for each $i \in [5]$.

We now connect these paths as follows. Let C_i be any of the regions. If the region shares a border with C_j, excluding the bottom border of C_6, then we connect the points from C_i and C_j closest to the border. Observe that this is exactly the process we used to form P_6. Again using Lemma 5, the tour T so formed is noncrossing.

To bound the length of T from below, we consider the length of the paths P_{S_i}, $i \in [k]$. The sum of these lengths is clearly a lower bound for $\ell(T)$. We thus have

$$\mathbb{E}(\ell(T)) \geq \sum_{i=1}^{k} \mathbb{E}(\ell(P_{S_i})),$$

by linearity of expectation. Moreover, observe that $\ell(P_{S_i}) = 0$ if fewer than 2 points are placed in S_i. Thus, we find by the law of total expectation

$$\mathbb{E}(\ell(P_{S_i})) = \mathbb{E}(\ell(P_{S_i}) \mid |S_i \cap X| \geq 2) \cdot \mathbb{P}(|S_i \cap X| \geq 2).$$

Assume the event $|S_i \cap X| \geq 2$ occurs. Let x, y be any two points in $S_i \cap X$. Then by the triangle inequality, $\ell(P_i) \geq d(x, y) \geq d_v(x, y)$, where d_v denotes the vertical distance between x and y. Observe that the vertical coordinates of x and y are independent and uniformly distributed over $[c, 1]$. This implies by Lemma 8 that

$$\mathbb{E}(\ell(P_i) \mid |S_i \cap X| \geq 2) \geq \frac{1-c}{3}.$$

Using Lemma 3 to bound $\mathbb{P}(|S_i \cap X| \geq 2)$ from below, we have

$$\mathbb{E}(\ell(T)) \geq \left\lfloor \frac{n}{\alpha} \right\rfloor \cdot \frac{1-c}{3} \cdot \left(1 - e^{1 - \frac{1}{2}\alpha(1-2c)(1-c)} \cdot e^{-\frac{1}{2\alpha(1-2c)(1-c)}} \right),$$

where we use the fact that $\text{area}(S_i) = \alpha(1-c)(1-2c)/n$.

It remains to fix values for c and α such that this expectation is nontrivial; for instance, $\alpha = 10$ and $c = 0.1$ suffice. We then find $\mathbb{E}(\ell(T)) = \Omega(n)$. Since there exists a tour of length $O(\sqrt{n})$ in the Euclidean TSP with high probability [7], we are done. □

4 Practical Performance of Uncrossing Tours

In this section, we show that in practical instances there is a large gap with the results suggested by Theorem 9. We generate instances with n points sampled from the uniform distribution over $[0,1]^2$, and run X-opt on these instances. As a starting tour, we pick a tour from the uniform distribution on all tours. We compute the lengths of the locally optimal tours obtained from our implementation of X-opt, and average them for each fixed value of n we evaluate. We consider the simplest possible pivot rule: starting from an arbitrary edge e in the tour, we check whether e intersects with any other edge, performing an exchange when we find the first such edge. If we do not find such an intersecting edge, we move on to the next edge in the tour and repeat the process. By "next", we mean that we order the edges of a tour according to the permutation on the vertices by which we represent the tour.

Since the optimal tour length is $\Theta(\sqrt{n})$ with high probability [7], we compare the length of the tours we obtain with this function. Their ratio then serves as a proxy for the approximation ratio of X-opt. We perform this procedure for $n \in \{100, \ldots, 1000, 2000, \ldots, 10000\}$. For each value of n, we take $N = 16,000$ samples. The results are shown in Fig. 4. To the precision we are able to obtain, we cannot distinguish the approximation ratio from constant.

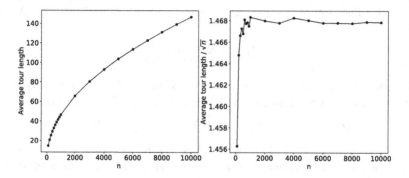

Fig. 4. Numerical evaluation of the average-case performance of X-opt.

5   Discussion

Although the results we presented in Theorem 1 and Theorem 9 are rather negative for X-opt, the numerical experiments of Sect. 4 paint a much more optimistic picture. The heuristic appears to be much more efficient in practice than our lower bounds suggest. Indeed, while Theorem 9 suggests an approximation ratio of $\Omega(\sqrt{n})$, the numerical experiments in Sect. 4 suggest a constant approximation ratio.

One possible explanation for this discrepancy is that we compare the optimal solution on any instance to local optima specifically constructed to be bad. This is a rather standard approach, and it is not surprising that it gives pessimistic results. However, the results in this case are especially pessimistic, considering that we can show a tight lower bound for the expected tour length in the average case.

We consider this to be an indication that this approach is incapable of explaining the practical approximation performance of local search heuristics. In order to more closely model the true behavior of heuristics, it seems one must analyze the landscape of local optima, and the probability of reaching different local optima. We stress that this discrepancy cannot be resolved by other standard methods of probabilistic analysis. In particular, smoothed analysis [10] cannot help, since the smoothed approximation ratio of an algorithm is bounded from below by the average-case approximation ratio.

References

1. Aarts, E., Lenstra, J.K. (eds.): Local Search in Combinatorial Optimization. Princeton University Press, Princeton (2003). https://doi.org/10.2307/j.ctv346t9c
2. Biniaz, A., Maheshwari, A., Smid, M.: Flip distance to some plane configurations (2019). https://doi.org/10.48550/arXiv.1905.00791
3. Brodowsky, U.A., Hougardy, S., Zhong, X.: The approximation ratio of the k-opt heuristic for the Euclidean traveling salesman problem (2021). https://doi.org/10.48550/arXiv.2109.00069
4. Chandra, B., Karloff, H., Tovey, C.: New results on the old k-opt algorithm for the traveling salesman problem. SIAM J. Comput. **28**(6), 1998–2029 (1999). https://doi.org/10.1137/S0097539793251244
5. da Fonseca, G.D., Gerard, Y., Rivier, B.: On the longest flip sequence to untangle segments in the plane (2022). https://doi.org/10.48550/arXiv.2210.12036
6. Englert, M., Röglin, H., Vöcking, B.: Smoothed analysis of the 2-opt algorithm for the general TSP. ACM Trans. Algorithms **13**(1), 10:1–10:15 (2016). https://doi.org/10.1145/2972953
7. Frieze, A.M., Yukich, J.E.: Probabilistic analysis of the TSP. In: Gutin, G., Punnen, A.P. (eds.) The Traveling Salesman Problem and Its Variations. Combinatorial Optimization, vol. 12, pp. 257–307. Springer, Boston (2007). https://doi.org/10.1007/0-306-48213-4_7
8. Korte, B., Vygen, J.: Combinatorial Optimization: Theory and Algorithms. Algorithms and Combinatorics. Springer, Heidelberg (2000). https://doi.org/10.1007/978-3-662-21708-5
9. Papadimitriou, C.H.: The Euclidean travelling salesman problem is NP-complete. Theoret. Comput. Sci. **4**(3), 237–244 (1977). https://doi.org/10.1016/0304-3975(77)90012-3
10. Spielman, D.A., Teng, S.H.: Smoothed analysis of algorithms: why the simplex algorithm usually takes polynomial time. J. ACM **51**(3), 385–463 (2004). https://doi.org/10.1145/990308.990310
11. Van Leeuwen, J., Schoone, A.: Untangling a traveling salesman tour in the plane. In: Proceedings of the 7th International Workshop Graph-Theoretical Concepts in Computer Science (1980)

A Frequency-Competitive Query Strategy for Maintaining Low Collision Potential Among Moving Entities

William Evans$^{(\boxtimes)}$ (iD) and David Kirkpatrick$^{(\boxtimes)}$ (iD)

Computer Science, University of British Columbia, Vancouver, Canada
{will,kirk}@cs.ubc.ca

Abstract. Consider a collection of entities moving with bounded speed, but otherwise unpredictably, in some low-dimensional space. Two such entities encroach upon one another at a fixed time if their separation is less than some specified threshold. Encroachment, of concern in many settings such as collision avoidance, may be unavoidable. However, the associated difficulties are compounded if there is uncertainty about the precise location of entities, giving rise to potential encroachment and, more generally, potential congestion within the full collection.

We adopt a model in which entities can be queried for their current location (at some cost) and the uncertainty region associated with an entity grows in proportion to the time since that entity was last queried. The goal is to maintain low potential congestion, measured in terms of the (dynamic) intersection graph of uncertainty regions, using the lowest possible query cost. Previous work, in the same uncertainty model, described query schemes that minimize several measures of congestion potential for *point entities*, using location queries of *some fixed frequency*. These schemes were shown to be $O(1)$-competitive, with other, even clairvoyant query schemes (that know the trajectories of all entities), subject to the same bound on query frequency.

In this paper we design a scheme that is competitive in terms of its query *granularity* (minimum spacing between queries), over all sufficiently large time intervals, while guaranteeing a *fixed bound on collision potential* (defined as the maximum degree of the intersection graph of uncertainty regions), for *entities with positive extent*. Our complementary optimization objective necessitates surprisingly different algorithms and analyses from that in previous work. Nevertheless, we also show that the competitive factor of our scheme is best possible, up to a constant factor, in the worst case.

Keywords: data in motion · uncertain inputs · collision avoidance · online algorithms · competitive analysis

This work was funded in part by Discovery Grants from the Natural Sciences and Engineering Research Council of Canada.

© The Author(s), under exclusive license to Springer Nature Switzerland AG 2023
J. Byrka and A. Wiese (Eds.): WAOA 2023, LNCS 14297, pp. 14–28, 2023.
https://doi.org/10.1007/978-3-031-49815-2_2

1   Introduction

Imagine a set of entities (robots) moving continuously in Euclidean space with bounded speed but following unpredictable trajectories. If we do not continuously monitor the positions of the entities, we risk the possibility of their collision. In our model, we can query an entity to obtain its current location but such queries are expensive. Between queries, the set of possible locations, called the *uncertainty region*, of each entity grows. How infrequently can we query in order to avoid potential collisions, where a potential collision is an intersection of the uncertainty regions of two entities? We must make the decision of which entity to query next (and when) knowing only the uncertainty regions of the entities that we have established from previous queries. This is an online task, but one in which we can choose the next piece of data (which entity's location) we obtain. For some sets of trajectories, there are no good sequences of choices that prevent potential collisions without incurring high cost; for example, imagine robots converging on a single point. For such trajectories, we cannot expect our (or any) query scheme to do well. Instead, we would like to design query schemes that have a cost that is not much worse than the smallest cost for a given set of trajectories. That is, we compare our scheme's performance against a *clairvoyant* query scheme that knows the trajectories but must also perform queries to keep uncertainty regions from intersecting.

Our primary goal is to formulate efficient query schemes that, for all possible collections of moving entities, maintain fixed bounds on the maximum number of entities that might collide with any one entity. Naturally for many such collections the required query frequency changes over time as entities cluster and spread, so efficient query schemes need to adapt to changes in the configuration of entities. While such changes are continuous and bounded in rate, they are only discernible through queries to individual entities, so entity configurations are *never* known precisely; future configurations are of course entirely hidden. In this latter respect our schemes and the competitive analysis of their efficiency, using as a benchmark a clairvoyant scheme that bases its queries on full knowledge of all entity trajectories (and hence all future configurations), resemble familiar scenarios that arise in the design and analysis of on-line algorithms.

1.1   The Query Model

To facilitate comparisons with earlier results, we adopt similar notation to that used by Evans et al. [9] and Busto et al. [3]. Let $\mathcal{E}$ be a set $\{e_1, e_2, \ldots, e_n\}$ of (mobile) entities. Each entity e_i is modelled as a d-dimensional closed ball with fixed extent and bounded speed, whose position (centre location) at any time is specified by the (unknown) function z_i from $[0, \infty)$ (time) to $\mathbb{R}^d$. We take the entity radius to be our *unit of distance*, and take the time for an entity moving at maximum speed to move a unit distance to be our *unit of time*.

The n-tuple $Z(t) = (z_1(t), z_2(t), \ldots, z_n(t))$ is called the $\mathcal{E}$-*configuration* at time t. Entities e_i and e_j are said to *encroach* upon one another at time t if the distance between their centres $\|z_i(t) - z_j(t)\|$ is less than some fixed *encroachment threshold* Ξ. For simplicity we assume to start that the distance between entity centres is always at least 2 (i.e. $\|z_i(t) - z_j(t)\| - 2$, the *separation* between entities e_i and e_j at time t, is always at least zero—so entities never properly intersect), and that the encroachment threshold is exactly 2 (i.e. we are only concerned with avoiding entity contact). The concluding section considers a relaxation (and decoupling) of these assumptions in which $\|z_i(t) - z_j(t)\|$, is always at least some positive constant ρ_0 (possibly less than 2), and the encroachment threshold Ξ is some constant at least ρ_0.

We wish to maintain knowledge of the positions of the entities over time by making location queries to individual entities, each of which returns the exact position[1] of the entity at the time of the query. A *(query) scheme* $\mathbb{S}$ is just an assignment of location queries to time instances. We measure the performance of a scheme in terms of its minimum query *granularity* (the time between consecutive queries) over a specified time interval T.

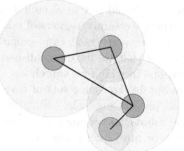

Fig. 1. Potential encroachment graph of four unit-radius entities (dark grey) with uncertainty regions (light grey) after queries four, three, two and one time unit in the past.

At any time $t \geq 0$, let $p_i^{\mathbb{S}}(t)$ denote the time, prior to t, that entity e_i was last queried; we define $p_i^{\mathbb{S}}(0) = -\infty$. The *uncertainty region* of e_i at time t, denoted $u_i^{\mathbb{S}}(t)$, is defined as the ball with centre $z_i(p_i^{\mathbb{S}}(t))$ and radius $1 + t - p_i^{\mathbb{S}}(t)$ (cf. Fig. 1); note that $u_i^{\mathbb{S}}(0)$ is unbounded. We omit $\mathbb{S}$ when it is understood and the dependence on t when t is fixed.

The set $U(t) = \{u_1(t), \ldots, u_n(t)\}$ is called the *(uncertainty) configuration* at time t. Entity e_i is said to *potentially encroach* upon entity e_j in configuration $U(t)$ if $u_i(t) \cap u_j(t) \neq \emptyset$ (that is, there are potential locations for e_i and e_j at time t such that $e_i \cap e_j \neq \emptyset$). In this way, any configuration U gives rise to an associated (symmetric) *potential encroachment graph* PE^U on the set $\mathcal{E}$. Note that, by our assumptions above, the potential encroachment graph associated with the initial uncertainty configuration $U(0)$ is complete. As in [3], notions of *congestion potential* are expressed as properties of the graph PE^U.

In this paper, we focus on the *max-degree* (hereafter *(uncertainty) degree*) of the graph PE^U, which is defined as the maximum, over entities e_i, of the number of entities e_j, *including* e_i, that potentially encroach upon e_i in configuration U. We also refer to this as the *collision potential* of U.

The assumption that entities never properly intersect is helpful since it means that for x larger than some dimension-dependent sphere-packing constant, it is always possible to maintain uncertainty degree at most x, using sufficiently high query frequency.

---

[1] The concluding discussion describes how this exactness condition can be relaxed.

1.2   Related Work

One of the most widely-studied approaches to computing functions of moving entities uses the *kinetic data structure* model which assumes precise information about the future trajectories of the moving entities and relies on elementary geometric relations among their locations along those trajectories to certify that a combinatorial structure of interest, such as their convex hull, remains essentially the same. The algorithm can anticipate when a relation will fail, or is informed if a trajectory changes, and the goal is to update the structure efficiently in response to these events [2,11,12]. Another less common model assumes that the precise location of *every* entity is given to the algorithm periodically. The goal again is to update the structure efficiently when this occurs [4–6].

More similar to ours is the framework introduced by Kahan [14,15] for studying data in motion problems that require repeated computation of a function (geometric structure) of data that is moving continuously in space where data acquisition via queries is costly. There, location queries occur simultaneously in batches, triggered by requests (Kahan refers to these as "queries") to compute the function at the time of the request rather than by a requirement to maintain a structure or property at all times. Performance is compared to a "lucky" algorithm that queries the minimum amount to calculate the function. Kahan's model and use of competitive evaluation is common to much of the work on query algorithms for uncertain inputs (see Erlebach and Hoffmann's survey [7]).

As mentioned, our model is essentially the same as the one studied by Evans et al. [9] and by Busto et al. [3], both of which focus on point entities. Paper [9] contains strategies whose goal is to guarantee competitively low congestion potential, compared to any other (even clairvoyant) scheme, at one specified target time. It provides precise descriptions of the impact on this guarantee for several measures of initial location knowledge and available lead time before the target. The other paper [3] contains a scheme for guaranteeing competitively low congestion potential at *all* times. For this more challenging task the scheme maintains congestion potential over time that is within a constant factor of that maintained by any other scheme over modest-sized time intervals. All of these results deal with the *optimization of congestion potential measures subject to fixed query frequency*.

In this paper, we consider a dual problem, for entities of bounded extent: *optimizing query frequency required to guarantee fixed bounds on collision potential*. This complementary optimization is fundamentally different: being able to minimize collision potential using fixed query frequency provides little insight into how to minimize query frequency to maintain a fixed bound on collision potential. In particular, even for stationary entities, a small change in the collision potential bound can lead to an arbitrarily large change in the required query frequency. Our schemes optimize query frequency in a strong sense: maximizing the minimum query granularity over all sufficiently large time intervals.

1.3   Our Results

At first, we imagine that entities remain stationary, even though they have the potential to move, since this avoids complications arising from entities changing location and nearby neighbours. We define a natural stationary frequency demand that serves to lower bound the number of queries (and hence upper bound the query granularity) required by any, even clairvoyant, scheme to maintain uncertainty degree x over any modest-length time interval T. This is complemented by a scheme that uses a constant times that number of (well-spaced) queries to achieve uncertainty degree x over *all* such time intervals.

When the entities are mobile, the stationary frequency demand changes over time as entities cluster and separate and the analysis becomes significantly more involved. However, by integrating the stationary measure over any modest-length time interval T we can again derive an expression that serves to lower bound the number of queries (and hence upper bound the query granularity) required by any, even clairvoyant, scheme to maintain uncertainty degree x over T. A query scheme is described that meets this bound on query granularity to within a factor $\Theta(x)$, over *all* such time intervals T. The competitive factor $\Theta(x)$ drops to $\Theta(\frac{x}{1+\Delta})$ when we relax the degree bound to $x + \Delta$, for some $\Delta \le x$ (i.e. we are willing to accept an approximation $x + \Delta$ of the degree bound x). We give a family of examples showing that the competitive factor cannot be improved by more than a constant factor in the worst case.

In the concluding section we describe how our results can be applied to other measures of congestion potential. We also discuss several other modifications to the underlying model that make our query optimization framework even more broadly applicable.

Returning to the problem concerning collision avoidance mentioned in the introduction, it follows from our results that, by maintaining uncertainty degree at most x, we maintain for each entity e_i a certificate identifying the, at most $x - 1$, other entities that could potentially encroach upon e_i (those warranting more careful local monitoring).

2   Geometric Preliminaries

In any $\mathcal{E}$-configuration $Z(t) = (z_1(t), z_2(t), \ldots, z_n(t))$ and for any positive integer x, we call the separation between e_i and its xth closest neighbour (not including e_i) its *x-separation*, and denote it by $\sigma_i^Z(x, t)$. We call the closed ball with radius $\sigma_i^Z(x, t)+1$ and centre $z_i(t)$, the *x-ball* of e_i, and denote it by $B_i^Z(x, t)$ (cf. Fig. 2.) We will omit Z when the configuration is understood. Note that, for all entities e_i and e_j,

$$\sigma_j(x, t) \le \|z_j(t) - z_i(t)\| + \sigma_i(x, t), \tag{1}$$

since the ball with radius $\|z_j(t) - z_i(t)\| + \sigma_i(x, t)$ centred at $z_j(t)$ contains the ball with radius $\sigma_i(x, t)$ centred at $z_i(t)$ (by the triangle inequality).

We first observe that, at any time, individual entities have bounded overlap with the $\hat{x}$-balls of other entities, for any positive integer $\hat{x}$. We use "$\hat{x}$" since Lemma 1 may be applied in situations where $\hat{x} \neq x$. See [8, Lemma 1] for the proof of the following:

Lemma 1. *In any $\mathcal{E}$-configuration $Z(t)$, any entity e_* intersects the $\hat{x}$-balls of at most $5^d\hat{x}$ entities.*

We have assumed that entities do not properly intersect. Define $c_{d,x}$ to be the smallest constant such that a unit-radius d-dimensional ball B can have x disjoint unit-radius d-dimensional balls (not including itself) with separation from B at most $c_{d,x}$. Thus, in any $\mathcal{E}$-configuration $Z(t)$, $\sigma_i(x,t) \geq c_{d,x}$ and hence

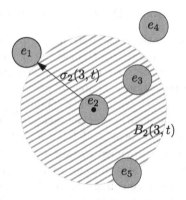

Fig. 2. A configuration of five unit-radius entities at time t. $B_2(3,t)$, the 3-ball of entity e_2, is shown shaded.

$$\frac{1 - \lambda_{d,x}}{\lambda_{d,x}} \sigma_i(x,t) \geq 1, \text{ where } \lambda_{d,x} = \frac{c_{d,x}}{1 + c_{d,x}}. \tag{2}$$

Since x unit-radius balls with separation at most ξ from B must all fit within a ball of radius $3 + \xi$ concentric with B, a straightforward volume argument shows that $c_{d,x} \geq \xi$ provided $x \geq (3 + \xi)^d$. Thus $\lambda_{d,x} \geq 1/2$ if $x \geq 4^d$.

Let X_d be the largest value of x for which $c_{d,x} = 0$ (e.g., $X_2 = 6$). Clearly, if $x \leq X_d$ there are entity configurations $Z(t)$ with $\sigma_i^Z(x,t) = 0$. Thus, for such x, maintaining uncertainty degree at most x might be impossible in general, for any query scheme. On the other hand, if $x > X_d$, then $\lambda_{d,x} > 0$. Thus:

Remark 1. Hereafter *we will assume that x, our bound on congestion potential, is greater than X_d.* The constants X_d and $\lambda_{d,x}$ will play a role in both the formulation and analysis of our query schemes in (arbitrary, but fixed) dimension d. If the reader prefers to focus on dimension 2, then assuming $x \geq 16$ ensures that $\lambda_{2,x} \geq 1/2$.

3   Query Optimization for Mobile Entities

Several factors contribute to the difficulty of maintaining low uncertainty degree in our model. One is that the increase in uncertainty over time arises from the entities' potential movement, whether or not they actually move. Another is that movement creates the potential for the formation of entity clusters whose duration, position and constituents change. To help isolate the first of these factors we first look at the special case of our problem where all entities are stationary.

3.1   Query Optimization for Stationary Entities

When entities are stationary, the entity configuration $Z(t)$ at time t, and hence the x-separation of entity e_i at time t, is the same for all t. Accordingly, we will use the notation $\sigma_i(x, *)$, to highlight this time invariance. If the radius of the uncertainty region u_i of $e_i \in \mathcal{E}$ exceeds $1 + \sigma_i(x, *)$ then u_i intersects at least $x + 1$ entities in $\mathcal{E}$. Thus, $\sigma_i(x, *)$ is an upper bound on the amount of uncertainty whose avoidance guarantees that the uncertainty degree of entity e_i remains at most x. It follows that $\phi_x = \sum_{e_i \in \mathcal{E}} \frac{1}{\sigma_i(x,*)}$, the *stationary frequency demand*, provides a lower bound on the total query frequency (measured as queries per unit of time) to avoid uncertainty degree greater than x, with entities in stationary configuration Z.[2]

Though necessary, it is not sufficient to query each entity e_i with frequency $\frac{1}{\sigma_i(x,*)}$, in order to avoid uncertainty degree greater than x, simply because the uncertainty regions of all entities, not just e_i, grow with time. Nevertheless, a constant multiple of the stationary frequency demand does suffice to keep the uncertainty degree of all entities from exceeding x.

The *Frequency-Weighted Round-Robin* scheme for maintaining uncertainty degree at most x, denoted FWRR[x], queries each entity e_i once every $t_i = 2^{g + \lfloor \lg[\sigma_i(x,*)\lambda_{d,x}/(\lambda_{d,x}+2)] \rfloor}$ time steps of size (granularity) $1/2^g$, where $g = \lceil \lg(\frac{\lambda_{d,x}+2}{\lambda_{d,x}} \phi_x) \rceil + 1$. The schedule repeats after $\max\{t_i\}$ steps.

Lemma 2. *The query scheme* FWRR[x] *maintains uncertainty configurations with uncertainty degree at most x at all times, and has an implementation using minimum query granularity at least $\frac{\lambda_{d,x}}{4(\lambda_{d,x}+2)} \frac{1}{\phi_x}$.*

Proof. Since $t_i/2^g \leq \frac{\lambda_{d,x}}{\lambda_{d,x}+2}\sigma_i(x, *)$, it follows that FWRR[$x$] will query every entity e_i with at most $\frac{\lambda_{d,x}}{\lambda_{d,x}+2}\sigma_i(x, *)$ time between queries. Equation (1) implies that any entity e_j whose separation from e_i is $s \geq \sigma_i(x, *)$ has the property that $\sigma_j(x, *) \leq s + 2 + \sigma_i(x, *) \leq 2s + 2$. Hence, using Eq. (2),

$$\sigma_i(x, *) + \sigma_j(x, *) \leq 3s + 2 \leq (3 + 2\frac{1 - \lambda_{d,x}}{\lambda_{d,x}})s = \frac{\lambda_{d,x} + 2}{\lambda_{d,x}}s,$$

and $\frac{\lambda_{d,x}}{\lambda_{d,x}+2}\sigma_i(x, *) + \frac{\lambda_{d,x}}{\lambda_{d,x}+2}\sigma_j(x, *) \leq s$. So the uncertainty regions of e_i and e_j never properly intersect, and thus the uncertainty degree of e_i remains at most x over time.

Since $\sum_{e_i \in \mathcal{E}} \frac{1}{t_i} < \frac{\lambda_{d,x}+2}{\lambda_{d,x} 2^{g-1}} \phi_x \leq 1$, it follows from a result of Anily et al. [1] (see Lemma 6.2) that a query schedule exists with at most one query for every slot of size $1/2^g$. Hence, the FWRR query scheme can be implemented with query granularity at least $\frac{\lambda_{d,x}}{4(\lambda_{d,x}+2)} \frac{1}{\phi_x}$.   □

---

[2] This frequency demand appears in work on scheduling jobs according to a vector v of periods, where job j must be scheduled at least once in every time interval of length v_j [10,13].

3.2   Query Optimization for General Mobile Entities

While the case of stationary entities exhibits some of the difficulties in maintaining uncertainty regions with low congestion, mobile entities add an additional level of complexity. As we have seen, when entities are stationary, the expression $|T|\phi_x = \sum_{e_i \in \mathcal{E}} \frac{|T|}{\sigma_i(x,*)}$, the *stationary query demand over time interval* T, plays a central role in characterizing the unavoidable number of queries needed to avoid uncertainty degree greater than x. For mobile entities, the more general expression $\phi_x(T) = \sum_{e_i \in \mathcal{E}} \int_T \frac{dt}{\sigma_i(x,t)}$ plays a similar role.

It follows from earlier work on the optimization of query degree using fixed query frequency [3] that a high stationary frequency demand *at one instant in time* does not necessarily imply that uncertainty degree at most x is unsustainable at a significantly lower query frequency. Nevertheless, as the following lemma demonstrates, high stationary query demand, sustained over a sufficiently large time interval T, does imply a lower bound on the number of queries over T or some very small shift of T.

Lemma 3. *Let $0 \leq \Delta \leq x$ and let T be a time interval for which $\phi_{x+\Delta}(T) \geq 1010|\mathcal{E}|$. Define $\overrightarrow{T}$ to be the interval T shifted by $|T|/3^{55}$ and $T^+ = T \cup \overrightarrow{T}$. Then any, even clairvoyant, query scheme that ensures maximum uncertainty degree at most x over T^+ must make $\Omega(\frac{1+\Delta}{x}\phi_{x+\Delta}(T))$ queries in total over either T or $\overrightarrow{T}$.*

Remark 2. Three features of this lower bound are worth highlighting: (i) It shows that all, even clairvoyant, query schemes that ensure uncertainty degree at most x over the time interval T^+, must have minimum query granularity $O(\frac{x}{1+\Delta}\frac{1}{\phi_{x+\Delta}(T)})$ over T^+, even if the granularity is arbitrarily smaller outside of that interval. (ii) We will describe a non-clairvoyant query scheme that uses minimum query granularity $\Omega(\frac{1}{\phi_{x+\Delta}(T)})$ to maintain maximum uncertainty degree $x + \Delta$ over *all* sufficiently large time intervals T. The $O(\frac{x}{1+\Delta}\frac{1}{\phi_{x+\Delta}(T)})$ granularity bound from Lemma 3 relates the granularity of our query scheme to that required for maintaining uncertainty degree at most x, highlighting the impact of allowing an approximation $x + \Delta$ of the uncertainty degree objective x. (iii) It can be shown that the appearance in the query lower bound of $\Omega(\frac{1+\Delta}{x})$, which does not appear in the stationary lower bound, is unavoidable, even for entity sets whose locations change only by some global translation.

Proof. (Sketch; see [8, Lemma 3 and Corollary 7] for full proof.) It helps to imagine first that the entities are stationary. For each $e_i \in \mathcal{E}$ partition T into $\lfloor \frac{|T|}{\sigma_i(x+\Delta,*)} \rfloor \geq \frac{|T|}{2\sigma_i(x+\Delta,*)}$ sub-intervals each of length at least $\sigma_i(x + \Delta, *)$. Let E_i denote the set of (at least) $x + \Delta$ entities in $\mathcal{E} \setminus e_i$ that intersect the $(x + \Delta)$-ball of entity e_i. We say that entity e_i is *satisfied* in a sub-interval if (i) e_i is queried in that sub-interval (in which case we say *directly satisfied*), or (ii) at least $1 + \Delta$ of the entities in E_i are queried in that sub-interval (in which case we say *indirectly satisfied*).

If e_i is not satisfied in a given sub-interval then at the end of the sub-interval at least x of the $x + \Delta$ entities in E_i must have uncertainty regions of radius at least $1 + \sigma_i(x + \Delta, *)$, all of which intersect the uncertainty region u_i of e_i, which means the uncertainty degree of e_i would be at least $x + 1$. Thus, to avoid uncertainty degree greater than x throughout T, every entity e_i must be satisfied in each of its sub-intervals. Since there are at least $\sum_{e_i \in \mathcal{E}} \frac{|T|}{2\sigma_i(x + \Delta, *)}$ sub-intervals in total, we can assume that at least half of these are satisfied indirectly (otherwise, the directly satisfied entity sub-intervals alone account for at least $\sum_{e_i \in \mathcal{E}} \frac{|T|}{4\sigma_i(x + \Delta, *)}$ queries.

By Lemma 1 entity e_j can intersect the $(x + \Delta)$-ball of at most $5^d(x + \Delta)$ entities, and so a query to e_j can help to indirectly satisfy at most $5^d(x + \Delta)$ other entity sub-intervals. Since there are at least $\sum_{e_i \in \mathcal{E}} \frac{|T|}{4\sigma_i(x + \Delta, *)}$ entity sub-intervals in total that are satisfied indirectly, it must be that there are at least $\frac{(1 + \Delta)}{4 \cdot 5^d(x + \Delta)} \sum_{e_i \in \mathcal{E}} \frac{|T|}{\sigma_i(x + \Delta, *)}$ queries in total over T.

At a high level the proof when entities are not necessarily stationary parallels that of the stationary case above. However, in the general case the $(x + \Delta)$-separation, and indeed the $(x + \Delta)$-neighbourhood of each entity, changes over time. A reasonable hope is that the *integral* of the entity's inverse $(x + \Delta)$-separation over T, summed over all entities, provides a similar basis for a lower bound.

As in the stationary case, for each entity e_i we partition T into sub-intervals; the length of a sub-interval starting at t_j is just $\sigma_i(x, t_j)$. Certainly, each entity e_i must either be queried, or have all but at most $x - 1$ of its $(x + \Delta)$-neighbours queried, in each of its sub-intervals. The difficulty is that one mobile entity can be the $(x + \Delta)$-neighbour of many entities over time so one query to that entity can partially satisfy the demands of many sub-intervals. (In the stationary case, one query can help satisfy the demands of at most $\Theta(x + \Delta)$ entities since this is the maximum number of stationary $(x + \Delta)$-neighbourhoods a stationary entity can be in.) However, if we restrict our attention to sub-intervals of an entity during which the entity's $(x + \Delta)$-separation remains approximately the same size, we can apply something similar to the stationary case argument. The challenge is to show that such sub-intervals, that are not simultaneously partially satisfied along with a large number of other sub-intervals, cover a substantial fraction of T for many entities.   □

Perception Versus Reality. For any query scheme, the true location of a moving entity e_i at time t, $z_i(t)$, may differ from its *perceived location*, $z_i(p_i(t))$, its location at the time of its most recent query. Let $N_i(x, t)$ be e_i plus the set of x entities whose perceived locations at time t are closest to the perceived location of e_i at time t. The *perceived x-separation* of e_i at time t, denoted $\tilde{\sigma}_i(x, t)$, is the separation between e_i and its perceived xth-nearest-neighbour at time t, i.e., $\tilde{\sigma}_i(x, t) = \max_{e_j \in N_i(x, t)} \| z_i(p_i(t)) - z_j(p_j(t)) \| - 2$.

Since a scheme only knows the perceived locations of the entities, it is important that each entity e_i be probed sufficiently often that its perceived x-

separation $\tilde{\sigma}_i(x,t)$ closely approximates its true x-separation $\sigma_i(x,t)$ at all times t. The following technical lemma asserts that once a close relationship between perception and reality has been established, it can be sustained by ensuring that the time between queries to an entity is bounded by some small fraction of its perceived x-separation. See [8, Lemma 8] for the proof.

Lemma 4. *Suppose that for some t_0 and for all entities e_i,*

(i) $\sigma_i(x, p_i(t_0))/2 \leq \tilde{\sigma}_i(x, p_i(t_0)) \leq 3\sigma_i(x, p_i(t_0))/2$, [perception is close to reality for e_i at time $p_i(t_0)$] and

(ii) for any $t \geq t_0$, $t - p_i(t) \leq \lambda_{d,x}\tilde{\sigma}_i(x, p_i(t))/12$ [all queries are done promptly based on perception].

Then for all entities e_i, $\sigma_i(x,t)/2 \leq \tilde{\sigma}_i(x,t) \leq 3\sigma_i(x,t)/2$, for all $t \geq p_i(t_0)$.

To obtain the preconditions of Lemma 4, we could assume that all entities are queried very quickly using low granularity for a short initialization phase. We show how to obtain these preconditions using granularity that is competitive with any scheme that guarantees uncertainty degree at most x from time t_0 onward (see [8, Lemma 9] for the proof). This establishes:

Lemma 5. *For any Δ, $0 \leq \Delta \leq x$, and any target time $t_0 \geq 0$, there exists an initialization scheme that guarantees*

(i) $\sigma_i(x + \Delta, t_0)/2 \leq \tilde{\sigma}_i(x + \Delta, t_0) \leq 3\sigma_i(x + \Delta, t_0)/2$, and
(ii) $t_0 - p_i(t_0) \leq \lambda_{d,x}\tilde{\sigma}_i(x + \Delta, p_i(t_0))/12$.

using minimum query granularity over the interval $[0, t_0]$ that is at most $\Theta(\frac{x}{1+\Delta})$ smaller than the minimum query granularity, over the interval $[0, (a+1)t_0]$, used by any other scheme that guarantees uncertainty degree at most x in the interval $[t_0, (a + 1)t_0]$, where $a = 64/(5\lambda_{d,x})$.

A Scheme to Maintain Low Degree for General Mobile Entities. A *bucket* is a set of entities and an associated time interval whose length (the bucket's *length*) is a power of two. The ith bucket B of length 2^b has time interval $T_B = [i2^b, (i + 1)2^b)$, for integers i and b. The time intervals of buckets of the same length partition $[0, \infty)$, and a bucket of length 2^b spans exactly 2^s *sub-buckets* of length 2^{b-s}.

Entities are assigned to exactly one bucket at any moment in time. Membership of entity e_j in a given bucket B implies a commitment to query e_j within the interval T_B. The *basic* version of the BucketScheme (see Algorithm 1) fulfills these commitments by scheduling a query to e_j at anytime within that time interval. That is, any version of Schedule(e_j, B) that allocates a query for e_j at some time within T_B satisfies the basic BucketScheme. After an entity e_j is queried, it is reassigned to a future bucket in a way that preserves (via Lemma 4) the following invariants: for all $t' \in T_B$, (i) $\sigma_j(x,t')/2 \leq \tilde{\sigma}_j(x,t') \leq 3\sigma_j(x,t')/2$; and (ii) $\sigma_j(x,t') = \Theta(2^b)$, so $\int_{T_B} \frac{dt}{\sigma_i(x,t)} = \Theta(1)$.

Algorithm 1. BucketScheme[x]

1: Assume perception-reality precondition properties hold at time t_0. ▷ See Lemma 5
2: **for all** entities e_j **do**   ▷ make initial query-time assignments
3:     Assign e_j to the first bucket B of length 2^b starting after time t_0, where $b = \lfloor \lg[(\lambda_{d,x}/24)\widetilde{\sigma}_j(x,t_0)] \rfloor$
4:     Schedule(e_j, B)   ▷ Assign e_j a query time in interval of bucket B
5: **repeat**
6:     Advance t to the next query time (say to entity e_j)
7:     Query e_j
8:     Assign e_j to the next bucket B of length 2^b starting after time $t + 2^b$, where $b = \lfloor \lg[(\lambda_{d,x}/24)\widetilde{\sigma}_j(x,t)] \rfloor$
9:     Schedule(e_j, B)   ▷ Assign e_j a query time in interval of bucket B

Theorem 1. *The basic BucketScheme[x] maintains uncertainty degree at most x indefinitely. Furthermore, over any time interval T in which the basic Bucket-Scheme[x] makes $3|\mathcal{E}|$ queries, $\phi_x(T) = \Omega(|\mathcal{E}|)$.*

Proof. It is straightforward to confirm that the assignment of entities to buckets (specified in line 7) ensures that the time between successive queries to any entity e_i satisfies precondition (ii) of Lemma 4. From the proof of Lemma 4 we see that this in turn implies that $t - p_i(t) \leq \frac{\lambda_{d,x}}{6}\sigma_i(x,t)$, for all entities e_i and all $t \geq t_0$. But $\frac{\lambda_{d,x}}{6}\sigma_i(x,t) \leq \frac{\lambda_{d,x}}{\lambda_{d,x}+2}\sigma_i(x,t)$, and so following the identical analysis used in the proof of Lemma 2, we conclude that uncertainty degree at most x is maintained indefinitely.

Since no entity has a query scheduled in overlapping buckets, it follows that if the basic BucketScheme[x] makes $3|\mathcal{E}|$ queries over T then, among these, it must make at least $|\mathcal{E}|$ queries to entities in buckets that are fully spanned by T. Since each entity in each fully spanned bucket contributes $\Theta(1)$ to $\phi_x(T)$, it follows that $\phi_x(T) = \Omega(|\mathcal{E}|)$.   □

A more fully specified implementation of the BucketScheme is not only competitive in terms of total queries over reasonably small intervals, but also competitive in terms of query granularity. The idea of this *refined* BucketScheme is to replace the simple scheduling policy Schedule of the basic BucketScheme with a recursive policy Schedule* that generates a refined reassignment of entities to buckets. Whenever a bucket B of length 2^b has been assigned two entities, these entities are immediately reassigned, one to each of the two sub-buckets of B of length 2^{b-1}. In this way, when all reassignments are finished, all of the entities are assigned to their own buckets. The entity associated with a bucket B has a *tentative next query time* at the midpoint of B. Tentative query times are updated of course when entities are reassigned (see Algorithm 2). At any point in time the next query is made to the entity with the earliest associated tentative next query time. Note that since distinct buckets have distinct midpoints, and no bucket has more than one associated entity, the current set of tentative next query times contains no duplicates. In fact, for any two tentative query times associated with entities in buckets B and B', it must be that either B and B' are disjoint, or the smaller bucket is a sub-bucket of one half of the larger bucket.

Recall from the invariant properties of bucket assignments in the basic BucketScheme that if e_i is assigned to bucket B, then $\sigma_i(x,t) = \Theta(|T_B|)$ (i.e. $1/\sigma_i(x,t) = \Theta(1/|T_B|)$), for $t \in T_B$. In the refined BucketScheme this property is generalized to: (i) if e_i is assigned to bucket B, then there is a subset of entities S_B, including e_i, such that $\sum_{e_j \in S_B} 1/\sigma_j(x,t) = \Theta(1/|T_B|)$, for $t \in T_B$, and (ii) if $T_B \cap T_{B'} \neq \emptyset$ then $S_B \cup S_{B'} = \emptyset$. It is straightforward to confirm that this property is preserved by the reassignment of entities in the bucket structure.

Since the gap between successive queries contains half of the smaller of the two buckets containing the two entities, it follows that every gap between queries has an associated integral of $\sum_{e_j \in \mathcal{E}} 1/\sigma_j(x,t)$ that is $\Theta(1)$. It follows from this that the stationary frequency demand is inversely proportional to the instantaneous granularity at the time of every query.

Algorithm 2. Schedule*(e_j, B) ▷ used by the refined BucketScheme

---

1: **if** bucket B already contains an entity e_i **then**   ▷ B contains at most one
2:     Unassign e_i from B
3:     Schedule*(e_i, B_{first}) ▷ B_{first} spans the first half-interval of B
4:     Schedule*(e_j, B_{second}) ▷ B_{second} spans the second half-interval of B
5: **else**
6:     Assign e_j to bucket B with query time at the midpoint of B.

We summarize with:

Lemma 6. *Over any time interval T in which the refined BucketScheme[x] makes $3|\mathcal{E}|$ queries, $\phi_x(T) = \Omega(|\mathcal{E}|)$. Furthermore, at any time the query granularity is inversely proportional to the stationary query demand.*

Combining Lemma 6 and Lemma 3, we reach our main result:

Theorem 2. *For any Δ, $0 \leq \Delta \leq x$, the refined BucketScheme[x + Δ] maintains uncertainty degree at most $x + \Delta$ and, over all sufficiently large time intervals T, is competitive, in terms of total queries over T or some small shift $\overrightarrow{T}$ of T, with any (even clairvoyant) query scheme that maintains uncertainty degree at most x over $T \cup \overrightarrow{T}$. The competitive factor is $O(\frac{x}{1+\Delta})$. Furthermore, at all times it uses query granularity that is inversely proportional to the stationary frequency demand.*

Remark 3. Observe that the competitive bound here is particularly strong: BucketScheme[x + Δ] maintains uncertainty degree at most $x + \Delta$ over *all* time and, for all time intervals T of sufficient length, is competitive with any (even clairvoyant) scheme that is designed to minimize queries on T, using arbitrarily high query frequency elsewhere.

It turns out that the competitive factor realized by the refined BucketScheme cannot be improved by more than a constant factor in general. This is demonstrated by a example in which two entity clusters, each with $(x + 1 + \Delta)/2$ entities, are situated in the plane, just above the x-axis on opposite sides of the y-axis, and move horizontally at maximum speed. A special subset (of size $1 + \Delta$) of each cluster, the *special entities*, move away from the opposite cluster, while all non-special entities move towards the opposite cluster. The initial cluster separation is chosen in such a way that (i) in the absence of queries in a specified time interval T the uncertainty regions will all intersect the origin at the end of T, but (ii) uniformly spaced queries to the special entities over T (and thereafter) will guarantee that no entity has uncertainty degree exceeding x in T (or thereafter). A non-clairvoyant scheme can be forced, by an adversarial scheduler, to query $\Theta(x)$ entities over T in order to ensure that sufficiently many of the special entities are queried to avoid uncertainty degree exceeding $x + \Delta$ over T.

4   Discussion

4.1   Other Measures of Congestion Potential

We have focused on collision potential but our results have immediate implications for other notions of congestion potential as well. Note that the maximum degree of vertices in the potential encroachment graph PE^U associated with an uncertainty configuration U, provides an upper bound on the maximum number of uncertainty regions in U that intersect in a common point (called the *uncertainty ply* of U in [3]), as well as the chromatic number of PE^U (called the *uncertainty thickness* of U in [3]). Thus our strategy for ensuring a fixed bound on the uncertainty degree of U also serves to maintain the same bound on uncertainty ply and thickness of U. It turns out that the query frequency lower bound of Lemma 3 holds, with only a small change in the constant, for query schemes that ensure maximum uncertainy ply at most x. So our refined BucketScheme, without change, is competitive among (even clairvoyant) query schemes for maintaining low uncertainty ply and thickness as well.

4.2   Generalizations of Our Model and Analysis

We describe below several modifications to our model and analyses that make our query optimization framework more broadly applicable.

Relaxing the Assumption on the Encroachment Threshold and Entity Disjointness. Without changing the units of distance and time, we can model a collection of unit-radius entities, any pair of which possibly intersect but whose centres always maintain distance at least some positive constant $\rho_0 < 2$, by simply scaling the constant $c_{d,x}$ by $\rho_0/2$ (and the constant $\lambda_{d,x}$ accordingly).

Similarly (and simultaneously), we can model a collection of unit-radius entities with encroachment threshold $\Xi > 2$ by (i) changing the *basic uncertainty radius* (the radius of the uncertainty region of an entity immediately after it has been queried) to $\Xi/2$ (thereby ensuring that entities with disjoint uncertainty regions do not encroach one another), and (ii) changing X_d to be the largest x such that $c_{d,x} \geq \Xi - 2$ (since for x exceeding this changed X_d there can be at most $x - 1$ entities that are within the encroachment threshold of any fixed entity). Note that this modification also allows us to relax the assumption that location queries are answered *exactly*: if location queries are answered to within some error ε then it suffices to set the encroachment threshold Ξ to $2(1 + \varepsilon)$.

Relaxing the Assumption of Uniform Entity Extent and Speed. We have assumed that all entities are d-dimensional balls with the same extent (radius). Completely relaxing this assumption would invalidate some of our packing arguments. Nevertheless, if entity extents differ by at most a constant factor, it is straightforward to modify the constants $\lambda_{d,x}$ and X_d, so that all of our results continue to hold. Similarly, the reader will not be surprised by the fact that our results are essentially unchanged if our assumption that all entities have the same (unit) bound on their maximum speed is relaxed to allow speed bounds that differ by at most a constant factor.

4.3   Motivating Applications

We return briefly to the motivating application mentioned in the introduction. Recall that by maintaining uncertainty degree at most x, we maintain (using optimal query frequency) for each robot (entity) e_i a certificate identifying the, at most $x-1$, other robots that could potentially collide with e_i (those warranting more careful local monitoring). This application becomes even more compelling if we take into consideration the more general notion of encroachment described in the preceding subsection.

An additional application, considered in [3], concerns entities that are mobile transmission sources, with associated broadcast ranges, where the goal is to minimize the number of broadcast channels so as to eliminate potential transmission interference. In this case, maintaining the uncertainty thickness to be at most x, using minimum query frequency, serves to maintain a fixed bound on the number of broadcast channels, an objective that seems to be at least as well motivated as optimizing the number of channels for a fixed query frequency (the objective in [3]). Our query scheme guarantees uncertainty degree $x + \Delta$ using a query frequency that is (up to a constant factor) optimally competitive with that required of any scheme to maintain uncertainty ply (which bounds from above the number of broadcast channels used to avoid potential broadcast interference) at most x. As we described in the previous subsection, our assumption of disjoint entities (i.e. broadcast ranges) is easily relaxed to permit intersections as long as the broadcast centres remain separated by at least some fixed positive distance.

References

1. Anily, S., Glass, C.A., Hassin, R.: The scheduling of maintenance service. Discret. Appl. Math. **82**, 27–42 (1998)
2. Basch, J., Guibas, L.J., Hershberger, J.: Data structures for mobile data. J. Algorithms **31**(1), 1–28 (1999)
3. Busto, D., Evans, W., Kirkpatrick, D.: Minimizing interference potential among moving entities. In: Proceedings of the ACM-SIAM Symposium on Discrete Algorithms (SODA), pp. 2400–2418 (2019)
4. de Berg, M., Roeloffzen, M., Speckmann, B.: Kinetic compressed quadtrees in the black-box model with applications to collision detection for low-density scenes. In: Epstein, L., Ferragina, P. (eds.) ESA 2012. LNCS, vol. 7501, pp. 383–394. Springer, Heidelberg (2012). https://doi.org/10.1007/978-3-642-33090-2_34
5. de Berg, M., Roeloffzen, M., Speckmann, B.: Kinetic convex hulls, Delaunay triangulations and connectivity structures in the black-box model. J. Comput. Geom. **3**(1), 222–249 (2012)
6. de Berg, M., Roeloffzen, M., Speckmann, B.: Kinetic 2-centers in the black-box model. In: Symposium on Computational Geometry, pp. 145–154 (2013)
7. Erlebach, T., Hoffmann, M.: Query-competitive algorithms for computing with uncertainty. Bull. Eur. Assoc. Theor. Comput. Sci. **2**(116) (2015)
8. Evans, W., Kirkpatrick, D.: Frequency-competitive query strategies to maintain low congestion potential among moving entities (2023). arXiv:2205.09243
9. Evans, W., Kirkpatrick, D., Löffler, M., Staals, F.: Minimizing co-location potential of moving entities. SIAM J. Comput. **45**(5), 1870–1893 (2016)
10. Fishburn, P.C., Lagarias, J.C.: Pinwheel scheduling: achievable densities. Algorithmica **34**(1), 14–38 (2002)
11. Guibas, L.J.: Kinetic data structures: a state of the art report. In: Proceedings of the Third Workshop on the Algorithmic Foundations of Robotics on Robotics: The Algorithmic Perspective, WAFR 1998, USA, pp. 191–209. A. K. Peters Ltd. (1998)
12. Guibas, L.J., Roeloffzen, M.: Modeling motion. In: Toth, C.D., O'Rourke, J., Goodman, J.E. (eds.) Handbook of Discrete and Computational Geometry, chap. 53, pp. 1401–1420. CRC Press (2017)
13. Holte, R., Mok, A., Rosier, L., Tulchinsky, I., Varvel, D.: The pinwheel: a real-time scheduling problem. In: Proceedings of the Twenty-Second Annual Hawaii International Conference on System Sciences. Volume II: Software Track, pp. 693–702 (1989)
14. Kahan, S.: A model for data in motion. In: Twenty-third Annual ACM Symposium on Theory of Computing, STOC 1991, pp. 265–277 (1991)
15. Kahan, S.: Real-time processing of moving data. Ph.D. thesis, University of Washington (1991)

Approximating Maximum Edge 2-Coloring by Normalizing Graphs

Tobias Mömke[1] , Alexandru Popa[2] , Aida Roshany-Tabrizi[1] ,
Michael Ruderer[1]([✉]) , and Roland Vincze[1]

[1] University of Augsburg, Augsburg, Germany
moemke@informatik.uni-augsburg.de,
{aida.roshany.tabrizi,michael.ruderer}@uni-a.de
[2] University of Bucharest, Bucharest, Romania
alexandru.popa@fmi.unibuc.ro

Abstract. In a simple, undirected graph G, an edge 2-coloring is a coloring of the edges such that no vertex is incident to edges with more than 2 distinct colors. The problem maximum edge 2-coloring (ME2C) is to find an edge 2-coloring in a graph G with the goal to *maximize* the number of colors. For a relevant graph class, ME2C models anti-Ramsey numbers and it was considered in network applications. For the problem a 2-approximation algorithm is known, and if the input graph has a perfect matching, the same algorithm has been shown to have a performance guarantee of $5/3 \approx 1.667$. It is known that ME2C is APX-hard and that it is UG-hard to obtain an approximation ratio better than 1.5. We show that if the input graph has a perfect matching, there is a polynomial time 1.625-approximation and if the graph is claw-free or if the maximum degree of the input graph is at most three (i.e., the graph is subcubic), there is a polynomial time 1.5-approximation algorithm for ME2C.

Keywords: Approximation Algorithms · Edge 2-Coloring · Matchings

1 Introduction

In a simple, undirected graph G, an edge 2-coloring is a coloring of the edges such that no vertex is incident to edges with more than 2 distinct colors. The problem maximum edge 2-coloring (ME2C) is to find an edge 2-coloring in G that uses a *maximal* number of colors. Formally, we aim to compute a coloring $\chi \colon E(G) \to \mathbb{N}$ that maximizes $|\{c \in \mathbb{N} \mid \chi(e) = c \text{ for an } e \in E(G)\}|$, such that for each vertex $v \in V(G)$, $|\{c \in \mathbb{N} \mid \chi(e) = c \text{ for an } e \text{ incident to } v\}| \leq 2$ holds.

Partially supported by DFG Grant 439522729 (Heisenberg-Grant) and DFG Grant 439637648 (Sachbeihilfe). Partially supported by a grant of the Ministry of Research, Innovation and Digitization, CNCS - UEFISCDI, project number PN-III-P1-1.1-TE-2021-0253, within PNCDI III.

© The Author(s), under exclusive license to Springer Nature Switzerland AG 2023
J. Byrka and A. Wiese (Eds.): WAOA 2023, LNCS 14297, pp. 29–44, 2023.
https://doi.org/10.1007/978-3-031-49815-2_3

Maximum edge 2-coloring is a particular case of anti-Ramsey numbers and has been considered in combinatorics. For given graphs G and H, the *anti-Ramsey number* $ar(G, H)$ is defined to be the maximum number of colors in an edge-coloring that does not produce a rainbow copy of H in G, i.e., a copy of H in G with every edge of H having a unique color. Classically, the graph G is a large complete graph and H is from a particular graph class. If H is a star with three leaves and G is an arbitrary graph, the anti-Ramsey number is precisely the maximum number of colors in an edge 2-coloring.

The study of anti-Ramsey numbers was initiated by Erdős, Simonovits and Sós in 1975 [8]. Since then, there have been a large number of results on the topic including the cases where $G = K_n$ and H is a cycle [3,8,19], tree [14,15], clique [4,8,12], matching [7,13,23] or a member of some other class of graphs [2, 8].

The main application of ME2C comes from wireless mesh networks. Raniwala et al. [21,22] proposed a wireless architecture in which each computer uses two network interface cards (NICs) compared to classical architectures that use only one NIC. In this model, each computer can communicate with the other computers in the network using two channels. Raniwala et al. [21,22] showed that using such an architecture can increase the throughput by a factor of 6. In order to minimize the interference, it is desirable to maximize the number of distinct channels used in the network. In ME2C computers correspond to nodes in the graph, while colors correspond to channels.

1.1   Previous Work

The problem of finding a maximum edge 2-coloring of a given graph has been first studied by Feng et al. [9–11]. They provided a 2-approximation algorithm for ME2C and show that ME2C is solvable in polynomial time for trees and complete graphs, but they left the complexity for general graphs as an open problem. The authors also studied a generalization of ME2C, the maximum edge q-coloring, where each vertex is allowed to be incident to at most q edges with distinct colors. For the maximum edge q-coloring they showed a $(1 + \frac{4q-2}{3q^2-5q+2})$-approximation for $q > 2$.

Later, Adamaszek and Popa [1] showed that the problem is APX-hard and proved that the algorithm above provides a 5/3-approximation for graphs which have a perfect matching. The APX-hardness is achieved via a reduction from the Maximum Independent Set problem and states that maximum edge 2-coloring problem is UG-hard to approximate within a factor better than $1.5 - \epsilon$, for some $\epsilon > 0$. Chandran et al. [6] showed that the matching-based algorithm of [11] yields a $(1 + \frac{2}{\delta})$-approximation for graphs with minimum degree δ and a perfect matching. If additionally the graph is triangle-free, the ratio improves to $(1 + \frac{1}{\delta-1})$. Recently, Chandran et al. [5] improved the analysis of the achieved approximation ratio for triangle-free graphs with perfect matching to 8/5. They also showed that the algorithm cannot achieve a factor better than 58/37 on triangle free graphs that have a perfect matching.

Larjomaa and Popa [16] introduced and studied the min-max edge 2-coloring problem a variant of the ME2C problem, where the goal is to find an edge 2-coloring that minimizes the largest color class. Mincu and Popa [18] introduced several heuristic algorithms for the min-max edge 2-coloring problem.

1.2   Our Results

Our core algorithm, Algorithm 1, is the 2-approximation algorithm for general graphs of Feng et al. [11]. The algorithm simply finds a maximum matching, colors each edge of the matching with a distinct color, removes the edges of the matching and finally, colors each connected component of the remaining graph with a distinct color.

Directly applying the algorithm, however, cannot provide an approximation ratio better than 2 in the general case [11] and not better than $5/3 \approx 1.667$ for graphs with perfect matchings [1]. To overcome this difficulty, we introduce a preprocessing phase which considerably simplifies the instance. The simplifications both improve the quality of the solution provided by the algorithm *and* lead to an improved upper bound on the size of an optimal solution. A graph is called *normalized* if no more preprocessing steps can be performed on it.

We first show that Algorithm 1 is a 1.5-approximation algorithm if *after* the normalization, the graph contains a perfect matching (Lemma 14). We can ensure this property if we normalize a subcubic graph[1], even if before applying the normalization it did not have a perfect matching.

Theorem 1. *ME2C   in   subcubic   graphs   has   a   polynomial-time 1.5-approximation algorithm.*

It has been shown that claw-free graphs contain a perfect matching [17,24]. Some preprocessing steps might introduce claws which worsen the quality of the solution provided by Algorithm 1, therefore we do not immediately obtain a 1.5-approximation. However, we develop a bookkeeping technique to counteract this effect.

Theorem 2. *There is a polynomial-time 1.5-approximation for claw-free graphs.*

In the more general case of graphs with perfect matchings, the effect of introduced unmatched vertices is more severe. We use a sophisticated accounting technique to quantify the effects of the appearing unmatched vertices on the quality of both the optimal solution and the solution given by Algorithm 1. As a result we obtain a weaker but improved approximation algorithm for graphs containing a perfect matching:

Theorem 3. *There is a 1.625-approximation for graphs that contain a perfect matching.*

---

[1] Recall that a graph is subcubic if no vertex has a degree larger than three.

Let us now elaborate on the key ideas behind our results. After the pre-processing phase we obtain a normalized graph via a series of *modifications*. Intuitively, the modifications achieve the following: 1) *Avoid leaves with equal neighborhoods;* 2) *Avoid degree-2 vertices;* 3) *Avoid a specific class of triangular cacti.*

A triangular cactus is a connected graph such that two cycles have at most one vertex in common and each edge is contained in a 3-cycle. For our purposes, we additionally require that no vertex of the cactus is incident to more than one edge not in the cactus.

While the three modifications are relatively simple, proving that they are approximation-preserving is non-trivial. If none of these modifications can be applied (anymore), we call the graph normalized. Our key insight is that the number of colors in an optimal solution of a normalized graph can be bounded from above, this is stated as Lemma 1 and shown in Sect. 3.

Lemma 1. *Let G be a normalized connected graph with $n \geq 3$ vertices and ℓ leaves. Then there is no feasible coloring χ with more than $3n/4 - \ell/4$ colors.*

We note that without normalization, an optimal solution can have n colors (e.g., if G is an n-cycle). In order to prove Lemma 1, we use the notion of *character graphs* introduced by Feng et al. [11]. A character graph of an edge 2-coloring is a graph that contains exactly one edge from each color class. We first show that for a normalized graph we can ensure the existence of a *nice* character graph, which is a character graph with several useful properties. These properties allow for a counting argument with respect to the number of components in the character graph, which allows us to prove the bound in Lemma 1.

For general graphs, the best result is still the known 2-approximation. There is a family of bipartite triangle free 2-connected graphs with minimum degree 3 which certifies this lower bound for our algorithm.

The rest of the paper is organized as follows. In Sect. 2 we describe the three modifications performed on the input graph before applying the algorithm. Then, in Sect. 3, we prove the upper bound on the optimal solution on normalized instances. In Sect. 4 we combine the results from Sects. 2 and 3 to prove Theorems 1 and 2, and finally, in Sect. 5 we prove Theorem 3.

Due to space constraints most of the proofs are moved into the journal version.

2   The Algorithm

Let G be a graph and χ a feasible 2-coloring of the edges. Recall that $\chi(e)$ marks the color of the edge e in χ. With a slight abuse of notation, let us denote the set of all colors of the edges of G by $\chi(G)$ and the colors incident to a vertex v by $\chi(v) := \{c \in \mathbb{N} \mid \exists u \in V(G) : \chi(uv) = c\}$. If a vertex v is incident to an edge colored c, we say that vertex v *sees* c. We also denote the number of colors in a coloring χ by $|\chi|$.

For a color c, $E(c)$ denotes the set of edges with color c, that is, $E(c) := \{e \in E(G) \mid \chi(e) = c\}$. We refer to $E(c)$ as the *color class* of c. Furthermore, we define by $V(c) := \{v \in V(G) \mid c \in \chi(v)\}$ the *color class* of c, i.e. the vertices that see the color c. Finally, $G(c) := (V(c), E(c))$ is the subgraph of G whose edges have color c. We call a cycle on 3 vertices a *3-cycle* or *triangle*. The term *pendant vertex* or *leaf* is used for degree-1 vertices, while the term *pendant edge* marks the edge incident to a pendant vertex.

Algorithm 1. The basic algorithm.

    Input: A simple undirected graph $G = (V, E)$.
    Output: An edge 2-coloring χ on the edges of G.
1: Calculate a maximum cardinality matching M in G.
2: Assign a distinct color in χ for every edge of M.
3: Assign a distinct color in χ for every nontrivial connected component of $E \setminus E(M)$.

While Algorithm 1 is well studied (cf. [1,5,11]), we apply some preprocessing steps to each problem instance G, before applying Algorithm 1 to the resulting graph G'. This preprocessing gives the graph G' more structure, which will help us to prove better approximation guarantees.

These preprocessing steps consist of different *modifications*, which will be defined throughout the paper. We note that modifications can increase the size of the maximum matching M, and therefore are not only for the analysis, but they change the instance in order to obtain stronger results.

Intuitively, a valid modification is a modification such that the number of colors in an optimal solution does not change and we can transform a solution for the modified instance to a solution for the original instance. Formally, we define the following equivalence relation. For a graph G let $\mathsf{opt}(G)$ denote the number of colors in an optimal edge 2-coloring of G.

Definition 1. *Two graphs $G = (V, E)$ and $G' = (V', E')$ form an* equivalent pair *with respect to edge 2-coloring, denoted by $\langle G, G' \rangle$, if*

1. *an optimal edge 2-coloring of G' uses the same number of colors as an optimal edge 2-coloring of G, i.e., $\mathsf{opt}(G) = \mathsf{opt}(G')$.*
2. *For every edge 2-coloring χ' of G' one can in polynomial time compute an edge 2-coloring χ for G that uses the same number of colors as χ', i.e., $|\chi| = |\chi'|$.*

To show $\langle G, G' \rangle$, it is sufficient to show $\mathsf{opt}(G) \leq \mathsf{opt}(G')$ for Condition 1 and $|\chi| \geq |\chi'|$ for Condition 2: For Condition 1, we use that there is a coloring χ' for G' and a coloring χ for G such that $\mathsf{opt}(G) \leq \mathsf{opt}(G') = |\chi'| \leq |\chi| \leq \mathsf{opt}(G)$ and thus all inequalities have to be satisfied with equality. For Condition 2, we note that it is always possible to reduce the number of colors.

Definition 2. *A* valid modification *is a sequence of vertex/edge alterations (additions or deletions), that result in a graph G' such that $\langle G, G' \rangle$ is an equivalent pair.*

All modifications that will be introduced in the following are indeed *valid* modifications. For proofs of their validity, we refer to the full version of this article.

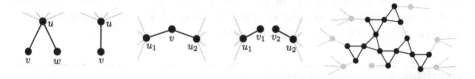

Fig. 1. Modifications 1 and 2, and a simple cactus (Modification 3) as a subgraph of G.

Lemma 2. *Let G be a graph and let G' be the graph obtained from G via a valid modification from Definition 2. Given a polynomial time α-approximation algorithm for G' we can obtain a polynomial time α-approximation algorithm for G.*

Modification 1: Avoid Pendant Vertices with Equal Neighborhoods. The first modification is to remove a leaf w from G, if there is another leaf v, such that both are incident to the same vertex u (see Fig. 1). Formally, we require that the following two conditions are simultaneously satisfied: (i) the degree of v in G is one; and (ii) there is a vertex $w \neq v$ of degree one and a vertex u of degree at least three adjacent to both v and w.

We note that we require the degree constraint on u to obtain a cleaner proof. If the degree equals 2, we will see that the following Modification 2 applies.

Modification 2: Avoid Degree-2 Vertices. Given a vertex v of degree 2, we break it into two vertices v_1 and v_2 with degree 1 each. More precisely, let u_1 and u_2 be the two vertices adjacent to v. We replace the edges u_1v and u_2v by the edges u_1v_1 and u_2v_2, replacing v by two new vertices v_1 and v_2.

Modification 3: Remove Triangular Cacti. Recall that a triangular cactus is a connected graph such that two cycles have at most one vertex in common and each edge is contained in a 3-cycle.[2] In other words it is a 'tree' of triangles where triangle pairs are joined by a single common vertex. Let C be a subgraph of G. For the modification we require that C is a triangular cactus such that for each vertex $v \in V(C)$, the degree of v in G is 3 or 4, and v is incident to at most one edge from $E(G) \setminus E(C)$. In the following, we call such a cactus a *simple cactus*, and we call an edge $e \in E(G) \setminus E(C)$ a *needle* of the cactus C. Note that we do not require C to be an *induced* subgraph of G. Indeed, two triangles can share a needle, as illustrated in Fig. 1.

---

[2] Note that our definition of a cactus is stricter than usual: we do not allow cut-edges.

Modification 3 replaces each 3-cycle by a single edge as follows:

Let $T_1, T_2, \ldots, T_\ell$ be the triangles that comprise the simple cactus C. For each T_i with vertices $\{u_i, v_i, w_i\}$, it removes the edges $u_i v_i, v_i w_i, w_i u_i$ and introduces two new vertices x_i and y_i with an edge $x_i y_i$. Finally it discards all isolated vertices (for an illustration, see the the full version of the article). We now show that the graph obtained this way is equivalent to the original graph.

Normalized Graph. Some of our claims rely on the problem instance G being a graph, such that none of Modifications 1–3 can be applied on G anymore. We define such a graph G to be *normalized.* Below we show that one can efficiently compute a normalized graph G' for every problem instance G, and that we can use the notion of normalized graphs to design approximation algorithms for the maximum edge 2-coloring problem.

Lemma 3. *Given a graph G, in polynomial time we can compute a normalized graph G' such that $\langle G, G' \rangle$ is an equivalent pair.*

Let C be a component of G such that C is an isolated vertex or C has two vertices connected by an edge. Then we say that C is a *trivial component.* Otherwise, the component is called *non-trivial.*

Lemma 4. *Suppose there is an α-approximation algorithm A for each non-trivial component of a normalized graph G' such that $\langle G, G' \rangle$ is an equivalent pair. Then there is an α-approximation algorithm for G.*

Due to Lemmas 3 and 4, from now on we can assume that the problem instance is a normalized connected graph G with more than two vertices. We now have a closer look at leaves.

Lemma 5. *Let G be a graph where no two leaves share a neighbor and let χ be an edge 2-coloring of G. Given G and χ, we can efficiently compute an edge 2-coloring $\hat{\chi}$ of G that uses at least as many colors as χ and assigns each pendant edge of G a unique color. In particular, in a normalized graph there is an optimal edge-2-coloring that assigns a unique color to each pendant edge.*

3   An Upper Bound on the Optimal Solution

To show Lemma 1, we analyze the character graph of the given instance.

3.1   Preparing the Character Graph

Intuitively, a character graph is an edge-induced subgraph with exactly one representative edge for each color.

Definition 3 (Character Graph). *Given an optimal solution χ for a graph G, a character graph of (G, χ) is a subgraph H with vertex set $V(G)$ and coloring $\chi_{|E(H)}$ such that (i) for each $e, f \in E(H)$ with $e \neq f$, $\chi(e) \neq \chi(f)$ and (ii) for each edge $e \in E(G)$ there is an edge $f \in E(H)$ with $\chi(e) = \chi(f)$.*

For ease of notation, in the following we write χ instead of $\chi_{|E(H)}$. Observe that in a character graph H, no vertex can have a degree larger than 2 since otherwise there would be a vertex with three incident colors. Thus H is a collection of isolated vertices, paths and cycles. We call a vertex in H a *free vertex* if its degree is zero, an *end vertex* if its degree is one and an *inner vertex* if its degree is two. We frequently use the following known simple but powerful lemma.

Lemma 6 (Feng et al. [11]). *Let χ be a feasible 2-edge coloring of a graph G and let $u \neq v$ be two vertices in $V(G)$. If $|\chi(u) \cup \chi(v)| \geq 4$, u and v are not adjacent in G. In particular, if H is a character graph of (G, χ) and $u \neq v$ are two inner vertices that are not neighbors in H then $uv \notin E(G)$.*

Based on Lemma 6, we can avoid cycles within a character graph.

Lemma 7 *Let G be a normalized graph, and χ a coloring of G. Then there is a character graph H of (G, χ) such that H is cycle-free.*

To further structure the character graph, we introduce a reachability measure.

Definition 4. *Let v be a vertex of a character graph H of (G, χ). The scope of v (scope(v)) is the set of vertices defined inductively as follows within G:*

(i) $v \in$ scope(v).
(ii) If $u \in$ scope(v) and there is an edge $e = uu' \in E(H)$ for an inner vertex u', then $V(\chi(e)) \subseteq$ scope(v) (i.e., we include the color class of $\chi(e)$).

We may choose a total ordering of the vertices and in (ii), we always choose the smallest vertex that satisfies the properties. Let $\kappa \geq 0$ be an integer which is at most the number of color classes added and let c_i be the color of the i-th color class added, for $1 \leq i \leq \kappa$. The scope graph of v and κ for a given ordering is the graph (scope(v), F), where $F := \bigcup_{i=1}^{\kappa} E(c_i)$. We skip the ordering and say that subgraph of G is a scope graph of v and κ if there exists an ordering for which it is a scope graph of v and κ.

Note that the (total) scope of a vertex does not depend on the chosen ordering. The scope of a vertex captures a natural sequence of dependencies between edge colors. In the following lemma, we show how to avoid free vertices in the scope of vertices, a property which is important in the proof of Lemma 1.

Lemma 8. *Let v be an inner vertex such that $vv' \in E(H)$ for an inner vertex v'. Each character graph H of (G, χ) can be transformed into a character graph H' of (G, χ) such that there is no free vertex in scope(v).*

We further extend the notion of scope to a set of vertices, which is not merely the union of scopes.

Definition 5. *Let S be a set of vertices of a character graph H of (G, χ). If $S = \{v\}$, we define the scope as scope(v) and define the scope graph accordingly. For $|S| > 1$, the scope of S (scope(S)) is the set of vertices defined inductively as follows within G:*

(i) $\text{scope}(v) \subseteq \text{scope}(S)$ *for each* $v \in S$.

(ii) *Let* $uu' \in E(H)$ *be the only edge of a path in* H. *Let* κ, κ' *be numbers and* $U, U' \subset S$ *sets with* $U \cap U' = \emptyset$. *If* u *is in a scope graph of* U *for* κ *and* u' *is in a scope graph of* U' *for* κ' *such that the colors of the two graphs are disjoint, then* $V(\chi(uu')) \subseteq \text{scope}(U \cup U')$.

(iii) *If* $u \in \text{scope}(U)$ *for* $U \subseteq S$ *and there is an edge* $e = uu' \in E(H)$ *for an inner vertex* u', *then* $V(\chi(e)) \subseteq \text{scope}(U)$.

If T *is a scope graph for* U *and* κ *and the color class of* c *is added, then* $(V(T) \cup V(\chi(c)), E(T) \cup E(\chi(c)))$ *is a scope graph for* U *and* $\kappa + 1$. *If additionally* T' *is a scope graph of* U' *and (ii) applies with respect to the two scope graphs,* $((V(T) \cup V(T') \cup V(\chi(uu')), E(T) \cup E(T') \cup E(\chi(uu')))$ *is a scope graph for* $U \cup U'$ *and* $\kappa + \kappa' + 1$.

We now strengthen Lemma 8.

Lemma 9. *Let* S *be a set of vertices such that each* $v \in S$ *is an inner vertex such that* $vv' \in E(H)$ *for an inner vertex* v'. *Each character graph* H *of* (G, χ) *can be transformed into a character graph* H' *of* (G, χ) *such that there is no free vertex in* $\text{scope}(S)$.

We say that a character graph is *nice* if it is (i) cycle-free and (ii) there is no free vertex in $\text{scope}(S)$ for an arbitrary set $S \subseteq V(G)$ such that $vv' \in E(H)$ for $v \in S$ and an inner vertex v'. The following lemma gives some guarantees for the existence of such a character subgraph.

Lemma 10. *Each normalized connected graph* G *with optimal coloring* χ *has a nice character graph* H.

The lemma follows directly from Lemma 7 and Lemma 9, noting that in the proof of Lemma 9 we do not introduce new cycles. Furthermore, given a normalized graph G and the coloring provided by Lemma 5, we may assume that H is a nice character graph where all pendant vertices of G are endpoints of paths in H.

3.2   The Proof of Lemma 1

With the preparation of Sect. 3.1, in this section we prove our main lemma.

Lemma 1. *Let* G *be a normalized connected graph with* $n \geq 3$ *vertices and* ℓ *leaves. Then there is no feasible coloring* χ *with more than* $3n/4 - \ell/4$ *colors.*

Let G be a normalized problem instance and H a nice character graph of (G, χ). Let F be the set of free vertices, T the set of end vertices and I the set of inner vertices of H. We define $n := |V(H)|$, $i := |I|$, $f := |F|$ and $t := |T|$, and clearly $n = i + t + f$. We show that there is a mapping ι from the set I of inner vertices to $T \cup F$ with the property that for each vertex v from $T \cup F$ there is at most one inner vertex mapped to v if $v \in T$ and at most two inner vertices are mapped to v if $v \in F$. Intuitively, we can see ι as an injective mapping, where each vertex in F is split into two vertices. The reason is that we can see a free vertex as a path of length zero and we count two end vertices for each path. Furthermore, we maintain that ι never maps an inner vertex to a pendant vertex of G. We first show that the mapping implies Lemma 1.

Lemma 11. *Suppose ι exists. Then no feasible coloring has more than $3n/4 - \ell/4$ colors.*

Proof. The mapping ι implies $2f + t - \ell \geq i = n - t - f$ and thus $3f + 2t \geq n + \ell$ since ℓ out of t end vertices cannot be used for the assignment. Each free vertex and each path is a component of the character graph H. Therefore the number of components is $f + t/2$, which is minimized if $f = 0$ and $t = (n + \ell)/2$. Thus there are at least $\frac{n+\ell}{4}$ components in H and the number of colors is at most $n - n/4 - \ell/4 = 3n/4 - \ell/4$, completing the proof.

We now construct the mapping ι. We associate the vertices with distinct natural numbers $\{1, 2, \ldots, n\}$ and define ι iteratively. An end vertex is *saturated* if there is an inner vertex mapped to it and a free vertex is saturated if there are *two* inner vertices mapped to it. While there are unassigned inner vertices, we continue the following process. Let $v \in I$ be the unassigned vertex with the smallest index. We define the set $U_v := \{u \in F \mid uv \in E(G) \setminus E(H) \text{ and } |\iota^{-1}(u)| \leq 1\} \cup \{u \in T \mid uv \in E(G) \setminus E(H) \text{ and } \iota^{-1}(u) = \emptyset\}$, that is, the set of unsaturated free- and end vertices adjacent to v via an edge *outside* of H. We remark that v and u are allowed to be in the same path of H, as long as they are not adjacent in H. If $U_v \neq \emptyset$, we set $\iota(v) := \min_{u \in U_v} u$. Clearly, if $U_v \neq \emptyset$, we find a valid mapping for v and can continue. If $U_v = \emptyset$, we add v to a set Q of postponed vertices and continue with the next vertex.

To finish the construction, we have to map the postponed vertices. Recall that if v is a vertex in Q, then $U_v = \emptyset$ holds. We then find a vertex u to map v to by growing a *plain* cactus: a plain cactus is a triangular cactus without needles that is a degree-4 bounded subgraph C of the problem instance G where each vertex of C that is connected to $G \setminus C$ can have any number of adjacent vertices in $V(G) \setminus V(C)$, as opposed to *one* in case of a *simple* cactus. In particular, we will be growing a plain cactus which is not a simple cactus. As a simple cactus, a plain cactus can have vertices adjacent in G. We call the edge between these vertices a *cactus chord*.

To gain some intuition, we first argue how to grow an initial triangle of the cactus (see also Fig. 2). There are no degree-two vertices in G, therefore v has a neighbor u' in G that is not a neighbor of v in H. Due to Lemma 6, u' is not an inner vertex. We note that u' cannot be a free vertex either: if u' was a free vertex and we could not map v to u', then u' would be saturated and v would be the third inner vertex adjacent to u'. By Lemma 9, however, that would mean u' seeing 3 colors, as each of the three inner vertices would be connected to u' with different colors, contradicting the feasibility of χ. Hence u' is an end vertex.

Since we cannot map v to u', there must be another vertex $\hat{v}$ already mapped to u'. Let P denote the path of u' and let u'' be the vertex adjacent to u' in

P. Observe that $u'' \notin \{v, \hat{v}\}$ because otherwise the edge $u'v$ or $u'\hat{v}$ would be contained in $E(H)$, but both have to be in $E(G) \backslash E(H)$ in order to be considered to map to v or $\hat{v}$, respectively. Since v and $\hat{v}$ are inner vertices, the colors of $u'v$ and $u'\hat{v}$ have to be from $\chi(v)$ and $\chi(\hat{v})$, respectively.

By Lemma 6, $\chi(u'v) \neq \chi(u'u'')$ and $\chi(u'\hat{v}) \neq \chi(u'u'')$. Therefore $c := \chi(u'v) = \chi(u'\hat{v})$ and, again by Lemma 6, v has an incident edge e and $\hat{v}$ has an incident edge $\hat{e}$ in H with $\chi(e) = \chi(\hat{e}) = c$, which implies $e = \hat{e}$. Therefore, v and $\hat{v}$ are neighbors in the same path P' of H; note that P' is not necessarily distinct from P.

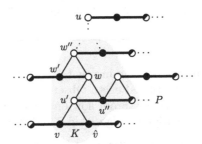

Fig. 2. End vertices are marked by hollow circles, inner vertices are marked by filled circles. A half-filled vertex can be either an inner- or an end vertex.

We observe that if there are only the edges from H and the edges from the triangle formed by $v, \hat{v}, u'$ incident to these three vertices, we have a simple cactus, i.e., a cactus of the form removed by Modification 3. Therefore there is another vertex adjacent to v, $\hat{v}$, or u'. We now argue that no matter which vertex has another adjacent vertex, we can either extend the plain triangular cactus or find the aimed-for vertex u.

More precisely, starting from $V(K) := \{v\}$ we grow a set of vertices $V(K)$; this process is shown in detail as Algorithm 2. Formally, we also grow an edge set $E(K)$ such that K is the aimed-for cactus. For a cactus K let $N(K) := \{\tilde{v} \in V(G) \backslash V(K) \mid$ there is a vertex $\tilde{u} \in V(K)$ with $\tilde{u}\tilde{v} \in E(G)\}$.

We will show that the cactus constructed by Algorithm 2 satisfies the following invariants.

1. The graph K is a plain cactus.
2. Each inner vertex of K except for v is mapped to an end vertex within $V(K)$.
3. For each end vertex of K, there is an inner vertex in $V(K)$ mapped to it.
4. All triangles of K are monochromatic.
5. For each color c in K, there is a color-c edge e in K that is also in $E(H)$.
6. $V(K) \subseteq \text{scope}(Q)$.
7. $V(K)$ does not include free vertices.
8. Each degree-2 vertex of K is incident to an edge from $E(H) \backslash E(K)$.

Algorithm 2. Mapping a vertex $v \in Q$.

1: Let $V(K) := \{v\}$ and $E(K) := \emptyset$;
2: **while** $\iota(v)$ is not yet determined **do**
3:     **if** $N(K)$ contains a not saturated free vertex or end vertex u **then**
4:         $\iota(v) := u$;
5:     **else**
6:         Find $w \in V(K)$ and $w', w'' \in V(G) \backslash V(K)$ with $w' \neq w''$, $ww' \in E(H), ww'' \in E(G)$;
7:         $V(K) := V(K) \cup \{w', w''\}$ and $E(K) := E(K) \cup \{ww', w'w'', w''w\}$;
8:         **if** $w' \in V(K')$ for a previously considered cactus K' **then**
9:             $V(K) := V(K) \cup V(K') \cup \{w''\}$ and $E(K) := E(K) \cup E(K') \cup \{ww', ww'', w'w''\}$;

We first show the invariants assuming that $V(K) \cup N(K)$ does not contain vertices from previously constructed cacti. Observe that the initial cactus with $V(K) = \{v\}$ satisfies all invariants. From now on we assume that K is a cactus constructed during the execution of Algorithm 2 and K satisfies all invariants.

Lemma 12. *In Algorithm 2, the vertices w, w', w'' exist, $\iota(w') = w''$, $\chi(ww') = \chi(ww'') = \chi(w'w'')$, and there is no other vertex $\bar{w} \in V \setminus V(K)$ adjacent to w.*

With Lemma 12 we know that adding the triangle $\{w, w', w''\}$ does not violate the conditions of plain cacti, i.e., adding it still satisfies Invariant 1. Since $\iota(w') = w''$, also Invariants 2 and 3 are satisfied. Invariant 4 follows directly from Lemma 12. Since $ww' \in E(H)$ by definition (within Algorithm 2), Invariant 5 follows. Invariant 6 follows by noting that $w \in \text{scope}(v)$, the edge ww' satisfies the conditions of the lemma, and thus the color class c is added to the scope. Invariant 7 is a direct consequence of Invariant 6. Invariant 8 requires additional arguments. Finally, we argue that all invariants are also preserved when merging two cacti. For a cactus K', let $v(K')$ be the vertex from Q mapped using cactus K'.

Lemma 13. *Let w', w'' be the vertices from Algorithm 2 and let $V(K')$ be the vertex set of a cactus constructed in a previous application of Algorithm 2. Then $w'' \notin V(K')$. Furthermore, if $w' \in V(K')$, $v(K')$ is mapped to w''.*

Then composing a new cactus from K, K', and the triangle formed by $\{w, w', w''\}$ satisfies all conditions: Since by induction they are satisfied by K and K', we only have to check the new triangle $\{w, w', w''\}$. We obtain a plain cactus since the degrees of w and w' are four and the degree of w'' is two. The inner vertex $v(K')$ is mapped to w'' which implies Invariant 2 and 3. The triangle $\{w, w', w''\}$ satisfies Invariant 4 by Lemma 12.

Since $ww' \in E(H)$, Invariant 5 is satisfied, both w and w' are in the scope of Q and K, K' provide disjoint scope graphs, the conditions of Definition 5 and therefore Invariant 6 are satisfied. Invariant 7 follows from Invariant 6 and that Q satisfies the conditions of Lemma 9. Finally, Invariant 8 follows since w'' is an end vertex and its incident edge from H is not in the constructed cactus.

4 Subcubic Graphs and Claw-Free Graphs

The upper bound shown in Sect. 3 directly gives the following result.

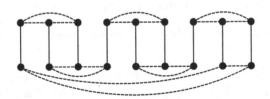

Fig. 3. Worst-case instance containing a perfect matching from [1]. Without applying modifications, Algorithm 1 may chose a perfect matching such that removing the matching leaves a connected graph – resulting in a $5/3 \approx 1.667$ approximation. An improved approximation ratio of 1.625 is possible due to Modification 3. In particular, Modification 3 replaces all simple cacti by independent edges, which results in a modified graph G' consisting only of independent edges, which means that Algorithm 1 actually computes an optimal solution.

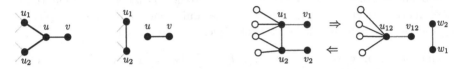

Fig. 4. Modification 4: Bridge removal, when $\deg(u) = 3$ and $\deg(v) = 1$, and Modification 5, with c_1, c_2 and c_3 marked with red, blue and black edges, respectively. (Color figure online)

Lemma 14. *There is a polynomial-time 1.5-approximation for normalized graphs that contain a perfect matching.*

In particular, our algorithm solves the tight worst-case instance depicted in Fig. 3 for the algorithm of Adamaszek and Popa [1] optimally. In order to handle subcubic graphs, we need to introduce additional modifications. The validity of these modifications follows from Lemma 5, which says that we can always assume that pendant edges are colored with a unique color.

Modification 4: Bridge Removal The next modification is only applied to subcubic graphs, after none of Modifications 1–3 can be applied anymore. Note that between two consecutive applications of Modification 4, we may have to update the graph by applying Modifications 1–3.

Suppose we have such a normalized, subcubic graph G with a cut of size one consisting of an edge uv, that is, the removal of uv disconnects G (see Fig. 4). Due to the definition of G, u and v are either degree-1 or degree-3 vertices: their degree cannot be more than 3, and if it was 2, Modification 2 could be applied.

Intuitively, Modification 4 disconnects G by removing all edges adjacent to uv and, if they exist, connects the former neighbors of u and v, respectively. Formally, let v, u_1, u_2 be the neighbors of u and u, v_1, v_2 the remaining neighbors of v, if they exist. We remove the edges u_1u, u_2u, v_1v, and v_2v. Then we add the edge u_1u_2 if u_1 and u_2 exist and v_1v_2 if v_1 and v_2 exist.

Theorem 1. *ME2C in subcubic graphs has a polynomial-time 1.5-approximation algorithm.*

Proofsketch. Let G be a subcubic graph. Apply Modifications 1–4, until they do not change the graph anymore, and denote the resulting graph by G'. We prove that all components of G' are either trivial or they contain only degree-3 vertices.

Observe that Modifications 1–4 cannot increase the degree of any vertex. Note that G' does not contain vertices of degree 2, due to Modification 2. Now consider vertices of degree 1. Since we remove pendant edges via Modification 4, the only vertices of degree 1 are the end vertices of independent edges, which are trivial components. Vertices of degree 0 are also trivial components.

After applying Modifications 1–4, each component is therefore either a trivial component or a bridgeless cubic graph. It is well-known that each bridgeless cubic graph has a perfect matching [20]. Using the equivalence of modified graphs due to Lemma 2, the claim follows as a consequence of Lemma 14.

In order to prove our result on claw-free graphs, we need to introduce a new modification that helps us control the number of claws.

Modification 5: Avoid neighboring pendant edges Suppose there are two adjacent vertices in G, denoted by u_1 and u_2, such that both of them have exactly one adjacent pendant vertex, denoted by v_1 and v_2, respectively (see Fig. 4). Modification 5 contracts the edge u_1u_2 into a new vertex u_{12} with exactly one pendant edge $u_{12}v_{12}$ incident. Then u 'inherits' all other neighbors of u_1 and u_2, without multiplicities. Furthermore, we introduce an isolated edge w_1w_2.

5   1.625-Approximation for Graphs with Perfect Matching

Definition 6. *A modification is* perfect matching preserving, *if for each graph G that has a perfect matching M, the modification generates a graph $G' \equiv G$ such that G' also has a perfect matching.*

Lemma 15. *Modification 1 and Modification 3 are perfect matching preserving.*

We remark that coming up with a perfect matching preserving modification that removes degree-2 vertices would yield a 1.5-approximation for graphs that contain a perfect matching. Indeed, in that case the arguments in the proof of Lemma 14 would give us the result, as we could apply Modifications 1–3 to a graph containing a perfect matching until we obtain a normalized graph with a perfect matching. As we do not have a perfect matching preserving Modification 2, we use another approach, yielding a slightly worse approximation factor.

Suppose G has a perfect matching, then Modification 2 introduces a new vertex, making the number of vertices odd, hence the resulting graph does not have a perfect matching anymore. This affects the approximation ratio of Algorithm 1: the number of vertices increases by one, the number of leaves by two, while the size of the maximum matching stays the same. One can show that although the approximation ratio can get worse, it does not get worse than $13/8 = 1.625$. Thus, we obtain Theorem 3.

References

1. Adamaszek, A., Popa, A.: Approximation and hardness results for the maximum edge Q-coloring problem. J. Discrete Algorithms **38–41**, 1–8 (2016)
2. Axenovich, M., Jiang, T.: Anti-Ramsey numbers for small complete bipartite graphs. Ars Comb. **73**, 311–318 (2004)
3. Axenovich, M., Jiang, T., Kündgen, A.: Bipartite anti-Ramsey numbers of cycles. J. Graph Theory **47**(1), 9–28 (2004)
4. Blokhuis, A., Faudree, R.J., Gyárfás, A., Ruszinkó, M.: Anti-Ramsey colorings in several rounds. J. Comb. Theory. Ser. B **82**(1), 1–18 (2001)
5. Chandran, L.S., Lahiri, A., Singh, N.: Improved approximation for maximum edge colouring problem. Discret. Appl. Math. **319**, 42–52 (2021)
6. Chandran, L.S., Hashim, T., Jacob, D., Mathew, R., Rajendraprasad, D., Singh, N.: New bounds on the anti-Ramsey numbers of star graphs (2023)
7. Chen, H., Li, X., Tu, J.: Complete solution for the rainbow numbers of matchings. Discrete Math. **309**(10), 3370–3380 (2009)
8. Erdős, P., Simonovits, M., Sós, V.T.: Anti-Ramsey theorems. Infinite and finite sets (Colloq., Keszthely, 1973; dedicated to P. Erdős on his 60th birthday), Vol. II, 633–643. In: Colloquia Mathematica Societatis János Bolyai, vol. 10 (1975)
9. Feng, W., Chen, P., Zhang, B.: Approximate maximum edge coloring within factor 2: a further analysis. In: ISORA, pp. 182–189 (2008)
10. Feng, W., Zhang, L., Qu, W., Wang, H.: Approximation algorithms for maximum edge coloring problem. In: Cai, J.-Y., Cooper, S.B., Zhu, H. (eds.) TAMC 2007. LNCS, vol. 4484, pp. 646–658. Springer, Heidelberg (2007). https://doi.org/10.1007/978-3-540-72504-6_59
11. Feng, W., Zhang, L., Wang, H.: Approximation algorithm for maximum edge coloring. Theor. Comput. Sci. **410**(11), 1022–1029 (2009)
12. Frieze, A., Reed, B.: Polychromatic Hamilton cycles. Discrete Math. **118**(1), 69–74 (1993)
13. Haas, R., Young, M.: The anti-Ramsey number of perfect matching. Discrete Math. **312**(5), 933–937 (2012)
14. Jiang, T.: Edge-colorings with no large polychromatic stars. Graphs Combin. **18**(2), 303–308 (2002)
15. Jiang, T., West, D.B.: Edge-colorings of complete graphs that avoid polychromatic trees. Discrete Math. **274**(1–3), 137–145 (2004)
16. Larjomaa, T., Popa, A.: The min-max edge Q-coloring problem. J. Graph Algorithms Appl. **19**(1), 507–528 (2015)
17. Las Vergnas, M.: A note on matchings in graphs. Cah. Cent. Étud. Rech. Opér. **17**, 257–260 (1975)

18. Mincu, R.S., Popa, A.: Heuristic algorithms for the min-max edge 2-coloring problem. In: Wang, L., Zhu, D. (eds.) COCOON 2018. LNCS, vol. 10976, pp. 662–674. Springer, Cham (2018). https://doi.org/10.1007/978-3-319-94776-1_55
19. Montellano-Ballesteros, J.J., Neumann-Lara, V.: An anti-Ramsey theorem on cycles. Graphs Combin. 21(3), 343–354 (2005)
20. Petersen, J.: Die theorie der regulären graphs. Acta Math. 15(1), 193–220 (1891)
21. Raniwala, A., Chiueh, T.: Architecture and algorithms for an IEEE 802.11-based multi-channel wireless mesh network. In: INFOCOM 2005, vol. 3, pp. 2223–2234 (2005)
22. Raniwala, A., Gopalan, K., Chiueh, T.: Centralized channel assignment and routing algorithms for multi-channel wireless mesh networks. Mob. Comput. Commun. Rev. 8(2), 50–65 (2004)
23. Schiermeyer, I.: Rainbow numbers for matchings and complete graphs. Discrete Math. 286(1–2), 157–162 (2004)
24. Sumner, D.P.: Graphs with 1-factors. Proc. Amer. Math. Soc. 42, 8–12 (1974)

An Improved Deterministic Algorithm for the Online Min-Sum Set Cover Problem

Mateusz Basiak[ID], Marcin Bienkowski[ID], and Agnieszka Tatarczuk[(✉)][ID]

University of Wrocław, Wrocław, Poland
`agnieszka.tatarczuk@uwr.edu.pl`

Abstract. We study the online variant of the Min-Sum Set Cover problem (Mssc), a generalization of the well-known list update problem. In the Mssc problem, an algorithm has to maintain the time-varying permutation of the list of n elements, and serve a sequence of requests $R_1, R_2, \ldots, R_t, \ldots$. Each R_t is a subset of elements of cardinality at most r. For a requested set R_t, an online algorithm has to pay the cost equal to the position of the first element from R_t on its list. Then, it may arbitrarily permute its list, paying the number of swapped adjacent element pairs.

We present the first *constructive* deterministic algorithm, whose competitive ratio does not depend on n. Our algorithm is $O(r^2)$-competitive, which beats both the *existential* upper bound of $O(r^4)$ by Bienkowski and Mucha [AAAI '23] and the previous constructive bound of $O(r^{3/2} \cdot \sqrt{n})$ by Fotakis et al. [ICALP '20]. Furthermore, we show that our algorithm attains an asymptotically optimal competitive ratio of $O(r)$ when compared to the best fixed permutation of elements.

Keywords: Min-sum set cover · Derandomization · Online algorithms · Competitive analysis

1 Introduction

In the online Min-Sum Set Cover (Mssc) problem [12,13], an algorithm has to maintain an ordered list of elements. During runtime, an online algorithm is given a sequence of requests $R_1, R_2, \ldots, R_t, \ldots$, each being a subset of elements. When a request R_t appears, an algorithm first pays the cost equal to the position of the first element of R_t in its list. Next, it may arbitrarily reorder the list, paying the number of swapped adjacent elements.

The online Mssc finds applications in e-commerce, for maintaining an ordered (ranked) list of all shop items (elements) to be presented to new shop customers [9]. This so-called cold-start list can be updated to reflect the preferences of already known users (where R_t corresponds to the set of items that

Supported by Polish National Science Centre grant 2022/45/B/ST6/00559. Full version available at https://arxiv.org/abs/2306.17755.

© The Author(s), under exclusive license to Springer Nature Switzerland AG 2023
J. Byrka and A. Wiese (Eds.): WAOA 2023, LNCS 14297, pp. 45–58, 2023.
https://doi.org/10.1007/978-3-031-49815-2_4

are interesting for user t). Each user should be able to find at least one interesting item close to the list beginning, as otherwise, they must scroll down, which could degrade their overall experience. Similar phenomena occur also in ordering results from a web search for a given keyword [1,10], or ordering news and advertisements.

The problem is also theoretically appealing as the natural generalization of the well-known list update problem [16], where all R_t are singletons (see [14] and references therein).

1.1 Model and Notation

For any integer ℓ, let $[\ell] = \{1,\ldots,\ell\}$. We use $\mathcal{U}$ to denote the universe of n elements. By permutation π of $\mathcal{U}$, we understand a mapping $\mathcal{U} \to [n]$ (from items to their list positions). Thus, for an element $z \in \mathcal{U}$, $\pi(z)$ is its position on the list.

An input $\mathcal{I}$ to the online MSSC problem consists of an initial permutation π_0 of $\mathcal{U}$ and a sequence of m sets $R_1, R_2, \ldots, R_m$. In step t, an online algorithm ALG is presented a request R_t and is charged the *access cost* $\min_{z \in R_t} \pi_{t-1}(z)$. Then, ALG chooses a new permutation π_t (possibly $\pi_t = \pi_{t-1}$) paying *reordering cost* $d(\pi_{t-1}, \pi_t)$, equal to the minimum number of swaps of adjacent elements necessary to change permutation π_{t-1} into π_t.[1]

We emphasize that the choice of π_t made by ALG has to be performed without the knowledge of future sets $R_{t+1}, R_{t+2}, \ldots$ and also without the knowledge of the sequence length m. We use r to denote the maximum cardinality of requested sets R_t.

1.2 Benchmarks

In the following, for an input $\mathcal{I}$ and an algorithm A, we use $A(\mathcal{I})$ to denote the total cost of A on $\mathcal{I}$. To measure the effectiveness of online algorithms, we use the standard notion of competitive ratio, but we generalize it slightly, for more streamlined definitions of particular scenarios.

We say that an online algorithm ALG is c-competitive *against class $\mathcal{C}$ of offline algorithms* if there exists a constant ξ, such that for any input $\mathcal{I}$ and any offline algorithm OFF $\in \mathcal{C}$, it holds that $\text{ALG}(\mathcal{I}) \leq c \cdot \text{OFF}(\mathcal{I}) + \xi$. If $\xi = 0$, then ALG is called *strictly* competitive. The competitive ratio of ALG against class $\mathcal{C}$ is the infimum of values of c, for which ALG is c-competitive against this class. For randomized algorithms, we replace the cost $\text{ALG}(\mathcal{I})$ with its expected value $\mathbf{E}[\text{ALG}(\mathcal{I})]$. We consider three scenarios.

Dynamic Scenario. In the dynamic scenario, the considered class $\mathcal{C}$ contains all possible offline algorithms, in particular those that adapt their permutation dynamically during runtime. This setting is equivalent to the traditional competitive ratio [8], where an online algorithm is compared to the optimal offline solution OPT. This scenario is the main focus of this paper.

---

[1] The value $d(\pi_{t-1}, \pi_t)$ is also equal to the number of inversions between π_{t-1} and π_t, i.e., number of unordered pairs (x,y) such that $\pi_{t-1}(x) < \pi_{t-1}(y)$ and $\pi_t(x) > \pi_t(y)$.

Static Scenario. Previous papers focused also on a simpler *static scenario*, where the considered class of algorithms FIXED contains all possible $n!$ fixed strategies: an algorithm from class FIXED starts with its list ordered according to a fixed permutation and never changes it [12]. (In this scenario, the starting permutation of an online algorithm and an offline solution are different.) Note that such an offline algorithm incurs no reordering cost, and pays access costs only. It is worth mentioning that there exist inputs $\mathcal{I}$, for which $\min_{A \in \text{FIXED}} A(\mathcal{I}) = \Omega(n) \cdot \text{OPT}(\mathcal{I})$ [12].

Learning Scenario. The static scenario can be simplified further, by assuming that reordering incurs no cost on ALG. We call such a setting *learning scenario*. Clearly, the competitive ratios achievable in the learning scenario are not larger than those for the static scenario, which are in turn not larger than those in the dynamic scenario.

1.3   Previous Results

Below, we discuss known results for the MSSC problem in the three scenarios described above (dynamic, static, and learning). Furthermore, we make a distinction between ratios achievable for polynomial-time algorithms and algorithms whose runtime per request is not restricted. The lower and upper bounds described below are also summarized in Table 1.

Lower Bounds. Feige et al. [11] studied the *offline* variant of MSSC (where all R_t's are given upfront and an algorithm has to compute a fixed permutation minimizing the access cost). They show that unless $\mathsf{P} = \mathsf{NP}$, no offline algorithm can achieve an approximation ratio better than 4. This result implies a lower bound of 4 on the competitive ratio of any polynomial-time online algorithm (assuming $\mathsf{P} \neq \mathsf{NP}$) as such a solution can be used to solve the offline variant as well. We note that 4-approximation algorithms for the offline variant are known as well [4,11].

The online version of MSSC was first studied by Fotakis et al. [12]. They show that no deterministic algorithm can achieve a ratio better than $\Omega(r)$ even in the learning scenario. This yields the same lower bound for the remaining scenarios as well.

Asymptotically Tight Upper Bounds. For the static scenario, the randomized $(1+\varepsilon)$-competitive solution (for any $\varepsilon > 0$) follows by combining multiplicative weight updates [2,15] with the techniques of Blum and Burch [7] designed for the metrical task systems. This approach has been successfully derandomized by Fotakis et al. [12], who gave a deterministic solution with an asymptotically optimal ratio of $O(r)$. These algorithms clearly also work in the learning scenario. However, in both scenarios, they require exponential time as they keep track of access costs for all possible $n!$ permutations.

Table 1. Known lower and upper bounds for the online MSSC problem for three scenarios (dynamic, static and learning), for polynomial-time and computationally unrestricted algorithms. Unreferenced results are trivial consequences of other results. The ratios proved in this paper are in bold.

		randomized		deterministic	
		LB	UB	LB	UB
learning	unrestr.	1	$1 + \varepsilon$	$\Omega(r)$ [12]	$O(r)$
	poly-time	4 [11]	11.713 [13]	$\Omega(r)$	$O(r)$ [13]
static	unrestr.	1	$1 + \varepsilon$ [7]	$\Omega(r)$	$O(r)$ [12]
	poly-time	4	$O(r^2)$ $O(r)$ **(Theorem 3)**	$\Omega(r)$	$\exp(O(\sqrt{\log n \cdot \log r}))$ [12] $O(r)$ **(Theorem 3)**
dynamic	unrestr.	1	$O(r^2)$	$\Omega(r)$	$O(r^4)$ [6] $O(r^2)$ **(Theorem 4)**
	poly-time	4	$O(r^2)$ [6]	$\Omega(r)$	$O(r^{3/2} \cdot \sqrt{n})$ [12] $O(r^2)$ **(Theorem 4)**

Fotakis et al. [13] showed that in the learning scenario, one can maintain a sparse representation of all permutations and achieve asymptotically optimal results that work in polynomial time: a randomized $O(1)$-competitive algorithm and a deterministic $O(r)$-competitive one.

Non-tight Upper Bounds. Much of the effort in the previous papers was devoted to creating algorithms for the dynamic scenario with low competitive ratios. For $r = 1$, a simple MOVE-TO-FRONT policy that moves the requested element to the first position is $O(1)$-competitive. Perhaps surprisingly, however, the competitive ratios of many of its natural generalizations were shown to be not better than $\Omega(n)$ [12].

Fotakis et al. [12] gave an online deterministic $O(r^{3/2} \cdot \sqrt{n})$-competitive algorithm MOVE-ALL-EQUALLY (MAE) and showed that its analysis is almost tight. They provided a better bound for the performance of MAE in the static scenario: in such a setting its competitive ratio is $\exp(O(\sqrt{\log n \cdot \log r}))$ [12].

For randomized solutions, this result was improved by Bienkowski and Mucha [6], who gave an $O(r^2)$-competitive *randomized* algorithm LMA (for the dynamic scenario). Their analysis holds also against so-called adaptive-online adversaries, and therefore, by the reduction of [5], it implies the *existence* of a deterministic $O(r^4)$-competitive algorithm. While using the techniques of Ben-David et al. [5], the construction of such an algorithm is possible, it was not done explicitly, and furthermore, a straightforward application of these techniques would lead to a huge running time.

1.4   Our Contribution

In Sect. 2, we present the first constructive deterministic algorithm whose competitive ratio in the dynamic scenario is a function of r only (and does not depend on n). Our algorithm, dubbed DETERMINISTIC-AND-LAZY-MOVE-TO-FRONT (DLM), runs in polynomial time, and we analyze its performance both in static and dynamic scenarios.

In the static scenario, studied in Sect. 4, DLM attains the optimal competitive ratio of $O(r)$, improving over the $\exp(O(\sqrt{\log n \cdot \log r}))$-competitive solution by Fotakis et al. [12] and matching $O(r)$ bound achieved by the exponential-time algorithm of [12].

In the dynamic scenario, studied in Sect. 5, we show that DLM is $O(r^2)$-competitive. As $r \leq n$, this bound is always better than the existing non-constructive upper bound of $O(r^4)$ [6] and the polynomial-time upper bound of $O(r^{3/2} \cdot \sqrt{n})$ [12]. Our analysis is asymptotically tight: in the full version of the paper, we show that the ratio of $O(r^2)$ is best possible for the whole class of approaches that includes DLM.

Finally, as the learning scenario is not harder than the static one, DLM is $O(r)$-competitive also there. While an upper bound of $O(r)$ was already known for the learning scenario [13], our algorithm uses a vastly different, combinatorial approach and is also much faster.

Our deterministic solution is inspired by the ideas for the randomized algorithm LMA of [6] and can be seen as its derandomization, albeit with several crucial differences.

- We simplify their approach as we solve the MSSC problem directly, while they introduced an intermediate exponential caching problem.
- We update item budgets differently, which allows us to obtain an optimal ratio for the static scenario (LMA is not better in the static scenario than in the dynamic one).
- Most importantly, [6] uses randomization to argue that LMA makes bad choices with a small probability. In this context, bad choices mean moving elements that OPT has near the list front towards the list tail in the solution of LMA. In the deterministic approach, we obviously cannot prove the same claim, but we show that it holds on the average. Combining this claim with the amortized analysis, by "encoding" it in the additional potential function Ψ, is the main technical contribution of our paper.

2   Our Algorithm DLM

DLM maintains a budget $b(z)$ for any element $z \in \mathcal{U}$. At the beginning of an input sequence, all budgets are set to zero.

In the algorithm description, we skip step-related subscripts when it does not lead to ambiguity, and we simply use $\pi(z)$ to denote the *current* position of element z in the permutation of DLM.

At certain times, DLM moves an element z to the list front. It does so using a straightforward procedure FETCH(z) (cf. Routine 1). It uses $\pi(z) - 1$ swaps that move z to the first position, and increment the positions of all elements that preceded z. Next, it resets the budget of z to zero.

Routine 1: FETCH(z), where z is any element

---

1 **for** $i = \pi(z), \ldots, 3, 2$ **do**
2 | swap elements on positions i and $i - 1$
3 $b(z) \leftarrow 0$

Algorithm 2: A single step of DETERMINISTIC-LAZY-MOVE-ALL-TO-FRONT (DLM)

Input: request $R = \{x, y_1, y_2, \ldots, y_{s-1}\}$, where $s \le r$ and $\pi(x) \le \pi(y_i)$ for $i \in [s-1]$, current permutation π of elements

---

1 $\ell \leftarrow \pi(x)$
2 **execute** FETCH(x)
3 **for** $i = 1, 2, \ldots, s - 1$ **do**
4 | $b(y_i) \leftarrow b(y_i) + \ell/s$
5 **while** exists z such that $b(z) \ge \pi(z)$ **do**
6 | **execute** FETCH(z)

Assume now that DLM needs to serve a request $R = \{x, y_1, y_2, \ldots, y_{s-1}\}$ (where $s \le r$ and $\pi(x) < \pi(y_i)$ for all y_i). Let $\ell = \pi(x)$. DLM first executes routine FETCH(x). Afterward, it performs a lazy counterpart of moving elements y_i towards the front: it increases their budgets by ℓ/s. Once a budget of any element reaches or exceeds its current position, DLM fetches it to the list front. The pseudocode of DLM on request R is given in Algorithm 2.

3   Basic Properties and Analysis Framework

We start with some observations about elements' budgets; in particular, we show that DLM is well defined, i.e., it terminates.

Lemma 1. DLM *terminates after every request.*

Proof. Let $C = \{z \in \mathcal{U} \mid b(z) \ge \pi(z)\}$. It suffices to show that the cardinality of C decreases at each iteration of the while loop in Line 5 of Algorithm 2. To this end, observe that in each iteration, we execute operation FETCH(z) for some $z \in C$. In effect, the budget of z is set to 0, and thus z is removed from C. The positions of elements that preceded z are incremented without changing their budget: they may only be removed from C but not added to it.

Observation 1. *Once* DLM *finishes list reordering in a given step,* $b(z) < \pi(z)$ *for any element* $z \in \mathcal{U}$. *Moreover,* $b(z) < (3/2) \cdot \pi(z)$ *also during list reordering.*

Proof. Once the list reordering terminates, by Lemma 1 and the while loop in Lines 5–6 of Algorithm 2, $b(z) < \pi(z)$ for any element z.

Within a step, the budgets are increased only for elements $y_i \in R$, i.e., only when $s \ge 2$. The budget of such an element y_i is increased from at most $\pi(y_i)$ by $\pi(x)/s \le \pi(x)/2 < \pi(y_i)/2$, i.e., its resulting budget is smaller than $(3/2) \cdot \pi(y_i)$.

3.1   Amortized Analysis

In our analysis, we compare the cost in a single step of DLM to the corresponding cost of an offline solution OFF. For a more streamlined analysis that will yield the result both for the static and dynamic scenarios, we split each step into two stages. In the first stage, both DLM and OFF pay their access costs, and then DLM reorders its list according to its definition. In the second stage, OFF reorders its list. Note that the second stage exists only in the dynamic scenario.

We use π and π^* to denote the current permutation of DLM and OFF, respectively. We introduce two potential functions Φ and Ψ, whose values depend only on π and π^*.

In Sect. 4, we show that in the first stage of any step, it holds that

$$\Delta\text{DLM} + \Delta\Phi + \Delta\Psi \leq O(r) \cdot \Delta\text{OFF}. \tag{1}$$

where ΔDLM, ΔOFF, $\Delta\Phi$, and $\Delta\Psi$ denote increases of the costs of DLM and OFF and the increases of values of Φ and Ψ, respectively. Relation (1) summed over all m steps of the input sequence yields the competitive ratio of $O(r)$ of DLM in the static scenario (where only the first stage is present).

In Sect. 5, we analyze the performance of DLM in the dynamic scenario. We say that an offline algorithm OFF is *MTF-based* if, for any request, it moves one of the requested elements to the first position of the list and does not touch the remaining elements. We define a class MTFB of all MTF-based offline algorithms. We show that in the second stage of any step, it holds that

$$\Delta\text{DLM} + \Delta\Phi + \Delta\Psi \leq O(r^2) \cdot \Delta\text{OFF}. \tag{2}$$

for any OFF $\in$ MTFB. Now, summing relations (1) and (2) over all steps in the input yields that DLM is $O(r^2)$-competitive against the class MTFB. We conclude by arguing that there exists an MTF-based algorithm OFF* which is a 4-approximation of the optimal solution OPT.

3.2   Potential Function

To define potential functions, we first split $\pi(z)$ into two summands, $\pi(z) = 2^{p(z)} + q(z)$, such that $p(z)$ is a non-negative integer, and $q(z) \in \{0, \ldots, 2^{p(z)}-1\}$. We split $\pi^*(z)$ analogously as $\pi^*(z) = 2^{p^*(z)} + q^*(z)$.

We use the following parameters: $\alpha = 2$, $\gamma = 5r$, $\beta = 7.5r + 5$, and $\kappa = \lceil \log(6\beta) \rceil$. Our analysis does not depend on the specific values of these parameters, but we require that they satisfy the following relations.

Fact 2. *Parameters α, β and γ satisfy the following relations: $\alpha \geq 2$, $\gamma \geq (3+\alpha) \cdot r$, $\beta \geq 3 + \alpha + (3/2) \cdot \gamma$. Furthermore, κ is an integer satisfying $2^\kappa \geq 6\beta$.*

For any element z, we define its potentials

$$\Phi_z = \begin{cases} \alpha \cdot b(z) & \text{if } p(z) \leq p^*(z) + \kappa, \\ \beta \cdot \pi(z) - \gamma \cdot b(z) & \text{if } p(z) \geq p^*(z) + \kappa + 1. \end{cases} \tag{3}$$

$$\Psi_z = \begin{cases} 0 & \text{if } p(z) \leq p^*(z) + \kappa - 1, \\ 2\beta \cdot q(z) & \text{if } p(z) \geq p^*(z) + \kappa. \end{cases} \tag{4}$$

We define the total potentials as $\Phi = \sum_{z \in \mathcal{U}} \Phi_z$ and $\Psi = \sum_{z \in \mathcal{U}} \Psi_z$.

Lemma 2. *At any time and for any element z, $\Phi_z \geq 0$ and $\Psi_z \geq 0$.*

Proof. The relation $\Psi_z \geq 0$ follows trivially from (4). By Fact 2, $\beta \geq (3/2) \cdot \gamma$. This, together with Observation 1, implies that $\Phi_z \geq 0$.

3.3   Incrementing Elements Positions

We first argue that increments of elements' positions induce small changes in their potentials. Such increments occur for instance when DLM fetches an element z to the list front: all elements that preceded z are shifted by one position towards the list tail. We show this property for the elements on the list of DLM first and then for the list of OFF.

We say that an element w is *safe* if $p(w) \leq p^*(w) + \kappa - 1$ and *unsafe* otherwise. Note that for a safe element w, it holds that $\pi(w) \leq 2^{p(w)+1} \leq 2^{\kappa} \cdot 2^{p^*(w)} \leq 2^{\kappa} \cdot \pi^*(w) = O(r) \cdot \pi^*(w)$, i.e., its position on the list of DLM is at most $O(r)$ times greater than on the list of OFF.

Lemma 3. *Assume that the position of an element w on the list of DLM increases by 1. Then, $\Delta\Phi_w + \Delta\Psi_w \leq 0$ if w was safe before the movement and $\Delta\Phi_w + \Delta\Psi_w \leq 3\beta$ otherwise.*

Proof. By $\pi(w) = 2^{p(w)} + q(w)$ and $\pi'(w) = \pi(w) + 1 = 2^{p'(w)} + q'(w)$ we denote the positions of w before and after the movement, respectively.

Assume first that w was safe before the movement. As $p'(w) \leq p(w) + 1 \leq p^*(w) + \kappa$, $\Delta\Phi_z = \alpha \cdot b(z) - \alpha \cdot b(z) = 0$. Furthermore, either $p'(w) = p(w)$, and then $\Delta\Psi_w = 0$ trivially, or $p'(w) = p(w) + 1$, and then $q'(w) = 0$. In the latter case $\Delta\Psi_w = 2\beta \cdot q'(z) - 0 = 0$ as well. This shows the first part of the lemma.

Assume now that w was unsafe ($p(w) \geq p^*(w) + \kappa$) before the movement. We consider two cases.

- $p(w) = p^*(w) + \kappa$ and $p'(w) = p(w) + 1$.
  It means that $q(w) = 2^{p(w)} - 1$ and $q'(w) = 0$. Then,

$$\Delta\Phi_w = \beta \cdot \pi'(z) - \gamma \cdot \beta(z) - \alpha \cdot \beta(z) \leq \beta \cdot \pi'(z) = \beta \cdot 2^{p'(w)} = 2\beta \cdot 2^{p(w)},$$
$$\Delta\Psi_w = 2\beta \cdot q'(z) - 2\beta \cdot q(z) = -2\beta \cdot (2^{p(w)} - 1) = -2\beta \cdot 2^{p(w)} + 2\beta.$$

That is, the large growth of Φ_w is compensated by the drop of Ψ_w, i.e, $\Delta\Phi_w + \Psi_w \leq 2\beta$.

$- p(w) > p^*(w) + \kappa$ or $p'(w) = p(w)$.

In such case, there is no case change in the definition of Φ_w, i.e.,

$$
\Delta\Phi_w = \begin{cases} \alpha \cdot b(w) - \alpha \cdot b(w) = 0 & \text{if } p(w) \leq p^*(w) + \kappa, \\ (\beta \cdot \pi'(w) - \gamma \cdot b(w)) \\ \qquad - (\beta \cdot \pi(w) - \gamma \cdot b(w)) = \beta & \text{otherwise.} \end{cases}
$$

Furthermore, as $q'(w) \leq q(w) + 1$, $\Delta\Psi(z) = 2\beta \cdot q'(w) - 2\beta \cdot q(w) \leq 2\beta$. Together, $\Delta\Phi_w + \Delta\Psi_w \leq \beta + 2\beta = 3\beta$.   □

Lemma 4. *Assume that the position of an element w on the list of* OFF *increases by 1. Then, $\Delta\Phi_w \leq 0$ and $\Delta\Psi_w \leq 0$.*

Proof. Note that $p^*(w)$ may be either unchanged (in which case the values of Φ_w and Ψ_w remain intact) or it may be incremented. We analyze the latter case.

By (3), the definition of Φ_w, the value of Φ_w may change only if $p^*(w)$ is incremented from $p(w) + \kappa - 1$ to $p(w) + \kappa$. In such case,

$$
\begin{aligned}
\Delta\Psi_w &= \alpha \cdot b(w) - \beta \cdot \pi(w) + \gamma \cdot b(w) \\
&\leq (\alpha + \gamma - \beta) \cdot \pi(w) && \text{(by Observation 1)} \\
&\leq 0. && \text{(by Fact 2)}
\end{aligned}
$$

By (4), the definition of Ψ_w, the value of Ψ_w may change only if $p^*(w)$ is incremented from $p(w) + \kappa$ to $p(w) + \kappa + 1$. In such case, $\Delta\Psi_w = -2\beta \cdot q(w) \leq 0$.   □

4   Analysis in the Static Scenario

As described in Sect. 3.1, in this part, we focus on the amortized cost of DLM in the first stage of a step, i.e., where DLM and OFF both pay their access costs and then DLM reorders its list.

Lemma 5. *Whenever* DLM *executes operation* FETCH(z), *it holds that ΔDLM$+ \Delta\Psi + \sum_{w \neq z} \Delta\Phi_w \leq 2 \cdot \pi(z)$.*

Proof. As defined in Routine 1, the cost of operation FETCH(z) is ΔDLM $= \pi(z) - 1 < \pi(z)$. We first analyze the potential changes of elements from set K of $\pi(z) - 1$ elements that originally preceded z.

Let $K' = \{w \in K \mid \pi^*(w) \leq 2^{p(z)-\kappa+1}\}$. Observe that any $w \in K \setminus K'$ satisfies $\pi^*(w) > 2^{p(z)-\kappa+1}$, which implies $p^*(w) \geq p(z) - \kappa + 1 \geq p(w) - \kappa + 1$, and thus w is safe. Thus, among elements of K, only elements from K' can be unsafe. By Lemma 3,

$$
\begin{aligned}
\sum_{w \in K} (\Delta\Phi_w + \Delta\Psi_w) &\leq \sum_{w \in K'} (\Delta\Phi_w + \Delta\Psi_w) \\
&\leq 3\beta \cdot |K'| = 3\beta \cdot 2^{p(z)-\kappa+1} \\
&\leq 2^{p(z)} \leq \pi(z) && \text{(by Fact 2)}
\end{aligned}
$$

As the only elements that may change their budgets are z and elements from K, we have $\Delta\text{DLM} + \Delta\Psi + \sum_{w \neq z} \Delta\Phi_w = \Delta\text{DLM} + \sum_{w \in K} \Delta\Psi_w + \Delta\Psi_z + \sum_{w \in K} \Delta\Phi_w \leq 2 \cdot \pi(z) + \Delta\Psi_z \leq 2 \cdot \pi(z)$. The last inequality follows as Ψ_z drops to 0 when z is moved to the list front. $\square$

Now we may split the cost of DLM in a single step into parts incurred by Lines 1–4 and Lines 5–6, and bound them separately.

Lemma 6. *Whenever* DLM *executes Lines 5–6 of Algorithm 2,* $\Delta\text{DLM} + \Delta\Phi + \Delta\Psi \leq 0$.

Proof. Let z be the element moved in Line 6. Line 5 guarantees that $b(z) \geq \pi(z)$ and Observation 1 implies $b(z) \leq (3/2) \cdot \pi(z)$. The value of Φ_z before the movement is then

$$\Phi_z \geq \min\{\alpha \cdot b(z), \beta \cdot \pi(z) - \gamma \cdot b(z)\}$$
$$\geq \min\{\alpha, \beta - (3/2) \cdot \gamma\} \cdot \pi(z)$$
$$\geq 2 \cdot \pi(z). \qquad \text{(by Fact 2)}$$

When z is moved to the list front, potential Φ_z drops to 0, and thus $\Delta\Phi_z \leq -2 \cdot \pi(z)$. Hence, using Lemma 5, $\Delta\text{DLM} + \Delta\Phi + \Delta\Psi \leq 2 \cdot \pi(z) + \Delta\Phi_z \leq 0$.

Lemma 7. *Fix any step and consider its first part, where* DLM *pays for its access and movement costs, whereas* OFF *pays for its access cost. Then,* $\Delta\text{DLM} + \Delta\Phi + \Delta\Psi \leq (3 + \alpha) \cdot 2^{\kappa+1} \cdot \Delta\text{OFF} = O(r) \cdot \Delta\text{OFF}$.

Proof. Let $R = \{x, y_1, \ldots, y_{s-1}\}$ be the requested set, where $s \leq r$ and $\pi(x) < \pi(y_i)$ for any $i \in [s-1]$. Let Φ_x denote the value of the potential just before the request. It suffices to analyze the amortized cost of DLM in Lines 1–4 as the cost in the subsequent lines is at most 0 by Lemma 6. In these lines:

- DLM pays $\pi(x)$ for the access.
- DLM performs the operation FETCH(x), whose amortized cost is at most $2 \cdot \pi(x) - \Phi_x$ (by Lemma 5).
- The budget of y_i grows by $\Delta b(y_i) = \pi(x)/s$ for each $i \in [s-1]$. As these elements do not move (within Lines 1–4), $\Delta\Psi_{y_i} = 0$.

Thus, we obtain

$$\Delta\text{DLM} + \Delta\Phi + \Delta\Psi \leq 3 \cdot \pi(x) - \Phi_x + \sum_{i \in [s-1]} \Delta\Phi_{y_i}. \qquad (5)$$

As elements y_i do not move (within Lines 1–4)), the change in Φ_{y_i} can be induced only by the change in the budget of y_i. Let $u \in R$ be the element with the smallest position on the list of OFF, i.e., $\Delta\text{OFF} = \pi^*(u)$. We consider three cases.

- $p(x) \leq p^*(u) + \kappa$.
  Then $\pi(x) \leq 2^{p(x)+1} \leq 2^{\kappa+1} \cdot 2^{p^*(u)} \leq 2^{\kappa+1} \cdot \pi^*(u) = 2^{\kappa+1} \cdot \Delta\text{OFF}$. Note that $\sum_{i \in [s-1]} \Delta\Phi_{y_i} \leq \sum_{i \in [s-1]} \alpha \cdot \Delta b(y_i) = (s-1) \cdot \alpha \cdot \pi(x)/s < \alpha \cdot \pi(x)$. By Lemma 2, $\Phi_x \geq 0$, and thus using (5),

$$\Delta\text{DLM} + \Delta\Phi + \Delta\Psi < 3 \cdot \pi(x) + \alpha \cdot \pi(x) \leq (3 + \alpha) \cdot 2^{\kappa+1} \cdot \Delta\text{OFF}.$$

- $p(x) \geq p^*(u) + \kappa + 1$ and $u = x$.

  In this case, $\Phi_x \geq \beta \cdot \pi(x) - \gamma \cdot b(x) \geq (\beta - (3/2) \cdot \gamma) \cdot \pi(x)$ (cf. Observation 1). By plugging this bound to (5), we obtain

$$\Delta\text{DLM} + \Delta\Phi + \Delta\Psi \leq 3 \cdot \pi(x) + (\beta - (3/2) \cdot \gamma) \cdot \pi(x) + \alpha \cdot \pi(x) \leq 0,$$

  where the last inequality follows as $\beta \geq 3 + (3/2) \cdot \gamma + \alpha$ by Fact 2.

- $p(x) \geq p^*(u) + \kappa + 1$ and $u = y_j$ for some $j \in [s-1]$.

  Recall that $\pi(x) < \pi(y_i)$, and thus $p(y_j) \geq p(x)$. Hence, $p(y_j) \geq p^*(y_j) + \kappa + 1$. In such a case,

$$\sum_{i \in [s-1]} \Delta\Phi_{y_i} = \Delta\Phi_{y_j} + \sum_{i \in [s-1] \setminus \{j\}} \Delta\Phi_{y_i}$$

$$\leq -\gamma \cdot \Delta b(y_j) + \sum_{i \in [s-1] \setminus \{j\}} \alpha \cdot \Delta b(y_i)$$

$$= -\gamma \cdot \pi(x)/s + (s-2) \cdot \alpha \cdot \pi(x)/s$$

$$< (\alpha - \gamma/r) \cdot \pi(x). \qquad \text{(as } s \leq r)$$

Plugging the bound above and $\Phi_x \geq 0$ to (5) yields

$$\Delta\text{DLM} + \Delta\Phi + \Delta\Psi \leq (3 + \alpha - \gamma/r) \cdot \pi(x) \leq 0,$$

where the last inequality again follows by Fact 2. □

Theorem 3. DLM *is $O(r)$-competitive in the static scenario.*

Proof. Fix any input $\mathcal{I}$ ad any offline solution OFF that maintains a fixed permutation. For any step t, let Φ^t and Ψ^t denote the total potentials right after step t, while Φ^0 and Ψ^0 be the initial potentials. By Lemma 7,

$$\text{DLM}_t(\mathcal{I}) + \Phi^t + \Psi^t - \Phi^{t-1} - \Psi^{t-1} = O(r) \cdot \text{OFF}_t(\mathcal{I}), \qquad (6)$$

where $\text{DLM}_t(\mathcal{I})$ and $\text{OFF}_t(\mathcal{I})$ denote the costs of DLM and OFF in step t, respectively. By summing (6) over all m steps of the input, we obtain $\text{DLM}(\mathcal{I}) + \Phi^m + \Psi^m - \Phi^0 - \Psi^0 \leq O(r) \cdot \text{OFF}(\mathcal{I})$. As $\Phi^m + \Psi^m \geq 0$,

$$\text{DLM}(\mathcal{I}) \leq O(r) \cdot \text{OFF}(\mathcal{I}) + \Phi^0 + \Psi^0.$$

Note that the initial potentials might be non-zero as in the static scenario OFF starts in its permutation which might be different from π_0. That said, both initial potentials can be universally upper-bounded by the amount independent of $\mathcal{I}$, and thus DLM is $O(r)$-competitive.

5   Analysis in the Dynamic Scenario

To analyze DLM in the dynamic scenario, we first establish an offline approximation of OPT that could be handled using our potential functions.

We say that an algorithm is *move-to-front based* (MTF-*based*) if, in response to request R, it chooses exactly one of the elements from R, brings it to the list front, and does not perform any further actions. We denote the class of all such (offline) algorithms by MTFB.

Lemma 8. *For any input* $\mathcal{I}$, *there exists an (offline) algorithm* $\text{OFF}^* \in \text{MTFB}$, *such that* $\text{OFF}^*(\mathcal{I}) \leq 4 \cdot \text{OPT}(\mathcal{I})$.

Proof. Based on the actions of OPT on $\mathcal{I} = (\pi_0, R_1, \ldots, R_m)$, we may create an input $\mathcal{J} = (\pi_0, R'_1, \ldots, R'_m)$ where R'_i is a singleton set containing exactly the element from R_i that OPT has nearest to the list front.

Clearly, $\text{OPT}(\mathcal{J}) \leq \text{OPT}(\mathcal{I})$. Note that $\mathcal{J}$ is an instance of the list update problem. Thus, if we take an algorithm MTF for the list update problem (which brings the requested element to the list front), then $\text{MTF}(\mathcal{J}) \leq 4 \cdot \text{OPT}(\mathcal{J})$ [16]. (The result of [16] shows the competitive ratio of 2, but the list update model ignores reordering costs. However, for MTF the reordering costs are equal to access costs minus 1 and hence, taking them into account at most doubles the competitive ratio.)

Let now OFF^* be an offline algorithm that, on input $\mathcal{I}$, performs the same list reordering as $\text{MTF}(\mathcal{J})$. Clearly, $\text{OFF}^* \in \text{MTFB}$. While the reordering cost of OFF^* on $\mathcal{I}$ coincide with that of MTF on $\mathcal{J}$, its access cost can be only smaller. Thus, $\text{OFF}^*(\mathcal{I}) \leq \text{MTF}(\mathcal{J})$.

Summing up, we obtain $\text{OFF}^*(\mathcal{I}) \leq \text{MTF}(\mathcal{J}) \leq 4 \cdot \text{OPT}(\mathcal{J}) \leq 4 \cdot \text{OPT}(\mathcal{I})$, which concludes the proof.   □

We now analyze the second stage of a step, where an offline algorithm OFF from the class MTFB reorders its list.

Lemma 9. *Assume* $\text{OFF} \in \text{MTFB}$. *Fix any step and consider its second stage, where* OFF *moves some element* z *to the list front. Then,* $\Delta\Phi + \Delta\Psi = O(r^2) \cdot \Delta\text{OFF}$.

Proof. We may assume that initially $\pi^*(z) \geq 2$, as otherwise there is no change in the list of OFF and the lemma follows trivially.

Apart from element z, the only elements that change their positions are elements that originally preceded z: their positions are incremented. By Lemma 4, the potential change associated with these elements is non-positive.

Thus, $\Delta\Phi + \Delta\Psi \leq \Delta\Phi_z + \Delta\Psi_z$. Element z is transported by OFF from position $\pi^*(z)$ to position 1, i.e., $\Delta\text{OFF} = \pi^*(z) - 1 \geq \pi^*(z)/2$ as we assumed $\pi^*(z) \geq 2$. Thus, to show the lemma it suffices to show that $\Delta\Phi_z = O(r^2) \cdot \pi^*(z)$ and $\Delta\Psi_z = O(r^2) \cdot \pi^*(z)$. We bound them separately.

- Note that $p^*(z)$ may only decrease. If initially $p^*(z) \leq p(z) - \kappa - 1$, then $\Phi_z = \beta \cdot \pi(z) - \gamma \cdot b(z)$ before and after the movement of z, and thus $\Delta\Phi_z = 0$. Otherwise, $p^*(z) \geq p(z) - \kappa$, which implies $\pi(z) < 2^{p(z)+1} \leq 2^{\kappa+1} \cdot 2^{p^*(z)} \leq 2^{\kappa+1} \cdot \pi^*(z)$. In such a case,

$$\Delta\Phi_z \leq \beta \cdot \pi(z) - \gamma \cdot b(z) - \alpha \cdot b(z) \leq \beta \cdot \pi(z) \leq \beta \cdot 2^{\kappa+1} \cdot \pi^*(z) = O(r^2) \cdot \pi^*(z).$$

- Similarly, if initially $p^*(z) \leq p(z) - \kappa$, then $\Psi_z = 2\beta \cdot q(z)$ before and after the movement of z, and thus $\Delta\Psi_z = 0$. Otherwise, $p^*(z) \geq p(z) - \kappa + 1$, which implies $\pi(z) < 2^{p(z)+1} \leq 2^{\kappa} \cdot 2^{p^*(z)} \leq 2^{\kappa} \cdot \pi^*(z)$. In such a case

$$\Delta\Psi_z \leq 2\beta \cdot q(z) - 0 \leq 2\beta \cdot \pi(z) \leq 2\beta \cdot 2^{\kappa} \cdot \pi^*(z) = O(r^2) \cdot \pi^*(z).$$

In either case, the lemma follows.   □

Theorem 4. DLM *is strictly* $O(r^2)$-*competitive in the dynamic scenario.*

Proof. The argument here is the same as for Theorem 3, but this time we sum the guarantees provided for the first stage of a step (Lemma 7) and for the second stage of a step (Lemma 9). This shows that for any offline algorithm OFF ∈ MTFB and any input $\mathcal{I}$, it holds that

$$\text{DLM}(\mathcal{I}) \le O(r^2) \cdot \text{OFF}(\mathcal{I}) + \Phi^0 + \Psi^0. \tag{7}$$

For the dynamic scenario, the initial permutations of DLM and OFF are equal, and hence the initial potential $\Phi^0 + \Psi^0$ is zero. As (7) holds against arbitrary OFF ∈ MTFB, it holds also against OFF* which is the 4-approximation of OPT (cf. Lemma 8). This implies that

$$\text{DLM}(\mathcal{I}) \le O(r^2) \cdot \text{OFF}^*(\mathcal{I}) \le O(4 \cdot r^2) \cdot \text{OPT}(\mathcal{I}),$$

which concludes the proof.   □

6   Final Remarks

In this paper, we studied achievable competitive ratios for the online MSSC problem. We closed the gaps for deterministic polynomial-time static scenarios and tighten the gaps for deterministic dynamic scenarios. Still, some intriguing open questions remain, e.g., the best randomized algorithm for the dynamic scenario has a competitive ratio of $O(r^2)$, while the lower bound is merely a constant.

Another open question concerns a generalization of the MSSC problem where each set R_t comes with a covering requirement k_t and an algorithm is charged for the positions of the first k_t elements from R_t on the list (see, e.g., [3]). The only online results so far are achieved in the easiest, learning scenario [13].

References

1. Agichtein, E., Brill, E., Dumais, S.T.: Improving web search ranking by incorporating user behavior information. SIGIR Forum **52**(2), 11–18 (2018). https://doi.org/10.1145/3308774.3308778
2. Arora, S., Hazan, E., Kale, S.: The multiplicative weights update method: a meta-algorithm and applications. Theory Comput. Syst. **8**(1), 121–164 (2012). https://doi.org/10.4086/toc.2012.v008a006
3. Bansal, N., Batra, J., Farhadi, M., Tetali, P.: Improved approximations for min sum vertex cover and generalized min sum set cover. In: Proceedings of the ACM-SIAM Symposium on Discrete Algorithms (SODA), pp. 998–1005. SIAM (2021). https://doi.org/10.1137/1.9781611976465.62
4. Bar-Noy, A., Bellare, M., Halldórsson, M.M., Shachnai, H., Tamir, T.: On chromatic sums and distributed resource allocation. Inf. Comput. **140**(2), 183–202 (1998). https://doi.org/10.1006/inco.1997.2677

5. Ben-David, S., Borodin, A., Karp, R.M., Tardos, G., Wigderson, A.: On the power of randomization in online algorithms. Algorithmica **11**(1), 2–14 (1994). https://doi.org/10.1007/BF01294260

6. Bienkowski, M., Mucha, M.: An improved algorithm for online reranking. In: Proceedings of the 37th AAAI Conference on Artificial Intelligence, pp. 6815–6822 (2023). https://doi.org/10.1609/aaai.v37i6.25835

7. Blum, A., Burch, C.: On-line learning and the metrical task system problem. Mach. Learn. **39**(1), 35–58 (2000). https://doi.org/10.1023/A:1007621832648

8. Borodin, A., El-Yaniv, R.: Online Computation and Competitive Analysis. Cambridge University Press, Cambridge (1998)

9. Derakhshan, M., Golrezaei, N., Manshadi, V.H., Mirrokni, V.S.: Product ranking on online platforms. In: Proceedings of the 21st ACM Conference on Economics and Computation (EC), p. 459. ACM (2020). https://doi.org/10.1145/3391403.3399483

10. Dwork, C., Kumar, R., Naor, M., Sivakumar, D.: Rank aggregation methods for the web. In: Proceedings of the 10th International World Wide Web Conference (WWW), pp. 613–622. ACM (2001). https://doi.org/10.1145/371920.372165

11. Feige, U., Lovász, L., Tetali, P.: Approximating min sum set cover. Algorithmica **40**(4), 219–234 (2004). https://doi.org/10.1007/s00453-004-1110-5

12. Fotakis, D., Kavouras, L., Koumoutsos, G., Skoulakis, S., Vardas, M.: The online min-sum set cover problem. In: Proceedings of the 47th International Colloquium on Automata, Languages and Programming (ICALP), pp. 51:1–51:16 (2020). https://doi.org/10.4230/LIPIcs.ICALP.2020.51

13. Fotakis, D., Lianeas, T., Piliouras, G., Skoulakis, S.: Efficient online learning of optimal rankings: Dimensionality reduction via gradient descent. In: Proceedings of the 33rd Annual Conference on Neural Information Processing Systems (NeurIPS), pp. 7816–7827 (2020)

14. Kamali, S.: Online list update. In: Kao, M.Y. (ed.) Encyclopedia of Algorithms, pp. 1448–1451. Springer, New York (2016). https://doi.org/10.1007/978-1-4939-2864-4_266

15. Littlestone, N., Warmuth, M.K.: The weighted majority algorithm. Inf. Comput. **108**(2), 212–261 (1994). https://doi.org/10.1006/inco.1994.1009

16. Sleator, D.D., Tarjan, R.E.: Amortized efficiency of list update and paging rules. Commun. ACM **28**(2), 202–208 (1985). https://doi.org/10.1145/2786.2793

Greedy Minimum-Energy Scheduling

Gunther Bidlingmaier[(⊠)]

Department of Computer Science, Technical University of Munich, Munich, Germany
g.bidlingmaier@tum.de

Abstract. We consider the problem of energy-efficient scheduling across multiple processors with a power-down mechanism. In this setting a set of n jobs with individual release times, deadlines, and processing volumes must be scheduled across m parallel processors while minimizing the consumed energy. When idle, each processor can be turned off to save energy, while turning it on requires a fixed amount of energy. For the special case of a single processor, the greedy Left-to-Right algorithm [7] guarantees an approximation factor of 2. We generalize this simple greedy policy to the case of $m \geq 1$ processors running in parallel and show that the energy costs are still bounded by $2\,\mathrm{OPT} + P$, where OPT is the energy consumed by an optimal solution and $P < \mathrm{OPT}$ is the total processing volume. Our algorithm has a running time of $\mathcal{O}(nf \log d)$, where d is the difference between the last deadline and the earliest release time, and f is the running time of a maximum flow calculation in a network of $\mathcal{O}(n)$ nodes.

Keywords: Scheduling · Greedy Algorithms · Approximation Algorithms

1 Introduction

Energy-efficiency has become a major concern in most areas of computing for reasons that go beyond the apparent ecological ones. At the hardware level, excessive heat generation from power consumption has become one of the bottlenecks. For the billions of mobile battery-powered devices, power consumption determines the length of operation and hence their usefulness. On the level of data centers, electricity is often the largest cost factor and cooling one of the major design constraints. Algorithmic techniques for saving power in computing environments employ two fundamental mechanisms, first the option to power down idle devices, and second the option to trade performance for energy-efficiency by speed-scaling processors. In this paper we study the former, namely classical deadline based scheduling of jobs on parallel machines which can be powered down with the goal of minimizing the consumed energy.

This work was supported by the Research Training Network of the Deutsche Forschungsgemeinschaft (DFG) (378803395: ConVeY).

© The Author(s), under exclusive license to Springer Nature Switzerland AG 2023
J. Byrka and A. Wiese (Eds.): WAOA 2023, LNCS 14297, pp. 59–73, 2023.
https://doi.org/10.1007/978-3-031-49815-2_5

In our setting, a computing device or processor has two possible states, it can be either *on* or *off*. If a processor is on, it can perform computations while consuming energy at a fixed rate. If a processor is off, the energy consumed is negligible but it cannot perform computation. Turning on a processor, i.e. transitioning it from the off-state to on-state consumes additional energy. The problem we have to solve is to schedule a number of jobs or tasks, each with its own processing volume and interval during which it has to be executed. The goal is to complete every job within its execution interval using a limited number of processors while carefully planning idle times for powering off processors such that the consumed energy is minimized. Intuitively, one aims for long but few idle intervals, so that the energy required for transitioning between the states is low, while avoiding turned on processors being idle for too long.

Previous Work. This fundamental problem in power management was first considered by [7] for a single processor. In their paper, they devise arguably the simplest algorithm one can think of which goes beyond mere feasibility. Their greedy algorithm *Left-to-Right* (LTR) is a 2-approximation and proceeds as follows. If the processor is currently busy, i.e. working on a job, then LTR greedily keeps the processor busy for as long as possible, always working on the released job with the earliest deadline. Once there are no more released jobs to be worked on, the processor becomes idle and LTR keeps the processor idle for as long as possible such that all remaining jobs can still be feasibly completed. At this point, the processor becomes busy again and LTR proceeds recursively until all jobs are completed. For a single processor, [3] develop an optimal dynamic program for unit jobs. [4] generalize this to general job weights with a running time of $\mathcal{O}(n^5)$, while [5] generalize it to multiple processors but again only unit jobs, increasing the complexity to $\mathcal{O}(n^7 m^5)$.

Obtaining good solutions for the general case of multiple processors and general job weights is difficult because of the additional constraint that every job can be worked on by at most a single processor at the same time. It is a major open problem whether the general multi-processor setting is NP-hard. It took further thirteen years for the first non-trivial result on this general setting to be developed. In their breakthrough paper, [1] develop the first constant-factor approximation for the problem. Their algorithm guarantees an approximation factor of $3 + \epsilon$ by relaxing an Integer Programming formulation of the problem. For making the rounded LP-solution feasible, they develop an additional extension algorithm *EXT-ALG*. This approximation factor is improved to $2 + \epsilon$ in [2] by incorporating into the Linear Program additional constraints for the number of processors required during every possible time interval. They also develop a combinatorial 6-approximation for the problem. As presented in the papers, all three algorithms run in pseudo-polynomial time. By using techniques presented in [1], the number of time slots which have to be considered can be reduced from d to $\mathcal{O}(n \log d)$, allowing the algorithms to run in polynomial time. More specifically, the number of constraints and variables of the Linear Programs reduces to $\mathcal{O}(n^2 \log^2 d)$. The running time of the EXT-ALG used by all three approximation algorithms is reduced to $\mathcal{O}(Fmn^3 \log^3 d)$, where F refers to a maximum flow calculation in a network with $\mathcal{O}(n \log d)$ nodes.

Contribution. In this paper we develop a greedy algorithm which is simpler and faster than the previous algorithms. The initially described greedy algorithm Left-to-Right of [7] is arguably the simplest algorithm one can think of for a single processor. We naturally extend LTR to multiple processors and show that this generalization still guarantees a solution of costs at most $2\,\mathrm{OPT} + P$, where $P < \mathrm{OPT}$ is the total processing volume. Our simple greedy algorithm *Parallel Left-to-Right* (PLTR) is the combinatorial algorithm with the best approximation guarantee and does not rely on Linear Programming and the necessary rounding procedures of [1] and [2]. It also does not require the EXT-ALG, which all previous algorithms rely on to make their infeasible solutions feasible in an additional phase.

Indeed, PLTR only relies on the original greedy policy of Left-to-Right: just keep processors in their current state (busy or idle) for as long as feasibly possible. For a single processor, LTR ensures feasibility by scheduling jobs according to the policy Earliest-Deadline-First (EDF). For checking feasibility if multiple processors are available, a maximum flow calculation is required since EDF is not sufficient anymore. Correspondingly, our generalization PLTR uses such a flow calculation for checking feasibility.

While the PLTR algorithm we describe in Sect. 2 is very simple, the structure exhibited by the resulting schedules is surprisingly rich. This structure consists of *critical sets of time slots* during which PLTR only schedules the minimum amount of volume which is feasibly possible. In Sect. 3 we show that whenever PLTR requires an additional processor to become busy at some time slot t, there must exist a critical set of time slots containing t. This in turn gives a lower bound for the number of busy processors required by any solution.

Devising an approximation guarantee from this structure is however highly non-trivial and much more involved than the approximation proof of the single-processor LTR algorithm, because one has to deal with sets of time slots and not just intervals. Our main contribution in terms of techniques is a complex procedure which (for the sake of the analysis only) carefully realigns the jobs scheduled in between critical sets of time slots such that it is sufficient to consider intervals as in the single processor case, see Sect. 4 for details. Here, we also show that our greedy policy leads to a much faster algorithm than the previous ones, namely to a running time $\mathcal{O}(nf \log d)$, where d is the maximal deadline and f is the running time for checking feasibility by finding a maximum flow in a network with $\mathcal{O}(n)$ nodes.

Formal Problem Statement. Formally, a problem instance consists of a set J of jobs with an integer release time r_j, deadline d_j, and processing volume p_j for every job $j \in J$. Each job $j \in J$ has to be scheduled across $m \geq 1$ processors for p_j units of time in the execution interval $E_j := [r_j, d_j]$ between its release time and its deadline. Preemption of jobs and migration between processors is allowed at discrete times and occurs without delay, but no more than one processor may process any given job at the same time. Without loss of generality, we assume the earliest release time to be 0 and denote the last deadline by d. The set of discrete time slots is denoted by $T:=\{0,\ldots,d\}$. The total amount of processing volume is $P:=\sum_{j\in J}p_j$.

Every processor is either completely off or completely on in every discrete time slot $t \in T$. A processor can only work on some job in the time slot t if it is in the on-state. A processor can be turned on and off at discrete times without delay. All processors start in the off-state. The objective now is to find a feasible schedule which minimizes the expended energy E, which is defined as follows. Each processor consumes 1 unit of energy for every time slot it is in the on-state and 0 units of energy if it is in the off-state. Turning a processor on consumes a constant amount of energy $q \geq 0$, which is fixed by the problem instance. In Graham's notation [6], this setting can be denoted with $m \mid r_j; \overline{d_j}; \text{pmtn} \mid E$.

Busy and Idle Intervals. We say a processor is *busy* at time $t \in T$ if some job is scheduled for this processor at time t. Otherwise, the processor is *idle*. Clearly a processor cannot be busy and off at the same time. An interval $I \subseteq T$ is a (full) *busy interval* for processor $k \in [m]$ if I is inclusion maximal on the condition that processor k is busy in every $t \in I$. Correspondingly, an interval $I \subseteq T$ is a *partial busy interval* for processor k if I is not inclusion maximal on the condition that processor k is busy in very $t \in I$. We define (partial and full) *idle intervals, on intervals*, and *off intervals* of a processor analogously via inclusion maximality. Observe that if a processor is idle for more than q units of time, it is worth turning the processor off during the corresponding idle interval. Our algorithm will specify for each processor when it is busy and when it is idle. Each processor is then defined to be in the off-state during idle intervals of length greater than q and otherwise in the on-state. Accordingly, we can express the costs of a schedule S in terms of busy and idle intervals.

For a multi-processor schedule S, let S^k denote the schedule of processor k. Furthermore, for fixed k, let $\mathcal{N}, \mathcal{F}, \mathcal{B}, \mathcal{I}$ be the set of on, off, busy, and idle intervals of S^k. We partition the costs of processor k into the costs $\text{on}(S^k)$ for residing in the on-state and the costs $\text{off}(S^k)$ for transitioning between the off-state and the on-state, hence $\text{costs}(S^k) = \text{on}(S^k) + \text{off}(S^k) = \sum_{N\in\mathcal{N}} q + |N|$. Equivalently, we partition the costs of processor k into the costs $\text{idle}(S^k):=\sum_{I\in\mathcal{I}}\min\{|I|,q\}$ for being idle and the costs $\text{busy}(S^k):=\sum_{B\in\mathcal{B}}|B|$ for being busy. The total costs of a schedule S are the total costs across all processors, i.e. $\text{costs}(S) = \sum_{k=1}^{m}\text{costs}(S^k)$. Clearly we have $\sum_{k=1}^{m}\text{busy}(k) = P$, this means for an approximation guarantee the critical part is bounding the idle costs.

Lower and Upper Bounds for the Number of Busy Processors. We specify a generalization of our problem which we call *deadline-scheduling-with-processor-bounds*. Where in the original problem, for each time slot t, between 0 and m processors were allowed to be working on jobs, i.e. being busy, we now specify a lower bound $l_t \geq 0$ and an upper bound $m_t \leq m$. For a feasible solution to *deadline-scheduling-with-processor-bounds*, we require that in every time slot t, the number of busy processors, which we denote with $\mathsf{vol}(t)$, lies within the lower and upper bounds, i.e. $l_t \leq \mathsf{vol}(t) \leq m_t$. This will allow us to express the PLTR greedy policy of keeping processors idle or busy, respectively. Note that this generalizes the problem *deadline-scheduling-on-intervals* introduced by [1] by additionally introducing lower bounds.

Properties of an Optimal Schedule

Definition 1. *Given some arbitrary but fixed order on the number of processors, a schedule S fulfills the* **stair-property** *if it uses the lower numbered processors first, i.e. for every $t \in T$, if processor $k \in [m]$ is busy at t, then every processor $k' \leq k$ is busy at t. This symmetrically implies that if processor $k \in [m]$ is idle at t, then every processor $k' \geq k$ is idle at t.*

Lemma 1. *For every problem instance we can assume the existence of an optimal schedule S_{opt} which fulfills the stair-property.*

2   Algorithm

The *Parallel Left-to-Right* (PLTR) algorithm shown in Algorithm 1 iterates through the processors in some arbitrary but fixed order and keeps the current processor idle for as long as possible such that the scheduling instance remains feasible. Once the current processor cannot be kept idle for any longer, it becomes busy and PLTR keeps it and all lower-numbered processors busy for as long as possible while again maintaining feasibility. The algorithm enforces these restrictions on the busy processors by iteratively making the lower and upper bounds l_t, m_t of the corresponding instance of *deadline-scheduling-with-processor-bounds* more restrictive. Visually, when considering the time slots on an axis from left to right and when stacking the schedules of the individual processors on top of each other, this generalization of the single processor *Left-to-Right* algorithm hence proceeds *Top-Left-to-Bottom-Right*.

Once PLTR returns with the corresponding tight upper and lower bounds m_t, l_t, an actual schedule S_{pltr} can easily be constructed by running the flow-calculation used for the feasibility check depicted in Fig. 1 or just taking the result of the last flow-calculation performed during PLTR. The mapping from this flow to an actual assignment of jobs to processors and time slots can then be defined as described in Lemma 2, which also ensures that the resulting schedule fulfills the stair-property from Definition 1, i.e. that it always uses the lower-numbered processors first.

As stated in Lemma 2, the check for feasibility in subroutines `keepidle` and `keepbusy` can be performed by calculating a maximum α-ω flow in the flow network given in Fig. 1 with a node u_j for every job $j \in J$ and a node v_t for every time slot $t \in T$ including the corresponding incoming and outgoing edges.

Lemma 2. *There exists a feasible solution to an instance of deadline-scheduling-with-processor-bounds l_t, m_t if and only if the maximum α-ω flow in the corresponding flow network depicted in Fig. 1 has value P.*

Algorithm 1. Parallel Left-to-Right

$m_t \leftarrow m$ for all $t \in T$
$l_t \leftarrow 0$ for all $t \in T$
for $k \leftarrow m$ to 1 **do**
    $t \leftarrow 0$
    while $t \le d$ **do**
        $t \leftarrow$ KEEPIDLE(k, t)
        $t \leftarrow$ KEEPBUSY(k, t)

---

function KEEPIDLE(k, t)
    find maximal $t' > t$ s.t. $\exists$ feasible schedule with $m_{t''} = k - 1$ for all $t'' \in [t, t')$
    $m_{t''} \leftarrow k - 1$ for all $t'' \in [t, t')$
    return t'
function KEEPBUSY(k, t)
    find maximal $t' > t$ s.t. $\exists$ feasible schedule with $l_{t''} = \max\{k, l_{t''}\}$ for all $t'' \in [t, t')$
    $l_{t''} \leftarrow \max\{k, l_{t''}\}$ for all $t'' \in [t, t')$
    return t'

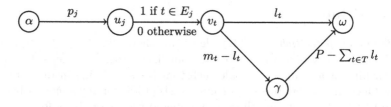

Fig. 1. The Flow-Network for checking feasibility of an instance of *deadline-scheduling-with-processor-bounds* l_t and m_t for the number of busy processors at $t \in T$. There are nodes u_j, v_t with the corresponding edges for every job $j \in J$ and for every time slot $t \in T$, respectively.

Theorem 1. *Given a feasible problem instance, algorithm PLTR constructs a feasible schedule.*

Proof. By definition of subroutines `keepidle` and `keepbusy`, PLTR only modifies the upper and lower bounds m_t, l_t for the number of busy processors such that the resulting instance of *deadline-scheduling-with-processor-bounds* remains feasible. The correctness of the algorithm then follows from the correctness of the flow-calculation for checking feasibility, which is implied by Lemma 2.

3 Structure of the PLTR-Schedule

3.1 Types of Volume

Definition 2. *For a schedule S, a job $j \in J$, and a set $Q \subseteq T$ of time slots, we define*

1. *the volume $vol_S(j, Q)$ as the number of time slots of Q for which j is scheduled by S,*

2. the forced volume $fv(j, Q)$ as the minimum number of time slots of Q for which j has to be scheduled in every feasible schedule, i.e. $fv(j, Q) := \max\{0; p_j - |E_j \setminus Q|\}$,

3. the unnecessary volume $uv_S(j, Q)$ as the amount of volume which does not have to scheduled during Q, i.e. $uv_S(j, Q) := vol_S(j, Q) - fv(j, Q)$,

4. the possible volume $pv(j, Q)$ as the maximum amount of volume which j can be feasibly scheduled in Q, i.e. $pv(j, Q) := \min\{p_j, |E_j \cap Q|\}$.

Since the corresponding schedule S will always be clear from context, we omit the subscript for vol and uv. We extend our volume definitions to sets $J' \subseteq J$ of jobs by summing over all $j \in J'$, i.e. $vol(J', Q) := \sum_{j \in J'} vol(j, Q)$. If the first parameter is omitted, we refer to the whole set J, i.e. $vol(Q) := vol(J, Q)$. For single time slots, we omit set notation, i.e. $vol(t) := vol(J, \{t\})$. Clearly we have for every feasible schedule, every $Q \subseteq T, j \in J$ that $fv(j, Q) \leq vol(j, Q) \leq pv(j, Q)$. The following definitions are closely related to these types of volume.

Definition 3. Let $Q \subseteq T$ be a set of time slots. We define

1. the density $\phi(Q) := fv(J, Q)/|Q|$ as the average amount of processing volume which has to be completed in every slot of Q,

2. the peak density $\hat{\phi}(Q) := \max_{Q' \subseteq Q} \phi(Q')$,

3. the deficiency $def(Q) := fv(Q) - \sum_{t \in Q} m_t$ as the difference between the amount of volume which has to be completed in Q and the processing capacity available in Q,

4. the excess $exc(Q) := \sum_{t \in Q} l_t - pv(Q)$ as the difference between the processor utilization required in Q and the amount of work available in Q.

If $\hat{\phi}(Q) > k - 1$, then clearly at least k processors are required in some time slot $t \in Q$ for every feasible schedule. If $def(Q) > 0$ or $exc(Q) > 0$ for some $Q \subseteq T$, then the problem instance is clearly infeasible.

3.2   Critical Sets of Time Slots

The following Lemma 5 provides the crucial structure required for the proof of the approximation guarantee. Intuitively, it states that whenever PLTR requires processor k to become busy at some time slot t, there must be some critical set $Q \subseteq T$ of time slots during which the volume scheduled by PLTR is minimal. This in turn implies that processor k needs to be busy at some point during Q in every feasible schedule. The auxiliary Lemmas 3 and 4 provide a necessary and more importantly also sufficient condition for the feasibility of an instance of *deadline-scheduling-with-processor-bounds* based on the excess $exc(Q)$ and the deficiency $def(Q)$ of sets $Q \subseteq T$. Lemmas 3 and 4 are again a generalization of the corresponding feasibility characterization in [1] for their problem deadline-scheduling-on-intervals, which only defines upper bounds.

Lemma 3. For every α-ω cut $(S, \bar{S})$ in the network given in Fig. 1 we have at least one of the following two lower bounds for the capacity $c(S)$ of the cut: $c(S) \geq P - def(Q(S))$ or $c(S) \geq P - exc(Q(\bar{S}))$, where $Q(S) := \{t \mid v_t \in S\}$.

Lemma 4. *An instance of deadline-scheduling-with-processor-bounds is feasible if and only if* $\mathsf{def}(Q) \leq 0$ *and* $\mathsf{exc}(Q) \leq 0$ *for every* $Q \subseteq T$.

Definition 4. *A time slot* $t \in T$ *is called an* engagement *of processor* k *if* $t = \min B$ *for some busy interval* B *on processor* k. *We say processor* k *is* engaged *at time* t *if* t *is an engagement of processor* k. *A time slot* $t \in T$ *is just called an* engagement *if it is an engagement of processor* k *for some* $k \in [m]$.

Lemma 5. *Let* $Q \subseteq T$ *be a set of time slots and* $t \in T$ *an engagement of processor* $k \in [m]$. *We call* Q *a* tight set *for engagement* t *of processor* k *if* $t \in Q$ *and*

$$\mathsf{fv}(Q) = \mathsf{vol}(Q),$$
$$\mathsf{vol}(t') \geq k - 1 \qquad\qquad \text{for all } t' \in Q \text{ , and}$$
$$\mathsf{vol}(t') \geq k \qquad\qquad \text{for all } t' \in Q \text{ with } t' \geq t.$$

For every engagement t *of some processor* $k \in [m]$ *in the schedule* S_{pltr} *constructed by PLTR, there exists a tight set* $Q_t \subseteq T$ *for engagement* t *of processor* k.

Proof. Suppose for contradiction that there is some engagement $t \in T$ of processor $k \in [m]$ and no such Q exists for t, i.e. every $Q \subseteq T$ containing t violates at least one of the three conditions in the Lemma. We show that PLTR would have extended the idle interval on processor k which ends at t. Consider the step in PLTR when t was the result of `keepidle` on processor k. Let $l_{t'}$, $m_{t'}$ be the lower and upper bounds for $t' \in T$ right after the calculation of t and the corresponding update of the bounds by `keepidle`. We modify the bounds by decreasing m_t by 1. Note that at this point $m_{t'} \geq k$ for every $t' > t$ and $m_{t'} \geq k - 1$ for every t'.

Consider $Q \subseteq T$ such that $t \in Q$ and $\mathsf{fv}(Q) < \mathsf{vol}(Q)$. Before our decrement of m_t we had $m_Q := \sum_{t' \in Q} m_{t'} \geq \mathsf{vol}(Q) > \mathsf{fv}(Q)$. The inequality $m_Q \geq \mathsf{vol}(Q)$ here follows since the upper bounds $m_{t'}$ are monotonically decreasing during PLTR. Since our modification decreases m_Q by at most 1, we hence still have $m_Q \geq \mathsf{fv}(Q)$ after the decrement of m_t. Consider $Q \subseteq T$ such that $t \in Q$ and $\mathsf{vol}(t') < k - 1$ for some t'. At the step in PLTR considered by us, i.e. when `keepidle` returned t on processor k, we hence have $m_{t'} \geq k - 1 > \mathsf{vol}(t')$. Before our decrement of m_t we therefore have $m_Q > \mathsf{vol}(Q) \geq \mathsf{fv}(Q)$, which implies $m_Q \geq \mathsf{fv}(Q)$ after the decrement. Finally, consider $Q \subseteq T$ such that $t \in Q$ and $\mathsf{vol}(t') < k$ for some $t' > t$. At the step in PLTR considered by us, we again have $m_{t'} \geq k > \mathsf{vol}(t')$, which implies $m_Q \geq \mathsf{fv}(Q)$ after our decrement of m_t. In summary, if for t no Q exists as characterized in the lemma, the engagement of processor k at t could not have been the result of `keepidle` on processor k.

Lemma 6. *We call a set* $C_k \subseteq T$ critical set *for processor* k *if* C_k *fulfills that*

- $C_k \supseteq C_{k'}$ *for every critical set for processor* $k' > k$,
- $t \in C_k$ *for every engagement* t *of processor* k,

- $fv(C_k) = vol(C_k)$,
- $vol(t) \geq k - 1$ for every $t \in C_k$, and
- $\phi(C_k)$ is maximal.

For every processor $k \in [m]$ of S_{pltr} which is not completely idle, there exists a critical set C_k for processor k.

Proof. We show the existence by induction over the processors $m, \ldots, 1$. For processor m, consider the union of all tight sets over engagements of processor m. This set fulfills all conditions necessary except for the maximality in regard to ϕ. Suppose that the critical sets $C_m, \ldots, C_{k+1}$ exist. Take $Q_k \subseteq T$ as the union of C_{k+1} and all tight sets over engagements of processor k. By definition of C_{k+1}, we have $Q_k \supseteq C_{k'}$ for all $k' > k$. By construction of Q_k, every engagement t of processor k is contained in Q_k. Finally, we have $fv(Q_k) = vol(Q_k)$ and $vol(t) \geq k - 1$ for every $t \in Q_k$ since all sets in the union fulfill these properties.

3.3   Definitions Based on Critical Sets

Definition 5. *For the critical set C_k of some processor $k \in [m]$, we define $crit(C_k):=k$. Let $\succeq$ be the total order on the set of critical sets C across all processors which corresponds to $crit$, i.e. $C \succeq C'$ if and only if $crit(C) \geq crit(C')$. Equality in regard to $\succeq$ is denoted with $\sim$. We extend the definition of $crit$ to general time slots $t \in T$ with $crit(t):= \max\{crit(C) \mid C$ is critical set, $t \in C\}$ if $t \in C$ for some critical set C and otherwise $crit(t):=0$. We further extend $crit$ to intervals $D \subseteq T$ with $crit(D):=\max\{crit(t) \mid t \in D\}$*

Definition 6. *A nonempty interval $V \subseteq T$ is a valley if V is inclusion maximal on the condition that $C \sim V$ for some fixed critical set C. Let $D_1, \ldots, D_l$ be the maximal intervals contained in a critical set C. A nonempty interval V is a valley of C if V is exactly the valley between D_a and D_{a+1} for some $a < l$, i.e. $V = [\max D_a + 1, \min D_{a+1} - 1]$. By the choice of C as a critical set (property 1), a valley of C is indeed a valley. We define the jobs $J(V) \subseteq J$ for a valley V as all jobs which are scheduled by S_{pltr} in every $t \in V$.*

Definition 7. *For a critical set C, an interval $D \subseteq T$ is a section of C if $D \cap C$ contains only full subintervals of C and at least one subinterval of C. For a critical set C and a section D of C, the left valley V_l is the valley of C ending at $\min(C \cap D) - 1$, if such a valley of C exists. Symmetrically, the right valley V_r is the valley of C starting at $\max(C \cap D) + 1$, if such a valley of C exists.*

Lemma 7. *For every critical set C, every section $D \subseteq T$ of C, we have: if $\phi(C \cap D) \leq crit(C) - \delta$ for some $\delta \in \mathbb{N}$, then the left valley V_l or the right valley V_r of C and D is defined and $|J(V_l)| + |J(V_r)| \geq \delta$. We take $|J(V)|:=0$ if V is not defined.*

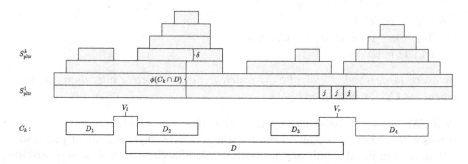

Fig. 2. The left and right valley V_l and V_r of the critical set C_k for processor k and a section D of C_k. Lemma 7 guarantees that δ jobs are scheduled at every slot of V_l or V_r.

Proof. Refer to Fig. 2 for a visual sketch of the lemma. By the choice of C as a critical set with $c:=\mathsf{crit}(C)$, we have $\mathsf{vol}(C \cap D) \geq (c-1) \cdot |C \cap D|$. If this inequality is fulfilled strictly, then with the premise $\mathsf{fv}(C \cap D)/|C \cap D| \leq c - \delta$ we directly get $\mathsf{uv}(C \cap D)/|C \cap D| > \delta - 1$. This implies that there are at least δ jobs j scheduled in $C \cap D$ with $\mathsf{uv}(j, C \cap D) > 0$. Such jobs can be scheduled in the part of C not contained in D, i.e. we must have $E_j \cap (C \setminus D) \neq \emptyset$ and hence the left valley V_l or the right valley V_r of C and D must be defined. Since these jobs j are scheduled in C only for the minimum amount possible, i.e. $\mathsf{vol}(j, C) = \mathsf{fv}(j, C) > 0$, they must be scheduled in every $t \in E_j \setminus C$ and are therefore contained in $J(V_l)$ or $J(V_r)$.

If on the other hand we have equality, i.e. $\mathsf{vol}(C \cap D) = (c-1) \cdot |C \cap D|$, then let t be an engagement of processor c. Since $\mathsf{vol}(t) > c - 1$, we must have $t \notin C \cap D$. By the same argument as before, we have that if $\mathsf{fv}(C \cap D)/|C \cap D| \leq c - \delta$, then $\mathsf{uv}(C \cap D)/|C \cap D| \geq \delta - 1$. Let $J':=\{j \in J \mid \mathsf{uv}(j, C \cap D) > 0\}$. Since $\mathsf{uv}(j, C \cap D) \leq |C \cap D|$ for every $j \in J$, we have $|J'| \geq \delta - 1$. If this lower bound is fulfilled with equality, then every $j \in J'$ must be scheduled in every time slot of $C \cap D$ and hence $\mathsf{fv}(J', C \setminus D) = \mathsf{vol}(J', C \setminus D)$. Now suppose for contradiction that all jobs j scheduled during $C \setminus D$ which are not contained in J' have $E_j \cap C \cap D = \emptyset$. Then $\mathsf{fv}(C \setminus D) = \mathsf{vol}(C \setminus D)$ and we get $\phi(C \setminus D) > \phi(C)$ since by case assumption $\mathsf{vol}(C \cap D)/|C \cap D| = (c-1) < \phi(C)$. With $\mathsf{vol}(t) \leq c-1$ for every $t \in C \cap D$, we know that $\mathsf{crit}(C \cap D) \leq c$ and therefore $C \setminus D$ is still a critical set for processor c but has higher density than C, contradicting the choice of C. Therefore, there must exist a job $j \notin J'$ scheduled in $C \setminus D$ with an execution interval intersecting $C \cap D$. In any case, we have at least δ jobs scheduled in C with an execution interval intersecting both $C \setminus D$ and $C \cap D$. This implies that the left valley V_l or the right valley V_r of C and D exists and that at least δ jobs are contained in $J(V_l)$ or $J(V_r)$.

4   Modification of the PLTR-Schedule for Analysis

In this section we modify the schedule S_{pltr} returned by PLTR in two steps. We stress that this is for the analysis only and not part of PLTR. The first step augments specific processors with auxiliary busy slots such that in every critical set C at least the first $\text{crit}(C)$ processors are busy all the time. For the single processor LTR algorithm, the crucial property for the approximation guarantee is that every idle interval of S_{opt} can intersect at most 2 distinct idle intervals of the schedule returned by LTR. The second modification step of S_{pltr} is more involved and establishes this crucial property on every processor $k \in [m]$ by making use of Lemma 7. More specifically, it will establish the stronger property that $\hat{\phi}(B) > k - 1$ for every busy interval B on processor k with $\text{crit}(B) \geq 2$, i.e. that every feasible schedule requires k busy processors at some point during B. Idle intervals surrounded by only busy intervals B with $\text{crit}(B) \leq 1$ are then handled in Lemma 12 with essentially the same argument as for the single processor LTR algorithm. By making sure that the modifications cannot decrease the costs of our schedule, we obtain an upper bound for the costs of S_{pltr}.

4.1   Augmentation and Realignment

We transform S_{pltr} into the *augmented schedule* S_{aug} by adding for every t with $k := \text{crit}(t) \geq 2$ and $\text{vol}(t) = k - 1$ an auxiliary busy slot on processor k. No job is scheduled in this auxiliary busy slot on processor k and it does also not count towards the volume of this slot. It merely forces processor k to be in the on-state at time k while allowing us to keep thinking in terms of idle and busy intervals in our analysis of the costs.

Lemma 8. *In S_{aug} processors $1, \ldots, \text{crit}(t)$ are busy in every slot $t \in T$ with $\text{crit}(t) \geq 2$.*

Proof. The property directly follows from our choice of the critical sets, the definition of $\text{crit}(t)$, and the construction of S_{aug}.

As a next step, we transform S_{aug} into the *realigned schedule* S_{real} using Algorithm 2. We briefly sketch the ideas behind this realignment. Lemma 8 guarantees us that every busy interval B on processor k is a section of the critical set C with $C \sim B$. It also guarantees that the left and right valley V_l, V_r of C and B do not end within an idle interval on processor k. Lemma 7 in turn implies that if the density of B is too small to guarantee that S_{opt} has to use processor k during B, i.e. if $\hat{\phi}(B) \leq k - 1$, then V_l or V_r is defined and there is some j scheduled in every slot of V_l or V_r. Let V be the corresponding left or right valley of C and D for which such a job j exists. Instead of scheduling j on the processors below k, we can schedule j on processor k in idle time slots during V. This merges the busy interval B with at least one neighbouring busy interval on processor k. In the definition of the realignment, we will call this process of filling the idle slots during V on processor k the *closing of valley V on processor k*. The corresponding subroutine is called $\texttt{close}(k, V)$.

The crucial part is ensuring that this merging of busy intervals by closing a valley continues to be possible throughout the realignment whenever we encounter a busy interval with a density too small. For this purpose, we go through the busy intervals on each processor in decreasing order of their criticality, i.e. in the order of $\succeq$. We also allow every busy slot to be used twice for the realignment (see variable $\sup_V$ in Algorithm 2) by introducing further auxiliary busy slots, since for a section D of the critical set C, both the right and the left valley might be closed on processor k in the worst case. This allows us to maintain the invariants stated in Lemma 9 during the realignment process, which correspond to the initial properties of Lemmas 7 and 8 for S_{aug}.

4.2   Invariants for Realignment

Lemma 9. *For an arbitrary step during the realignment of S_{aug} and a valley $V \subseteq T$, let the* critical processor k_V *for V be the highest processor such that*

- *processor k_V is not fully filled yet, i.e. $\texttt{fill}(k_V, T)$ has not yet returned,*
- *no $V' \supseteq V$ has been closed on k_V so far, and*
- *there is a (full) busy interval $B \subseteq V$ on processor k_V.*

We take $k_V := 0$ if no such processor exists. At every step in the realignment of S_{aug} the following invariants hold for every valley V, where C denotes the critical set with $C \sim V$.

1. *If $\phi(C \cap D) \leq k_V - \delta$ for some $\delta \in \mathbb{N}$, some section $D \subseteq V$ of C, then the left valley V_l or the right valley V_r of C, D exists and $\sup_{V_l} + \sup_{V_r} \geq 2\delta$.*
2. *For every $t \in C \cap V$, processors $1, \ldots, k_V$ are busy at t.*
3. *Every busy interval $B \subseteq V$ on processor k_V with $B \sim V$ is a section of C.*

Lemma 10. *The resulting schedule S_{real} of the realignment of S_{aug} is defined.*

Lemma 11. *For every processor $k \in [m]$ and every busy interval B on processor k in S_{real} with $\text{crit}(B) \geq 2$, we have $\hat{\phi}(B) > k - 1$.*

Proof. We show that $\texttt{fill}(k, T)$ establishes the property on processor k. The claim then follows since $\texttt{fill}(k, T)$ does not change the schedules of processors above k. We know that on processor k busy intervals are only extended, since in $\texttt{fill}(k, T)$ we only close valleys for busy intervals B on k which are a section of the corresponding critical set C. Let $B \subseteq V$ be a busy interval on processor k in S_{real} with $B \sim V$ and $\text{crit}(B) \geq 2$. No valley $W \supseteq V$ can have been closed on k since otherwise there would be no $B \subseteq V$ in S_{real}. Therefore, at some point $\texttt{fill}(k, V)$ must be called. Consider the point in $\texttt{fill}(k, V)$ when the while-loop terminates. Clearly at this point all busy intervals $B' \subseteq V$ with $B' \sim V$ on processor k have $\hat{\phi}(B') > k-1$. At this point there must also be at least one such B' for B to be a busy interval on k in S_{real} with $B \sim V$ and $B \subseteq V$. In particular, one such B' must have $B' \subseteq B$, which directly implies $\hat{\phi}(B) \geq \hat{\phi}(B') > k - 1$.

Algorithm 2. Realignment of S_{aug} for analysis only

$\sup_V \leftarrow 2|J(V)|$ for every valley V
for $k \leftarrow m$ to 1 **do**
   FILL(k, T)
   $\sup_V \leftarrow \sup_V -1$ for every V s.t. some V' with $V' \cap V \neq \emptyset$ was closed on proc. k

---

function FILL(k, V)
   if crit$(V) \leq 1$ **then**
      return
   let C be the critical set s.t. $C \sim V$
   while $\exists$ busy interval $B \subseteq V$ on processor k with $B \sim V$ and $\hat{\phi}(B) \leq k-1$ **do**
      let V_l, V_r be the left, right valley for C and B (given B is a section of C)
      if V_l exists and $\sup_{V_l} > 0$ **then**
         CLOSE(k, V_l)
      else if V_r exists and $\sup_{V_r} > 0$ **then**
         CLOSE(k, V_r)
   for every valley $V' \subseteq V$ of C which has not been closed on k **do**
      FILL(k, V')
function CLOSE(k, V)
   for every $t \in V$ which is idle on processor k **do**
      if processors $1, \ldots, k-1$ are idle at t **then**
         introduce new auxiliary busy slot on processor k at time t
      else
         move busy slot t of highest processor $\leq k-1$ to processor k

---

While with Lemma 11 we have our desired property for busy intervals B of crit$(B) \geq 2$, we still have to handle busy intervals of crit$(B) \leq 1$. To be precise, we have to handle idle intervals which are surrounded only by busy intervals B of crit$(B) \leq 1$. We will show that this constellation can only occur in S_{real} on processor 1 and that the realignment has not done any modifications in these intervals, i.e. S_{pltr} and S_{real} do not differ for these intervals. With the same argument as for the original single-processor Left-to-Right algorithm, we then get that at least one processor has to be busy in any schedule during these intervals.

Lemma 12. Let I be an idle interval in S_{real} on some processor k and let B_l, B_r be the busy intervals on k directly to the left and right of I with crit$(B_l) \leq 1$ and crit$(B_r) \leq 1$. Allow B_l to be empty, i.e. we might have $\min I = 0$, but B_r must be nonempty, i.e. $\max I < d$. Then we must have $k = 1$ and $\hat{\phi}(B_l \cup I \cup B_r) > 0$.

Lemma 13. For every processor k, every idle interval on processor k in S_{opt} intersects at most two distinct idle intervals of processor k in S_{real}.

Proof. Let I_{opt} be an idle interval in S_{opt} on processor k intersecting three distinct idle intervals of processor k in S_{real}. Let I be the middle one of these three idle intervals. Lemma 12 and Lemma 11 imply that k busy processors are required during I and its neighboring busy intervals. This makes it impossible for S_{opt} to be idle on processor k during the whole interval I_{opt}.

4.3   Approximation Guarantee and Running Time

Lemma 13 finally allows us to bound the costs of the schedule S_{real} with the same arguments as in the proof for the single-processor LTR algorithm of [7]. We complement this with an argument that the augmentation and realignment could have only increased the costs of S_{pltr} and that we have hence also bounded the costs of the schedule returned by our algorithm PLTR.

Theorem 2. *Algorithm* PLTR *constructs a schedule of costs at most* $2\,\text{OPT} + P$.

Proof. We begin by bounding $\text{costs}(S_{\text{real}})$ as in the lemma. First, we show that $\text{idle}(S_{\text{real}}^k) \leq 2\,\text{off}(S_{\text{opt}}^k) + \text{on}(S_{\text{opt}}^k)$ for every processor $k \in [m]$. Let $\mathcal{I}_1$ be the set of idle intervals on S_{real}^k which intersect some off interval of S_{opt}^k. Lemma 13 implies that $\mathcal{I}_1$ contains as most twice as many intervals as there are off intervals in S_{opt}^k. Since the costs of each idle interval are at most q, and the costs of each off interval are exactly q, the costs of all idle intervals in $\mathcal{I}_1$ is bounded by $2\,\text{off}(S_{\text{opt}}^k)$. Let $\mathcal{I}_2$ be the set of idle intervals on S_{real}^k which do not intersect any off interval in S_{opt}^k. The total length of these intervals is naturally bounded by $\text{on}(S_{\text{opt}}^k)$.

We continue by showing that $\text{busy}(S_{\text{real}}) \leq 2P$. By construction of S_{aug} and the definition of $\sup_V$ and close, we introduce at most as many auxiliary busy slots at every slot $t \in T$ as there are jobs scheduled at t in S_{pltr}. For S_{aug}, an auxiliary busy slot is only added for t with $\text{crit}(t) \geq 2$ and hence $\text{vol}(t) \geq 1$. Furthermore, initially $\sup_V = 2|J(V)|$ for every valley V and $\sup_V$ is decremented if some V' intersecting V is closed during $\text{fill}(k, T)$. During $\text{fill}(k, T)$ at most a single V' containing t is closed for every $t \in T$. Finally, auxiliary busy slots introduced by S_{aug} are used in the subroutine close. This establishes the lower bound $\text{costs}(S_{\text{real}}) = \text{idle}(S_{\text{real}}) + \text{busy}(S_{\text{real}}) \leq 2\,\text{off}(S_{\text{opt}}) + \text{on}(S_{\text{opt}}) + 2P \leq 2\,\text{OPT} + P$ for our realigned schedule.

We complete the proof by arguing that $\text{costs}(S_{\text{pltr}}) \leq \text{costs}(S_{\text{real}})$ since transforming S_{real} back into S_{pltr} does not increase the costs of the schedule. Removing the auxiliary busy slots clearly cannot increase the costs. Since the realignment of S_{aug} only moves busy slots between processors, but not between different time slots, we can easily restore S_{pltr} (up to permutations of the jobs scheduled on the busy processors at the same time slot) by moving all busy slots back down to the lower numbered processors. By the same argument as in Lemma 1, this does not increase the total costs of the schedule.

Theorem 3. *Algorithm* PLTR *has a running time of* $\mathcal{O}(nf \log d)$ *where f denotes the time needed for finding a maximum flow in a network with $\mathcal{O}(n)$ nodes.*

Acknowledgement. A comprehensive version of this paper, including all proofs, is available on arXiv: https://arxiv.org/abs/2307.00949. Thanks to Prof. Dr. Susanne Albers for her supervision during my studies. The idea of generalizing the Left-to-Right algorithm emerged in discussions during this supervision.

References

1. Antoniadis, A., Garg, N., Kumar, G., Kumar, N.: Parallel machine scheduling to minimize energy consumption. In: Proceedings of the Thirty-First Annual ACM-SIAM Symposium on Discrete Algorithms, SODA 2020, pp. 2758–2769. Society for Industrial and Applied Mathematics, USA (2020)
2. Antoniadis, A., Kumar, G., Kumar, N.: Skeletons and minimum energy scheduling. In: Ahn, H.K., Sadakane, K. (eds.) 32nd International Symposium on Algorithms and Computation (ISAAC 2021). Leibniz International Proceedings in Informatics (LIPIcs), vol. 212, pp. 51:1–51:16. Schloss Dagstuhl - Leibniz-Zentrum für Informatik, Dagstuhl, Germany (2021). https://doi.org/10.4230/LIPIcs.ISAAC.2021.51. https://drops.dagstuhl.de/opus/volltexte/2021/15484
3. Baptiste, P.: Scheduling unit tasks to minimize the number of idle periods: a polynomial time algorithm for offline dynamic power management. In: Proceedings of the Seventeenth Annual ACM-SIAM Symposium on Discrete Algorithm, SODA 2006, pp. 364–367. Society for Industrial and Applied Mathematics, USA (2006)
4. Baptiste, P., Chrobak, M., Dürr, C.: Polynomial time algorithms for minimum energy scheduling. In: Arge, L., Hoffmann, M., Welzl, E. (eds.) ESA 2007. LNCS, vol. 4698, pp. 136–150. Springer, Heidelberg (2007). https://doi.org/10.1007/978-3-540-75520-3_14
5. Demaine, E.D., Ghodsi, M., Hajiaghayi, M.T., Sayedi-Roshkhar, A.S., Zadimoghaddam, M.: Scheduling to minimize gaps and power consumption. In: Proceedings of the Nineteenth Annual ACM Symposium on Parallel Algorithms and Architectures, SPAA 2007, pp. 46–54. Association for Computing Machinery, New York (2007). https://doi.org/10.1145/1248377.1248385
6. Graham, R., Lawler, E., Lenstra, J., Rinnooy Kan, A.: Optimization and approximation in deterministic sequencing and scheduling: a survey. Ann. Discret. Math. **5**, 287–326 (1979). https://doi.org/10.1016/S0167-5060(08)70356-X
7. Irani, S., Shukla, S.K., Gupta, R.K.: Algorithms for power savings. In: Proceedings of the Fourteenth Annual ACM-SIAM Symposium on Discrete Algorithms, 12–14 January 2003, Baltimore, Maryland, USA, pp. 37–46. ACM/SIAM (2003). http://dl.acm.org/citation.cfm?id=644108.644115

Scheduling with Speed Predictions

Eric Balkanski$^{(\boxtimes)}$, Tingting Ou$^{(\boxtimes)}$, Clifford Stein$^{(\boxtimes)}$, and Hao-Ting Wei$^{(\boxtimes)}$

Department of Industrial Engineering and Operations Research,
Columbia University, New York, USA
{eb3224,to2372,hw2738}@columbia.edu, cliff@ieor.columbia.edu

Abstract. Algorithms with predictions is a recent framework that has been used to overcome pessimistic worst-case bounds in incomplete information settings. In the context of scheduling, very recent work has leveraged machine-learned predictions to design algorithms that achieve improved approximation ratios in settings where the processing times of the jobs are initially unknown. In this paper, we study the speed-robust scheduling problem where the speeds of the machines, instead of the processing times of the jobs, are unknown and augment this problem with predictions.

Our main result is an algorithm that achieves a $\min\{\eta^2(1+\alpha),(2+2/\alpha)\}$ approximation, for any $\alpha \in (0,1)$, where $\eta \geq 1$ is the prediction error. When the predictions are accurate, this approximation outperforms the best known approximation for speed-robust scheduling without predictions of $2-1/m$, where m is the number of machines, while simultaneously maintaining a worst-case approximation of $2+2/\alpha$ even when the predictions are arbitrarily wrong. In addition, we obtain improved approximations for three special cases: equal job sizes, infinitesimal job sizes, and binary machine speeds. We also complement our algorithmic results with lower bounds. Finally, we empirically evaluate our algorithm against existing algorithms for speed-robust scheduling. The full version of the paper can be referred to the following link https://arxiv.org/abs/2205.01247.

Keywords: Algorithms with prediction · Scheduling · Approximation algorithm

1 Introduction

In many optimization problems, the decision maker faces crucial information limitations due to the input not being completely known in advance. A natural goal in such settings is to find solutions that have a good worst-case performance over all potential input instances. However, even though worst-case analysis provides a useful measure for the robustness of an algorithm, it is also known to be a measure that often leads to needlessly pessimistic results.

© The Author(s), under exclusive license to Springer Nature Switzerland AG 2023
J. Byrka and A. Wiese (Eds.): WAOA 2023, LNCS 14297, pp. 74–89, 2023.
https://doi.org/10.1007/978-3-031-49815-2_6

A recent, yet extensive, line of work on *algorithms with predictions* models the partial information that is often available to the decision maker and overcomes worst-case bounds by leveraging machine-learned predictions about the inputs (see [22] for a survey of the early work in this area). In this line of work, the algorithm is given some type of prediction about the input, but the predictions are not necessarily accurate. The goal is to design algorithms that achieve stronger bounds when the provided predictions are accurate, which are called *consistency* bounds, but also maintain worst-case *robustness* bounds that hold even when the predictions are inaccurate. Optimization problems that have been studied under this framework include online paging [20], scheduling [23], secretary [10], covering [6], matching [8,9,17], knapsack [16], facility location [13], Nash social welfare [7], and graph [4] problems. Most of the work on scheduling in this model has considered predictions about the processing times of the jobs [2,3,5,15,18,21,23].

There is a large body of work considering uncertainty in the input to scheduling problems, including whole fields like stochastic scheduling. Most of it studies uncertainty in the jobs. A recent line of work considers scheduling problems where there is uncertainty surrounding the available machines (e.g. [1,11,12,24]). In particular, we emphasize *scheduling with an unknown number of parallel machines*, introduced in [24] where, given a set of jobs, there is a first *partitioning* stage where they must be partitioned into *bags* without knowing the number of machines available and then, in a second *scheduling* stage, the algorithm learns the number of machines and the bags must be scheduled on the machines without being split up. This problem was generalized to *speed-robust scheduling* [11] where there are m machines, but speeds of the machines are unknown in the partitioning stage and are revealed in the scheduling stage[1]. We will use the speed robust scheduling model in the rest of this paper, as it captures applications where partial packing decisions have to be made with only partial information about the machines. As discussed in [24], such applications include MapReduce computations in shared data centers where data is partitioned into groups by a mapping function that is designed without full information about the machines that will be available in the data center, or in a warehouse where items are grouped into boxes without full information about the trucks that will be available to ship the items.

In this paper, we introduce and study the problem of scheduling with machine-learned predictions about the speeds of the machines. In the two applications mentioned above, MapReduce computations and package shipping, it is natural to have some relevant historical data about the computing resources or the trucks that will be available, which can be used to obtain machine-learned predictions about these quantities. In the *scheduling with speed predictions* problem, we are given jobs and predictions about the speeds of the m machines. In the first, *partitioning stage*, jobs are partitioned into m bags, using only the predictions about the speeds of the machines. Then, in the second, *scheduling stage*,

---

[1] This problem strictly generalizes the first problem by setting speed to 1 for actual machines, and speed to 0 for the other (non)machines.

Table 1. Robustness of deterministic $1 + \alpha$ consistent algorithms, where $\alpha \in (0, 1)$ except for the $(4 - 2\alpha)/3$ lower bound, for which $\alpha \in (0, 1/2)$.

Job sizes	Speeds	Upper bound	Lower bound
General	General	$2 + 2/\alpha$ (Theorem 3)	$1 + (1 - \alpha)/2\alpha - O(1/m)$ (Theorem 1)
Equal-size	General	$2 + 1/\alpha$ (Theorem 4)	$1 + (1 - \alpha)/2\alpha - O(1/m)$ (Theorem 1)
Infinitesimal	General	$1 + 1/\alpha$ (Theorem 5)	$1 + (1 - \alpha)^2/4\alpha - O(1/m)$ (Theorem 1)
General	$\{0,1\}$	2 (Theorem 6)	$(4 - 2\alpha)/3$ (Theorem 7)

the true speeds of the machines are revealed, and the bags must be scheduled on the machines without being split up. The goal is to use the predictions to design algorithms that achieve improved guarantees for speed-robust scheduling. The fundamental question we ask is:

Can speed predictions be used to obtain both improved guarantees when the predictions are accurate and bounded guarantees when the prediction errors are arbitrarily large?

We focus on the classical makespan (completion time of the last completed job) minimization objective. Two main evaluation metrics for our problem, or for any algorithms with predictions problem, are robustness and consistency. The consistency of an algorithm is the approximation ratio it achieves when the speed predictions are equal to the true speeds of the machines, and its robustness is its worst-case approximation ratio over all possible machine speeds, i.e., when the predictions are arbitrarily wrong. The main focus of this paper is on general job processing times and machine speeds, but we also consider multiple special cases.

Without predictions, [11] achieves a $(2 - 1/m)$-approximation. Thus, if we do not trust the predictions, we can ignore them and use this algorithm to achieve a $2 - 1/m$ consistent and $2 - 1/m$ robust algorithm. On the other hand, if we fully trust the predictions, we can pretend that the predictions are correct and use a polynomial time approximation scheme (PTAS) for makespan minimization on related machines to obtain a $1 + \epsilon$ consistent algorithm, for any constant $\epsilon > 0$. However, as we show in Sect. 3, this approach would have unbounded robustness. Thus, the main challenge is to develop an algorithm that leverages predictions to improve over the best known $2 - 1/m$ approximation when the predictions are accurate, while maintaining bounded robustness guarantees even when the predictions are arbitrarily wrong.

1.1 Our Results

Our main result is a deterministic algorithm for minimizing makespan in the scheduling with speed predictions (SSP) model that is $1 + \alpha$ consistent and $2 + 2/\alpha$ robust, for any $\alpha \in (0, 1)$ (Theorem 2). When the predictions are accurate, the $1 + \alpha$ consistency outperforms the best-known approximation for speed-robust scheduling without predictions of $2 - 1/m$ [11], while maintaining a $2 + 2/\alpha$ robustness guarantee that holds even when the predictions are arbitrarily wrong. To obtain a polynomial time algorithm, the consistency and robustness both increase by a $1 + \epsilon$ factor, for any constant $\epsilon \in (0, 1)$, due to the PTAS for makespan minimization on related machines that we use as a subroutine [14].

We extend this result to obtain an approximation ratio that interpolates between $1 + \alpha$ and $2 + 2/\alpha$ as a function of the prediction error. More precisely, for any $\alpha \in (0, 1)$, our algorithm achieves an approximation of $\min\{\eta^2(1 + \alpha), (2 + 2/\alpha)\}$ (Theorem 3), where the prediction error $\eta := \max_{i \in [m]} \frac{\max\{\hat{s}_i, s_i\}}{\min\{\hat{s}_i, s_i\}}$ is the maximum ratio between the predicted speed $\hat{s}_i$ and the true speed s_i of the m machines. The following hardness result motivates this choice for the prediction error: for any $\alpha \in (0, 1)$, any deterministic $1 + \alpha$ consistent algorithm has robustness at least $1 + \frac{1-\alpha}{2\alpha} - O(\frac{1}{m})$, even when a single machine speed is incorrectly predicted (Theorem 1). Thus, a single incorrectly predicted machine speed can cause a strong lower bound on the approximation ratio. We also note that the maximum ratio over all the predictions is a common definition for the prediction error in scheduling with predictions (see, e.g., [18,19]). Additionally, we obtain the following results (summarized in Table 1):

- When the job processing times are equal or infinitesimal, the best-known approximations without predictions are 1.8 and $e/(e-1) \approx 1.58$ [11], respectively. For these cases, our $1 + \alpha$ consistent algorithm achieves a robustness of $2 + 1/\alpha$ (Theorem 4) and $1 + 1/\alpha$ (Theorem 5), respectively.
- When the machine speeds are either 0 or 1, which corresponds to the scenario where the number of machines is unknown, the best-known approximation without predictions is 5/3 [24]. We develop an algorithm that is 1 consistent and 2 robust (Theorem 6). We also show that, for any $\alpha \in [0, 1/2)$, any deterministic $1 + \alpha$ consistent algorithm has robustness at least $(4 - 2\alpha)/3$ (Theorem 7).
- Even when the prediction error is relatively large, our algorithm often empirically outperforms existing speed-robust algorithms that do not use predictions.

We note that, subsequent to our work, a scheduling with predictions problem where the machine speeds are unknown was also studied in [19], but in an incomparable online setting where the speeds can be job-dependent.

1.2 Technical Overview

We give an overview of the main technical ideas used to obtain our main result (Theorem 3). The second stage of the SSP problem corresponds to a standard makespan minimization problem in the full information setting, so the main problem is the first stage where jobs must be partitioned into bags given predictions about the speeds of the machines. At a high level, our partitioning algorithm initially creates a partition of the jobs in bags, and a tentative assignment of the bags to machines, assuming that the predictions are the true speeds of the machines. This tentative solution is optimal if the predictions are perfect, but as we discuss in Sect. 3, if the predictions are wrong, its makespan may be far from optimal. To address this concern, the algorithm iteratively moves away from the initial partition in order to obtain a more robust partitioning, while also maintaining that the bags can be scheduled to give a $(1+\alpha)$-approximation of the makespan if the predictions are correct. The parameter $\alpha \in (0, 1)$ is an input to the algorithm that controls the consistency-robustness trade-off, i.e., it controls how much the predictions should be trusted. Starting from a consistent solution and then robustifying has been used in some other algorithms with predictions. Our main technical contribution is in designing such a robustification algorithm for the SSP problem.

More concretely, let the total processing time of a bag be the sum of the processing time of the jobs in that bag. The partitioning algorithm always maintains a tentative assignment of bags to the machines. To robustify this assignment, the algorithm iteratively reassigns the bag with minimum total processing time to the machine that is assigned the bag with maximum total processing time. If there are now ℓ bags assigned to this machine, we break open these ℓ bags, and reassign the jobs to ℓ new bags using the Longest Processing Time first algorithm, which will roughly balance the size of the ℓ bags assigned to this machine. Thus, at every iteration, the bags that had the maximum and minimum total processing times at the beginning of that iteration end up with approximately equal total processing times, which improves the robustness of the partition. The algorithm terminates when the updated partition would not achieve a $1+\alpha$ consistency anymore.

The analysis of the $2 + 2/\alpha$ robustness consists of three main lemmas. The algorithm and analysis use a parameter β, which is the ratio of the maximum total processing time of a bag that contains at least two jobs to the minimum total processing time of a bag. We use this particular parameter partly to handle the case of very large jobs. Informally, both the algorithm and the adversary will need to put that one job in its own bag and on its own machine, so we can just "ignore" such jobs. We first show that if we can solve the second-stage scheduling problem optimally, then the robustness achieved by any partition is at most $\max\{2, \beta\}$. Then, we show that at each iteration, the minimum total processing time of a bag is non-decreasing. Finally, we use this monotonicity property to show that, for the partition returned by the algorithm, $\beta \leq 2 + 2/\alpha$. Together with the first lemma, this implies that the algorithm achieves a $2 + 2/\alpha$ robustness. The last lemma requires a careful argument to show that, if $\beta > 2 + 2/\alpha$, then an additional iteration of the algorithm does not break the $1+\alpha$ consistency achieved by the current partition. To obtain a polynomial-time algorithm, we pay an extra factor of $1 + \epsilon$ in the scheduling stage by using the PTAS of [14].

Finally, we provide an empirical evaluation of our algorithm that shows that, even when the prediction error is relatively large, it often outperforms existing speed-robust algorithms that do not use predictions.

2   Preliminaries

We first describe the speed-robust scheduling problem, which was introduced by [11] and builds on the scheduling with an unknown number of machines problem from [24]. There are n jobs with processing times $\mathbf{p} = (p_1, \ldots, p_n) \geq \mathbf{0}$ and m machines with speeds $\mathbf{s} = (s_1, \ldots, s_m) > \mathbf{0}$ such that the time needed to process job j on machine i is p_j/s_i.[2] The problem consists of the following two stages. In the first stage, called the partitioning stage, the speeds of the machines are unknown and the jobs must be partitioned into m (possibly empty) bags $B_1, \ldots, B_m$ such that $\cup_{i \in [m]} B_i = [n]$ (where $[n] = \{1, \ldots, n\}$) and $B_{i_1} \cap B_{i_2} = \emptyset$ for all $i_1, i_2 \in [m]$, $i_1 \neq i_2$. In the second stage, called the scheduling stage, the speeds $\mathbf{s}$ are revealed to the algorithm and each bag B_i created in the partitioning stage must be assigned, i.e., scheduled, on a machine without being split up.

The paper on speed-robust scheduling, [11], considers the classical makespan minimization objective. Let $\mathcal{M}_i$ be the bags assigned to machine i; the goal is to minimize $\max_{i \in [m]} (\sum_{B \in \mathcal{M}_i} \sum_{j \in B} p_j)/s_i$. An algorithm for speed-robust scheduling is β-robust if it achieves an approximation ratio of β compared to the optimal schedule that knows the speeds in advance, i.e., $\max_{\mathbf{p},\mathbf{s}} alg(\mathbf{p}, \mathbf{s})/opt(\mathbf{p}, \mathbf{s}) \leq \beta$ where $alg(\mathbf{p}, \mathbf{s})$ and $opt(\mathbf{p}, \mathbf{s})$ are the makespans of the schedule returned by the algorithm (that learns $\mathbf{s}$ in the second stage) and the optimal schedule (that knows $\mathbf{s}$ in the first stage).

We augment the speed-robust scheduling problem with predictions about the speeds of the machines and call this problem Scheduling with Speed Predictions (SSP). The difference between SSP and speed-robust scheduling is that, during the partitioning stage, the algorithm is now given access to, potentially incorrect, predictions $\hat{\mathbf{s}} = (\hat{s}_1, \ldots, \hat{s}_m) \geq \mathbf{0}$ about the speeds of the machines (see Appendix A.1 of the full version of the paper for additional discussion about how we learn the machine speeds and obtain $\hat{s}$). The true speeds of the machines $\mathbf{s}$ are revealed during the scheduling stage, as in the speed-robust scheduling problem. We also want to minimize the makespan.

Consistency and robustness are two standard measures in algorithms with predictions [20]. An algorithm is c-consistent if it achieves a c approximation ratio when the predictions are correct, i.e., if $\max_{\mathbf{p},\mathbf{s}} alg(\mathbf{p}, \mathbf{s})/opt(\mathbf{p}, \mathbf{s}) \leq c$ where $alg(\mathbf{p}, \hat{\mathbf{s}}, \mathbf{s})$ is the makespan of the schedule returned by the algorithm when it is given predictions $\hat{\mathbf{s}}$ in the first stage and speeds $\mathbf{s}$ in the second stage. An algorithm is β-robust if it achieves a β approximation ratio when the predictions can be arbitrarily wrong, i.e., if $\max_{\mathbf{p},\hat{\mathbf{s}},\mathbf{s}} alg(\mathbf{p}, \hat{\mathbf{s}}, \mathbf{s})/opt(\mathbf{p}, \mathbf{s}) \leq \beta$. We note that a β-robust algorithm for speed-robust scheduling is also a β-robust (and β-consistent) algorithm for SSP which ignores the speed predictions.

The main challenge in algorithms with predictions problems is to simultaneously achieve "good" consistency and robustness, which requires partially trusting the predictions (for consistency), but not trusting them too much (for

---

[2] The non-zero speed assumption is for ease of notation. Having a machine with speed $s_i = 0$ is equivalent to $s_i = \epsilon$ for ϵ arbitrarily small since in both cases no schedule can assign a job to i without the completion time of this job being arbitrarily large.

robustness). In particular, the goal is to obtain an algorithm that achieves a consistency that improves over the best known approximation without predictions $(2 - 1/m$ for speed-robust scheduling), ideally close to the best known approximation in the full information setting ($1+\epsilon$, for any constant $\epsilon > 0$, for makespan minimization on related machines), while also achieving bounded robustness.

Even though consistency and robustness capture the main trade-off in SSP, we are also interested in giving approximation ratios as a function of the prediction error. It is important, in any algorithms with predictions problem, to define the prediction error appropriately, so that it actually captures the proper notion of error in the objective. It might seem that, for example, L_1 distance between the predictions and data is natural, but for many problems, including this one, such a definition would mainly give vacuous results. We define the prediction error $\eta \geq 1$ to be the maximum ratio[3] between the true speeds $\mathbf{s}$ and the predicted speeds $\hat{\mathbf{s}}$, or vice versa, i.e., $\eta(\hat{\mathbf{s}}, \mathbf{s}) = \max_{i \in [m]} \frac{\max\{\hat{s}_i, s_i\}}{\min\{\hat{s}_i, s_i\}}$ (see Appendix A.2 of the full version for further discussion on the choice of error measure). Given a bound η on the prediction error, an algorithm achieves a $\gamma(\eta)$ approximation if $\max_{\mathbf{p}, \hat{\mathbf{s}}, \mathbf{s}: \eta(\hat{\mathbf{s}}, \mathbf{s}) \leq \eta} alg(\mathbf{p}, \hat{\mathbf{s}}, \mathbf{s})/opt(\mathbf{p}, \mathbf{s}) \leq \gamma(\eta)$.

Given arbitrary bags $B_1, \ldots, B_m$, the scheduling stage corresponds to a standard makespan minimization problem in the full information setting, for which polynomial-time approximation schemes (PTAS) are known [14]. Thus, the main challenge is the partitioning stage. We define the consistency and robustness of a partitioning algorithm $\mathcal{A}_P$ to be the consistency and robustness achieved by the two-stage algorithm that first runs $\mathcal{A}_P$ and then solves the scheduling stage optimally. If we want to require that algorithms be polynomial time, we may simply run the PTAS for makespan minimization in the scheduling stage, and the bounds increase by a $(1 + \epsilon)$ factor. We will not explicitly mention this in the remainder of the paper.

3 Consistent Algorithms are not Robust

A natural first question is whether there is an algorithm with optimal consistency that also achieves a good robustness. We answer this question negatively and show that there exists an instance for which any 1-consistent algorithm cannot be $o(n)$-robust. This impossibility result is information-theoretic and is not due to computational constraints. The proofs in this section can be found in Appendix B of the full version.

Proposition 1. *For any $n > m$, there is no algorithm that is 1-consistent and $\frac{n-m+1}{\lceil n/m \rceil}$-robust, even in the case of equal-size jobs. In particular, for $m = n/2$, there is no algorithm that is 1-consistent and $o(n)$-robust.*

More generally, we show that there is a necessary non-trivial trade-off between consistency and robustness for the SSP problem. In particular, the

---

[3] We scale $\mathbf{s}, \hat{\mathbf{s}}$ such that $\max_i s_i = \max_i \hat{s}_i$ before computing η, to make sure the speeds are on the same scale.

robustness of any deterministic algorithm for SSP must grow inversely proportional as a function of the consistency.

Theorem 1. *For any $\alpha \in (0,1)$, if a deterministic algorithm for SSP is $(1+\alpha)$-consistent, then its robustness is at least $1 + \frac{1-\alpha}{2\alpha} - O(\frac{1}{m})$, even in the case where the jobs have equal processing times. In the special case where the processing times are infinitesimal, the robustness of a deterministic $(1+\alpha)$-consistent algorithm is at least $1 + \frac{(1-\alpha)^2}{4\alpha} - O(\frac{1}{m})$.*

Recall that in the setting without predictions, the best known algorithm is $(2 - 1/m)$-robust (and thus also $(2 - 1/m)$-consistent) [11]. Since we have shown that algorithms with near-optimal consistency must have unbounded robustness, a main question is thus whether it is even possible to achieve a consistency that improves over $(2 - 1/m)$ while also obtaining bounded robustness. We note that the natural idea of randomly choosing to run the $(2 - 1/m)$-robust algorithm or an algorithm with near-optimal consistency (with unbounded robustness), aiming to hedge between robustness and consistency, does not work since the resulting algorithm would still have unbounded robustness due to SSP being a minimization problem.

4 The Algorithm

In this section, we give an algorithm for scheduling with speed predictions with arbitrary-sized jobs that achieves a $\min\{\eta^2(1+\epsilon)(1+\alpha), (1+\epsilon)(2+2/\alpha)\}$ approximation for any constant $\epsilon \in (0,1)$ and any $\alpha \in (0,1)$. The proofs in this section can be found in Appendix C of the full version.

4.1 Description of the Algorithm

Our algorithm, called IPR and formally described in Algorithm 1, takes as input the processing times of the jobs $\mathbf{p}$, the predicted speeds of the machines $\hat{s}$, an accuracy parameter ϵ, a consistency goal $1+\alpha$, and a parameter ρ that influences the ratio between the size of the smallest and largest bags. For general job processing times and machine speeds, we use $\rho = 4$. For some special cases in Sect. 5, we use $\rho = 2$. IPR first uses the PTAS for makespan minimization [14] to construct a partition of the jobs into bags $B_1, \ldots, B_m$ such that scheduling the jobs in B_i on machine i achieves a $1 + \epsilon$ approximation when the predictions are correct. In other words, it initially assumes that the predictions are correct and creates a $(1 + \epsilon)$-consistent partition of the jobs into bags. In addition, it also creates a tentative assignment $\mathcal{M}_1 = \{B_1\}, \ldots, \mathcal{M}_m = \{B_m\}$ of the bags $B_1, \ldots, B_m$ on the machines.

Even though this tentative assignment achieves a good consistency, its robustness is arbitrarily poor. To improve the robustness, IPR iteratively rebalances this partition while maintaining a $(1 + \epsilon)(1 + \alpha)$ bound on its consistency. The design and analysis of such an iterative rebalancing procedure is the main challenge.

At each iteration, the subroutine LPT-REBALANCE rebalances the bags and modifies $\mathcal{M}_1, \ldots, \mathcal{M}_m$. We define the processing time $p(B)$ of a bag to the total processing time of the jobs in that bag, i.e., $p(B) = \sum_{j \in B} p_j$. The algorithm terminates either when scheduling the bags in each $\mathcal{M}_i$ on machine i violates the desired $(1 + \epsilon)(1 + \alpha)$ consistency bound or when the ratio of the largest processing time of a bag containing at least two jobs to the smallest processing time of a bag is at most ρ. To verify the consistency bound, the algorithm compares the makespan of the new tentative assignment to the makespan $\overline{\mathrm{OPT}}_C$ of the initial assignment, assuming that the speed predictions are correct.

Algorithm 1. ITERATIVE-PARTIAL-REBALANCING (IPR)

Input: predicted machine speeds $\hat{s}_1 \geq \cdots \geq \hat{s}_m$, job processing times $p_1, \ldots, p_n$, consistency $1 + \alpha$, accuracy $\epsilon \in (0, 1)$, maximum bag size ratio $\rho \geq 1$.

1: $\{B_1, \ldots, B_m\} \leftarrow$ a $(1 + \epsilon)$-consistent partition such that $p(B_1) \geq \ldots \geq p(B_m)$
2: $\overline{\mathrm{OPT}}_C \leftarrow \max_{i \in [m]} p(B_i)/\hat{s}_i$
3: $\mathcal{M}_1, \ldots, \mathcal{M}_m \leftarrow \{B_1\}, \ldots, \{B_m\}$
4: **while** $\max_{B \in \cup_i \mathcal{M}_i, |B| \geq 2} p(B) > \rho \min_{B \in \cup_i \mathcal{M}_i} p(B)$:
5: $\quad \mathcal{M}'_1, \ldots, \mathcal{M}'_m \leftarrow$ LPT-REBALANCE$(\mathcal{M}_1, \ldots, \mathcal{M}_m)$
6: $\quad$ **if** $\max_{i \in [m]} \sum_{B \in \mathcal{M}'_i} p(B)/\hat{s}_i > (1 + \alpha)\overline{\mathrm{OPT}}_C$:
7: $\qquad \{B_1, \ldots, B_m\} \leftarrow \cup_{i \in [m]} \mathcal{M}_i$
8: $\qquad$ **return** $\{B_1, \ldots, B_m\}$
9: $\quad \mathcal{M}_1, \ldots, \mathcal{M}_m \leftarrow \mathcal{M}'_1, \ldots, \mathcal{M}'_m$
10: $\{B_1, \ldots, B_m\} \leftarrow \cup_{i \in [m]} \mathcal{M}_i$
11: **return** $\{B_1, \ldots, B_m\}$

The LPT-Rebalance Subroutine. This subroutine first moves the bag $B_{\min}$ with the smallest processing time to the collection of bags $\mathcal{M}_{\max}$ that contains the bag with the largest processing time among the bags that contain at least two jobs. Let ℓ be the number of bags in $\mathcal{M}_{\max}$, including $B_{\min}$. The subroutine then balances the processing time of the bags in $\mathcal{M}_{\max}$ by running the Longest Processing Time first (LPT) algorithm over all jobs in bags in $\mathcal{M}_{\max}$, i.e. jobs in $\cup_{B \in \mathcal{M}_{\max}} B$, to create ℓ new, balanced, bags that are placed in $\mathcal{M}_{\max}$. LPT-REBALANCE finally returns the updated assignment of bags to machines $\mathcal{M}_1, \ldots, \mathcal{M}_m$. We note that among these m collections of bags, only two, $\mathcal{M}_{\min}$ and $\mathcal{M}_{\max}$, are modified. We use Fig. 1 to illustrate this rebalancing procedure.

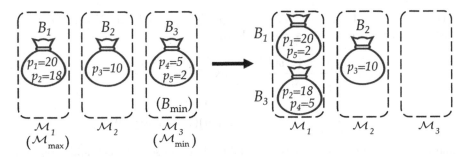

Fig. 1. Illustration of one iteration of the IPR algorithm on an example with $m = 3$ bags and machines and $n = 5$ jobs.

Algorithm 2. LPT-REBALANCE

Input: assignments of bags $\mathcal{M}_1, \ldots, \mathcal{M}_m$
1: $B_{\min} \leftarrow \operatorname{argmin}_{B \in \cup_i \mathcal{M}_i} p(B)$
2: $\mathcal{M}_{\min} \leftarrow$ the collection of bags $\mathcal{M}$ such that $B_{\min} \in \mathcal{M}$
3: $\mathcal{M}_{\max} \leftarrow \operatorname{argmax}_{\mathcal{M}_i : i \in [m]} \max_{B \in \mathcal{M}_i : |B| \geq 2} p(B)$
4: $\mathcal{M}_{\max} \leftarrow \mathcal{M}_{\max} \cup \{B_{\min}\}, \mathcal{M}_{\min} \leftarrow \mathcal{M}_{\min} \setminus \{B_{\min}\}$
5: $J_{\max} \leftarrow \cup_{B \in \mathcal{M}_{\max}} B, \ell \leftarrow |\mathcal{M}_{\max}|$
6: $B'_1, \ldots, B'_\ell \leftarrow \{\}, \ldots, \{\}$
7: **while** $|J_{\max}| > 0$ **do**
8:     $j' \leftarrow \operatorname{argmax}_{j \in J_{\max}} p_j$
9:     $B' \leftarrow \operatorname{argmin}_{B \in \{B'_1, \ldots, B'_\ell\}} p(B)$
10:    $B' \leftarrow B' \cup \{j'\}, J_{\max} \leftarrow J_{\max} \setminus \{j'\}$
11: $\mathcal{M}_{\max} \leftarrow \{B'_1, \ldots, B'_\ell\}$
12: **return** $\mathcal{M}_1, \ldots, \mathcal{M}_m$

4.2 Analysis of the Algorithm

We first show that IPR with parameter $\rho = 4$ in the general case is a $(1 + \epsilon)(1 + \alpha)$-consistent and $(2 + 2/\alpha)$-robust partitioning algorithm (Lemma 1 and Theorem 2). Then, we use these consistency and robustness guarantees to obtain the $\min\{\eta^2(1 + \epsilon)(1 + \alpha), (1 + \epsilon)(2 + 2/\alpha)\}$ approximation as a function of the prediction error η (Theorem 3). Finally, we analyze the running time (Lemma 5). The main challenge is to analyze IPR's robustness.

The Consistency. The consistency almost comes from the definition of IPR.

Lemma 1. *For any constants* $\alpha, \epsilon \in (0, 1)$, *IPR is a* $(1 + \epsilon)(1 + \alpha)$-*consistent partitioning algorithm.*

Proof. To prove the consistency, we consider the final tentative assignment of the bags on the machines $\mathcal{M}_1, \ldots, \mathcal{M}_m$ when IPR terminates. With true speeds s, the makespan of this schedule is $\max_{i\in[m]} \sum_{B\in\mathcal{M}_i} p(B)/s_i$. When the speed predictions are correct, i.e., $\mathbf{s} = \hat{\mathbf{s}}$, we have

$$\max_{i\in[m]} \frac{\sum_{B\in\mathcal{M}_i} p(B)}{s_i} = \max_{i\in[m]} \frac{\sum_{B\in\mathcal{M}_i} p(B)}{\hat{s}_i}$$
$$\leq (1+\alpha)\overline{\mathrm{OPT}}_C$$
$$\leq (1+\alpha)(1+\epsilon)opt(\mathbf{p}, \mathbf{s}).$$

Line 6 of IPR enforces the first inequality. For the second inequality, observe that when $\mathbf{s} = \hat{\mathbf{s}}$, $\overline{\mathrm{OPT}}_C$ is the makespan of the initial assignment, which is a $1+\epsilon$ approximation to the optimal makespan $opt(\mathbf{p}, \mathbf{s})$. Since there exists an assignment of the bags returned by IPR that achieves a $(1+\epsilon)(1+\alpha)$ approximation when $\mathbf{s} = \hat{\mathbf{s}}$, IPR is a $(1+\epsilon)(1+\alpha)$-consistent partitioning algorithm.

The Robustness. The main part of the analysis is to bound the algorithm's robustness. First, we show that the ratio $\beta(\mathcal{B}) = \frac{\max_{B\in\mathcal{B}, |B|\geq 2} p(B)}{\min_{B\in\mathcal{B}} p(B)}$ of the maximum total processing time of a bag containing at least two jobs to the minimum total processing time of a bag can be used to bound the robustness of any partition $\mathcal{B}$.

Lemma 2. *Let $\mathcal{B} = \{B_1, \cdots, B_m\}$ be a partition of n jobs with processing times $p_1, \ldots p_n$ into m bags. Then $\mathcal{B}$ is a $\max\{2, \beta(\mathcal{B})\}$-robust partition, where $\beta(\mathcal{B}) = \frac{\max_{B\in\mathcal{B}, |B|\geq 2} p(B)}{\min_{B\in\mathcal{B}} p(B)}$.*

By Lemma 2, it remains to bound the ratio β of the bags $\mathcal{B}$ returned by IPR. Let $\mathcal{B}^{(i)}$ denote the collection of all bags B at iteration i of the algorithm and define $b_{\min}^{(i)} = \min_{B\in\mathcal{B}^{(i)}} p(B)$ to be the minimum processing time of a bag at each iteration i. To bound the ratio β, we first show in Lemma 3 that $b_{\min}^{(i)}$ is non-decreasing in i.

Lemma 3. *At each iteration i of IPR with $\rho = 4$, $b_{\min}^{(i+1)} \geq b_{\min}^{(i)}$.*

Using Lemma 3, we bound the size ratio β needed for Lemma 2.

Lemma 4. *Let $\mathcal{B}_{\mathrm{IPR}} = \{B_1, \ldots, B_m\}$ be the partition of the n jobs returned by IPR with $\rho = 4$. Then, we have that $\beta(\mathcal{B}_{\mathrm{IPR}}) \leq 2 + 2/\alpha$.*

We are now ready to show the algorithm's robustness.

Theorem 2. *For any constants $\alpha, \epsilon \in (0, 1)$, IPR with $\rho = 4$ is a $(2 + 2/\alpha)$-robust partitioning algorithm.*

Proof. Let $\mathcal{B}_{\mathrm{IPR}} = \{B_1, \ldots, B_m\}$ be the partition of the n jobs returned by IPR with $\rho = 4$. By Lemma 4, we have that $\beta(\mathcal{B}_{\mathrm{IPR}}) \leq 2 + 2/\alpha$. Thus, by Lemma 2, the robustness of IPR with $\rho = 4$ is $2 + 2/\alpha$.

The Approximation as a Function of the Prediction Error. We extend the consistency and robustness results for IPR to obtain our main result. We show that for the SSP problem, the algorithm that runs IPR in the partitioning stage and then a PTAS in the scheduling stage achieves an approximation ratio that gracefully degrades as a function of the prediction error η from $(1+\epsilon)(1+\alpha)$ to $(1+\epsilon)(2+2/\alpha)$.

Theorem 3. *Consider the algorithm that runs IPR with $\rho = 4$ in partitioning stage and a PTAS for makespan minimization in scheduling stage. For any constant $\epsilon \in (0,1)$ and any $\alpha \in (0,1)$, this algorithm achieves a $\min\{\eta^2(1+\epsilon)(1+\alpha), (1+\epsilon)(2+2/\alpha)\}$ approximation for SSP where $\eta = \max_{i \in [m]} \frac{\max\{\hat{s}_i, s_i\}}{\min\{\hat{s}_i, s_i\}}$ is the prediction error.*

If we do not care about the computation runtime; that is, we can solve each scheduling problem optimally including the initial step of IPR in the partition stage and the scheduling stage, then our result improves to a $\min\{\eta^2(1+\alpha), (2+2/\alpha)\}$ approximation.

The Running Time of IPR. We show that the main algorithm performs $O(m^2)$ iterations, which implies that its running time is polynomial in n and m.

Lemma 5. *At most $O(m^2)$ iterations are needed for IPR with $\rho = 4$ to terminate.*

5   Improved Trade-Offs for Special Cases

When all job processing times are either equal or infinitesimal, t the IPR algorithm with $\rho = 2$ achieves an improved robustness. The proofs of this section can be found in Sect. 5 of the full version.

Theorem 4. *If $p_j = 1$ for all $j \in [n]$, then, for any constant $\epsilon \in (0,1)$ and any $\alpha \in (0,1)$, IPR with $\rho = 2$ is $(1+\epsilon)(1+\alpha)$-consistent and $(2+1/\alpha)$-robust.*

Theorem 5. *If all jobs are infinitesimal, then, for any constant $\epsilon \in (0,1)$ and any $\alpha \in (0,1)$, IPR with $\rho = 2$ is $(1+\epsilon)(1+\alpha)$-consistent and $(1+1/\alpha)$-robust.*

When the machine speeds are in $\{0,1\}$, we propose a different partitioning algorithm that is $(1+\epsilon)$-consistent and $2(1+\epsilon)$-robust for this special case.

Theorem 6. *For any constant $\epsilon > 0$, there is a $(1+\epsilon)$-consistent and $2(1+\epsilon)$-robust partitioning algorithm for the $\{0,1\}$-speed SSP problem.*

We also provide a robustness lower bound for $\{0,1\}$ speeds.

Theorem 7. *For any $\alpha \in [0,1/2)$, if a deterministic algorithm for the $\{0,1\}$-speed SSP problem is $(1+\alpha)$-consistent, then its robustness is at least $(4-2\alpha)/3$.*

6 Experiments

We empirically evaluate the performance of IPR on synthetic data against benchmarks that achieve either the best-known consistency or the best-known robustness for SSP.

6.1 Experiment Settings

Benchmarks. We compare three algorithms. **IPR** is Algorithm 1 with $\rho = 4$ and $\alpha = 0.5$. The Largest Processing Time first partitioning algorithm, which we call **LPT-Partition**, creates m bags by adding each job, in decreasing order of their processing time, to the bag with minimum total processing time. LPT-PARTITION is 2-robust (and 2-consistent since it ignores the predictions) [11]. The **1-consistent** algorithm completely trusts the prediction and generates a partition that is 1-consistent (but has arbitrarily poor robustness due to our lower bound in Proposition 1). In practice, PTAS algorithms for scheduling are extremely slow. Instead of using a PTAS for the scheduling stage, we give an advantage to the two benchmarks by solving their scheduling stage via integer programming (IP). However, since we want to ensure that our algorithm has a polynomial running time, we use the LPT algorithm to compute a schedule during both the partitioning and scheduling stage of IPR, instead of a PTAS or an IP and we use IP to compute the optimal solution.

Data Sets. In the first set of experiments, we generate synthetic datasets with $n = 50$ jobs and $m = 10$ machines and evaluate the performance of the different algorithms as a function of the standard deviation of the prediction error distribution. The job processing times p_j are generated i.i.d. either from $\mathcal{U}(0, 100)$, the uniform distribution in the interval $(0, 100)$, or $\mathcal{N}(50, 5)$, the normal distribution with mean $\mu_p = 50$ and standard deviation $\sigma_p = 5$. The machine speeds s_i are also generated i.i.d., either from $\mathcal{U}(0, 40)$ or $\mathcal{N}(20, 4)$. We evaluate the performance of the algorithms over each of the 4 possible combinations of job processing time and machine speed distributions. The prediction error $err(i) = \hat{s}_i - s_i$ of each machine is sampled i.i.d. from $\mathcal{N}(0, x)$ and we vary x from $x = 0$ to $x = \mu_s$ (the mean of machine speeds).

In the second set of experiments, we fix the distributions of the processing times, machine speeds, and prediction errors to be $\mathcal{N}(50, \sigma_p)$, $\mathcal{N}(20, \sigma_s)$, and $\mathcal{N}(0, 4)$ respectively, with default values of $\sigma_p = 5$ and $\sigma_s = 4$. We evaluate the algorithms' performance as a function of (1) the number n of jobs, (2) the number m of machines, (3) σ_p, and (4) σ_s. For each figure, the approximation ratio achieved by the different algorithms are averaged over 100 instances generated i.i.d. as described above. Additional details of the experiment setup are provided in Appendix D of the full version.

6.2   Experiment Results

Experiment Set 1. From the first row of Fig. 2, we observe that, in all four settings, when we vary the magnitude of the prediction error, IPR outperforms LPT-PARTITION when the error is small and outperforms 1-CONSISTENT when the error is large. Since LPT-PARTITION does not use the predictions, its performance remains constant as a function of the prediction errors. Since 1-CONSISTENT completely trusts the predictions, it is optimal when the predictions are exactly correct but its performance deteriorates quickly as the prediction errors increase.

IPR combines the advantages of LPT-PARTITION and 1-CONSISTENT: when the predictions are relatively accurate, it is able to take advantage of the predictions and outperform LPT-PARTITION. When the predictions are increasingly inaccurate, IPR has a slower deterioration rate compared to 1-CONSISTENT. It is noteworthy that, in some settings, IPR simultaneously outperforms both benchmarks for a wide range of values of the standard deviation σ_{err} of the prediction error distribution. When the distributions of job processing times and machine speeds are $\mathcal{N}(50,5)$ and $\mathcal{N}(20,4)$ respectively, IPR achieves the best performance when $\sigma_{err}/\mu_s \geq 0.2$. When they are $\mathcal{N}(50,5)$ and $\mathcal{U}(0,40)$, IPR outperforms both benchmarks when $\sigma_{err}/\mu_s \geq 0.4$.

Experiment Set 2. The number of jobs has almost no impact on the performance of any of the algorithms. However, the approximations achieved by the algorithms do improve as the number of machines m increases, especially for LPT-PARTITION. The reason is that m is also the number of bags, so when the number of bags increases, there is more flexibility in the scheduling stage, especially when the total processing times of the bags are balanced.

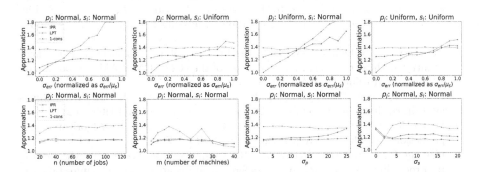

Fig. 2. The approximation ratio achieved by our algorithm, IPR, and the two benchmarks as a function of the standard deviation of the prediction error σ_{err} for different job processing time and true speed distributions (row 1) and as a function of the number of jobs n, the number of machines m, the standard deviation σ_p of the job processing time distribution, and the standard deviation σ_s of the true speed distribution (row 2).

IPR is the algorithm most sensitive to the standard deviation σ_p of the job processing times. It has performance close to that of 1-CONSISTENT when σ_p is small, and similar to LPT-PARTITION when σ_p is large. The approximation ratio of LPT-PARTITION increases as σ_s increases, while our algorithm and the 1-CONSISTENT partitioning algorithm are relatively insensitive to the change in σ_s. Since the LPT-PARTITION algorithm generates balanced bags of similar total processing times, it performs well when the machine speeds are all almost equal, but its performance then quickly degrades as σ_s increases. An additional set of experiments that studies the impact of the α parameter in the performance of the IPR algorithm can be found in Sect. 6 of the full version.

References

1. Albers, S., Schmidt, G.: Scheduling with unexpected machine breakdowns. Discret. Appl. Math. **110**(2–3), 85–99 (2001)
2. Azar, Y., Leonardi, S., Touitou, N.: Flow time scheduling with uncertain processing time. In: Proceedings of the 53rd Annual ACM SIGACT Symposium on Theory of Computing, pp. 1070–1080 (2021)
3. Azar, Y., Leonardi, S., Touitou, N.: Distortion-oblivious algorithms for minimizing flow time. In: Proceedings of the 2022 Annual ACM-SIAM Symposium on Discrete Algorithms (SODA), pp. 252–274. SIAM (2022)
4. Azar, Y., Panigrahi, D., Touitou, N.: Online graph algorithms with predictions. In: Proceedings of the Thirty-Third Annual ACM-SIAM Symposium on Discrete Algorithms (2022)
5. Bamas, E., Maggiori, A., Rohwedder, L., Svensson, O.: Learning augmented energy minimization via speed scaling. In: Larochelle, H., Ranzato, M., Hadsell, R., Balcan, M.F., Lin, H. (eds.) Advances in Neural Information Processing Systems, vol. 33, pp. 15350–15359. Curran Associates, Inc. (2020)
6. Bamas, E., Maggiori, A., Svensson, O.: The primal-dual method for learning augmented algorithms. In: Larochelle, H., Ranzato, M., Hadsell, R., Balcan, M.F., Lin, H. (eds.) Advances in Neural Information Processing Systems, pp. 20083–20094 (2020)
7. Banerjee, S., Gkatzelis, V., Gorokh, A., Jin, B.: Online nash social welfare maximization with predictions. In: Proceedings of the 2022 ACM-SIAM Symposium on Discrete Algorithms, SODA 2022. SIAM (2022)
8. Dinitz, M., Im, S., Lavastida, T., Moseley, B., Vassilvitskii, S.: Faster matchings via learned duals. Adv. Neural. Inf. Process. Syst. **34**, 10393–10406 (2021)
9. Dinitz, M., Im, S., Lavastida, T., Moseley, B., Vassilvitskii, S.: Algorithms with prediction portfolios. arXiv preprint arXiv:2210.12438 (2022)
10. Dütting, P., Lattanzi, S., Paes Leme, R., Vassilvitskii, S.: Secretaries with advice. In: Proceedings of the 22nd ACM Conference on Economics and Computation, pp. 409–429 (2021)
11. Eberle, F., Hoeksma, R., Megow, N., Nölke, L., Schewior, K., Simon, B.: Speed-robust scheduling - sand, bricks, and rocks. In: Integer Programming and Combinatorial Optimization - 22nd International Conference, IPCO 2021, Atlanta, GA, USA, 19–21 May 2021, Proceedings, pp. 283–296 (2021)
12. Epstein, L., et al.: Universal sequencing on an unreliable machine. SIAM J. Comput. **41**(3), 565–586 (2012)

13. Fotakis, D., Gergatsouli, E., Gouleakis, T., Patris, N.: Learning augmented online facility location. CoRR abs/2107.08277 (2021). https://arxiv.org/abs/2107.08277
14. Hochbaum, D.S., Shmoys, D.B.: A polynomial approximation scheme for scheduling on uniform processors: using the dual approximation approach. SIAM J. Comput. **17**(3), 539–551 (1988)
15. Im, S., Kumar, R., Montazer Qaem, M., Purohit, M.: Non-clairvoyant scheduling with predictions. In: Proceedings of the 33rd ACM Symposium on Parallelism in Algorithms and Architectures, pp. 285–294 (2021)
16. Im, S., Kumar, R., Montazer Qaem, M., Purohit, M.: Online knapsack with frequency predictions. In: Advances in Neural Information Processing Systems, vol. 34 (2021)
17. Jin, B., Ma, W.: Online bipartite matching with advice: Tight robustness-consistency tradeoffs for the two-stage model. arXiv preprint arXiv:2206.11397 (2022)
18. Lattanzi, S., Lavastida, T., Moseley, B., Vassilvitskii, S.: Online scheduling via learned weights. In: Proceedings of the 2020 ACM-SIAM Symposium on Discrete Algorithms (SODA), pp. 1859–1877 (2020)
19. Lindermayr, A., Megow, N., Rapp, M.: Speed-oblivious online scheduling: knowing (precise) speeds is not necessary. arXiv preprint arXiv:2302.00985 (2023)
20. Lykouris, T., Vassilvtiskii, S.: Competitive caching with machine learned advice. In: International Conference on Machine Learning, pp. 3296–3305. PMLR (2018)
21. Mitzenmacher, M.: Scheduling with Predictions and the Price of Misprediction. In: 11th Innovations in Theoretical Computer Science Conference (ITCS 2020). Leibniz International Proceedings in Informatics (LIPIcs), vol. 151, pp. 14:1–14:18 (2020)
22. Mitzenmacher, M., Vassilvitskii, S.: Algorithms with predictions. arXiv preprint arXiv:2006.09123 (2020)
23. Purohit, M., Svitkina, Z., Kumar, R.: Improving online algorithms via ml predictions. In: Bengio, S., Wallach, H., Larochelle, H., Grauman, K., Cesa-Bianchi, N., Garnett, R. (eds.) Advances in Neural Information Processing Systems. Curran Associates, Inc. (2018)
24. Stein, C., Zhong, M.: Scheduling when you do not know the number of machines. ACM Trans. Algorithms (2019)

The Power of Amortization on Scheduling with Explorable Uncertainty

Alison Hsiang-Hsuan Liu[1]([✉]) [ID], Fu-Hong Liu[1,2] [ID], Prudence W. H. Wong[2] [ID], and Xiao-Ou Zhang[1]

[1] Utrecht University, Utrecht, The Netherlands
alison.hhliu@gmail.com
[2] University of Liverpool, Liverpool, UK

Abstract. In this work, we study a scheduling problem with explorable uncertainty. Each job comes with an upper limit of its processing time, which could be potentially reduced by testing the job, which also takes time. The objective is to schedule all jobs on a single machine with a minimum total completion time. The challenge lies in deciding which jobs to test and the order of testing/processing jobs.

The online problem was first introduced with unit testing time [5,6] and later generalized to variable testing times [1]. For this general setting, the upper bounds of the competitive ratio are shown to be 4 and 3.3794 for deterministic and randomized online algorithms [1]; while the lower bounds for unit testing time stands [5,6], which are 1.8546 (deterministic) and 1.6257 (randomized).

We continue the study on variable testing times setting. We first enhance the analysis framework in [1] and improve the competitive ratio of the deterministic algorithm in [1] from 4 to $1 + \sqrt{2} \approx 2.4143$. Using the new analysis framework, we propose a new deterministic algorithm that further improves the competitive ratio to 2.316513. The new framework also enables us to develop a randomized algorithm improving the expected competitive ratio from 3.3794 to 2.152271.

Keywords: Explorable uncertainty · Online scheduling algorithms · Total completion time · Competitive analysis · Amortized analysis

1 Introduction

In this work, we study the single-machine *Scheduling with Uncertain Processing time* (SUP) problem with the minimized total completion time objective. We are given n jobs, where each job has a *testing time* t_j and an *upper limit* u_j of its *real processing time* $p_j \in [0, u_j]$. A job j can be executed (without testing), taking u_j time units. A job j can also be tested using t_j time units, and after it is tested, it takes p_j time to execute. Note that any algorithm needs to test a

P. W. H. Wong—The work is partially supported by University of Liverpool Covid Recovery Fund.

© The Author(s), under exclusive license to Springer Nature Switzerland AG 2023
J. Byrka and A. Wiese (Eds.): WAOA 2023, LNCS 14297, pp. 90–103, 2023.
https://doi.org/10.1007/978-3-031-49815-2_7

job j beforehand to run it in time p_j. The online algorithm does not know the exact value of p_j unless it tests the job. On the other hand, the optimal offline algorithm knows in advance each p_j even before testing. Therefore, the optimal strategy is to test job j if and only if $t_j + p_j \leq u_j$ and execute the shortest job first, where the processing time of a job j is $\min\{t_j + p_j, u_j\}$ [1,5,6]. However, since the online algorithm only learns about p_j after testing j, the challenge to the online algorithm is to decide which jobs to test and the order of tasks that could be testing, execution, or execution-untested.

It is typical to study uncertainty in scheduling problems, for example, in the worst case scenario for online or stochastic optimization. Kahan [15] has introduced a novel notion of explorable uncertainty where queries can be used to obtain additional information with a cost. The model of scheduling with explorable uncertainty studied in this paper was introduced by Dürr et al. recently [5,6]. In this model, job processing times are uncertain in the sense that only an upper limit of the processing time is known, and can be reduced potentially by testing the job, which takes a testing time that may vary according to the job. An online algorithm does not know the real processing time before testing the job, whereas an optimal offline algorithm has the full knowledge of the uncertain data.

One of the motivations to study scheduling with uncertain processing time is clinic scheduling [3,16]. Without a pre-diagnosis, it is safer to assign each treatment the maximum time it may need. With pre-diagnosis, the precise time a patient needs can be identified, which can improve the performance of the scheduling. Other applications are, as mentioned in [5,6], code optimization [2], compression for file transmission over network [20], fault diagnosis in maintenance environments [17]. Application in distributed databases with centralized master server [18] is also discussed in [1].

In addition to its practical motivations, the model of explorable uncertainty also blurs the line between offline and online problems by allowing a restricted uncertain input. It enables us to investigate how uncertainty influences online decision quality in a more quantitative way. The concept of exploring uncertainty has raised a lot of attention and has been studied on different problems, such as sorting [13], finding the median [11], identifying a set with the minimum-weight among a given collection of feasible sets [8], finding shortest paths [10], computing minimum spanning trees [14], etc. More recent work and a survey can be found in [7,10,12]. Note that in many of the works, the aim of the algorithm is to find the optimal solution with the minimum number of testings for the uncertain input, comparing against the optimal number of testings.

Another closely related model is Pandora's box problem [4,9,19], which was based on the secretary problem, that was first proposed by Weitzman [19]. In this problem, each candidate (that is, the box) has an independent probability distribution for the reward value. To know the exact reward a candidate can provide, one can open the box and learn its realized reward. More specifically, at any time, an algorithm can either open a box, or select a candidate and terminate the game. However, opening a box costs a price. The goal of the algorithm is to maximize the reward from the selected candidate minus the total cost of opening boxes. The Pandora's box problem is a foundational framework for studying how the cost of revealing uncertainty affects the decision quality. More importantly, it suggests what information to acquire next after gaining some pieces of information.

Previous Works. For the SUP problem, Dürr et al. studied the case where all jobs have the same testing time [5,6]. In the paper, the authors proposed a THRESHOLD algorithm for the special instances. For the competitive analysis, the authors proposed a delicate *instance-reduction* framework. Using this framework, the authors showed that the worst case instance of THRESHOLD has a special format. An upper bound of the competitive ratio of 2 of THRESHOLD is obtained by the ratio of the special format instance. Using the instance-reduction framework, the authors also showed that when all jobs have the same testing time and the same upper limit, there exists a 1.9338-competitive BEAT algorithm. The authors provided a lower bound of 1.8546 for any deterministic online algorithm. For randomized algorithms, the authors showed that the expected competitive ratio is between 1.6257 and 1.7453.

Later, Albers and Eckl studied a more general case where jobs have variable testing time [1]. In the paper, the authors proposed a classic and elegant framework where the completion time of an algorithm is divided into contribution segments by the jobs executed prior to it. For the jobs with "correct" execution order as they are in the optimal solution, their total contribution to the total completion time is charged to twice the optimal cost by the fact that the algorithm does not pay too much for wrong decisions of testing a job or not. For the jobs with "wrong" execution order, their total contribution to the total completion time is charged to another twice the optimal cost using a *comparison tree* method, which is bound with the proposed (α, β)-SORT algorithm. The authors also provide a preemptive 3.2361-competitive algorithm and an expected 3.3794-competitive randomized algorithm.

In the works [1,5,6], the objective of minimizing the maximum completion time on a single machine was also studied. For the uniform-testing-time setting, Dürr et al. [5,6] proposed a ϕ-competitive deterministic algorithm and a $\frac{4}{3}$-competitive randomized algorithm, where both algorithms are optimal. For a more general setting, Albers and Eckl [1] showed that variable testing time does not increase the competitive ratios of online algorithms.

Our Contribution. We first analyze the (α, β)-SORT algorithm proposed in the work [1] in a more amortized sense. Instead of charging the jobs in the correct order and in the wrong order to the optimal cost separately, we manage to partition the tasks into groups and charge the total cost in each of the groups to the optimal cost regarding the group. The introduction of amortization to the analysis creates room for improving the competitive ratio by adjusting the values of α and β. The possibility of picking $\alpha > 1$ helps balance the penalty incurred by making a wrong guess on testing a job or not. On the other hand, the room for different β values allows one to differently prioritize the tasks that provide extra information and the tasks that immediately decide a completion

Table 1. Summary of the results. The results from this work are bold and in red.

	Testing time	Upper limit	Upper Bound	Lower bound
Deterministic	1	Uniform	1.9338 [5,6]	1.8546 [5,6]
		Variable	2 [5,6]	
	Variable	Variable	4 [1] → 2.414 (Theorem 1)	
			2.316513 (Theorem 2)	
		(Prmp.)	3.2361 [1]	
			2.316513 (Theorem 2)	
Randomized	1	Variable	1.7453 [5,6]	1.6257 [5,6]
	Variable	Variable	3.3794 [1]	
			2.152271 (Theorem 3)	

time for a job. By this new analysis and the room of choosing different values of α and β, we improve the upper bound of the competitive ratio of (α, β)-SORT from 4 to $1 + \sqrt{2}$. With the power of amortization, we improve the algorithm by further prioritizing different tasks using different parameters. The new algorithm, $\mathrm{PCP}_{\alpha,\beta}$, is 2.316513-competitive. This algorithm is extended to a randomized version with an expected competitive ratio of 2.152271. Finally, we show that under the current problem setting, preempting the execution of jobs does not help in gaining a better algorithm. A summary of the results can be found in Table 1.

Paper Organization. In Sect. 2, we introduce the notation used in this paper. We also review the algorithm and analysis of the (α, β)-SORT algorithm proposed in the work [1]. In Sect. 3, we elaborate on how amortized analysis helps to improve the competitive analysis of (α, β)-SORT (Subsect. 3.1). Upon the new framework, we propose a better algorithm, $\mathrm{PCP}_{\alpha,\beta}$, in Subsect. 3.2. In Subsect. 3.3, we argue that the power of preemption is limited in the current model. Finally, we show how amortization helps to improve the performance of randomized algorithms. For the sake of the page limit, we leave the proofs in the full version.

2   Preliminary

Given n jobs $1, 2, \cdots, n$, each job j has a *testing time* t_j and an *upper limit* u_j of its *real processing time* $p_j \in [0, u_j]$. A job j can be executed-untested in u_j time units or be tested using t_j time units and then executed in p_j time units. Note that if a job is tested, it does not need to be executed immediately. That is, for a tested job, there can be tasks regarding other jobs between its testing and its execution.

We denote by p_j^A the time spent by an algorithm A on job j, i.e., $p_j^A = t_j + p_j$ if A tests j, and $p_j^A = u_j$ otherwise. Similarly, we denote by p_j^* the time spent by OPT, the optimal algorithm. Since OPT knows p_j in advance, it can decide optimally whether to test a job, i.e., $p_j^* = \min\{u_j, t_j + p_j\}$, and execute the jobs in the ascending order of p_j^*. We denote by $cost(A)$ the total completion time of any algorithm A.

The *tasks* regarding a job j are the testing, execution, or execution-untested of j (taking t_j, p_j, or u_j, respectively). We follow the notation in the work of Albers and Eckl [1] and denote $c(k, j)$ as the *contribution* of job k in the completion time of job j in the online schedule A. That is, $c(k, j)$ is the total time of the tasks regarding job k before the completion time of job j. The *completion time* of job j in the schedule A is then $\sum_{k=1}^{n} c(k, j)$. Similarly, we define $c^*(k, j)$ as the contribution of job k in the completion time of job j in the optimal schedule. As observed, OPT schedules in the order of p^*, $c^*(k, j) = 0$ if k is executed after j in the optimal schedule, and $c^*(k, j) = p_k^*$ otherwise.

We denote by $i <_o j$ if the optimal schedule executes job i before job j. We also define $i >_o j$ and $i =_o j$ similarly (in the latter case, job i and job j are the same job). The completion time of job j in the optimal schedule is denoted by $c_j^* = \sum_{i \leq_o j} p_i^*$. The total completion time of the optimal schedule is then $\sum_{j=1}^{n} c_j^*$. Note that there is an optimal strategy where $p_i^* \leq p_j^*$ if $i \leq_o j$.

2.1 Review (α, β)-SORT Algorithm [1]

For completeness, we summarise the (α, β)-SORT algorithm and its analysis proposed in the work of Albers and Eckl [1].

Intuitively, the algorithm tests a job j if and only if $u_j \geq \alpha \cdot t_j$. Depending on whether a job is tested or not, the job is transformed into one task (execution-untested task) or two tasks (testing task and execution task). These tasks are then maintained in a priority queue for the algorithm to decide their processing order. More specifically, a testing task has a weight of $\beta \cdot t_j$, an execution task has a weight of p_j, and an execution-untested task has a weight of u_j. (See Algorithm 1.) After the tasks regarding the jobs are inserted into the queue, the algorithm executes the tasks in the queue and deletes the executed tasks, starting from the task with the shortest (weighted) time. If the task is a testing of a job j, the resulting p_j is inserted into the queue after testing. (See Algorithm 2.) Intuitively, both α and β are at least 1. The precise values of α and β will be decided later based on the analysis.

Analysis [1]. Recall that $c(k, j)$ is the contribution of job k of the completion time of job j, and the completion time of job j is $c_j^A = \sum_{k=1}^{n} c(k, j)$. The key

---

Algorithm 1. (α, β)-SORT algorithm [1]

Initialize a priority queue Q
  for $j = 1, 2, 3, \cdots, n$ **do**
    if $u_j \geq \alpha \cdot t_j$ **then**
      Insert a testing task with weight $\beta \cdot t_j$ into Q
    else
      Insert an execution-untested task with weight u_j into Q
  Queue-Execution(Q) ▷ See Algorithm 2

Algorithm 2. Procedure **Queue-Execution** (Q)

---

procedure QUEUE-EXECUTION(Q)
    while Q is not empty **do**
        $x \leftarrow$ Extract the smallest-weight task in Q
        if x is a testing task for a job j **then**
            Test job j ▷ It takes t_j time
            Insert an execution task with weight p_j into Q
        else if x is an execution task for a job j **then**
            Execute (tested) job j ▷ It takes p_j time
        else   ▷ x is an execution-untested task for a job j
            Execute job j untested   ▷ It takes u_j time

---

idea of the analysis is that given job j, partitioning the jobs (say, k) that are executed before j into two groups, $k \leq_o j$ or $k >_o j$. Since the algorithm only tests a job j when $u_j \geq \alpha t_j$, $p_k^A \leq \max\{\alpha, 1 + \frac{1}{\alpha}\} \cdot p_k^*$. Therefore, the total cost incurred by the first group of jobs is at most $\max\{\alpha, 1 + \frac{1}{\alpha}\} \cdot cost(\text{OPT})$. Note that the ratio, in this case, reflects the penalty to the algorithm that makes a wrong guess on testing a job or not.

For the second group of jobs, the authors proposed a classic and elegant *comparison tree* framework to charge each $c(k, j)$ with $k >_o j$ to the time that the optimal schedule spends on job j. More specifically, $c(k, j) \leq \max\{(1 + \frac{1}{\beta})\alpha, 1 + \frac{1}{\alpha}, 1 + \beta\} \cdot p_j^*$ for any k and j. Hence, the total cost incurred by the second group of jobs can be charged to $\max\{(1 + \frac{1}{\beta})\alpha, 1 + \frac{1}{\alpha}, 1 + \beta\} \cdot cost(\text{OPT})$.

By summing up the $c(k, j)$ values for all pairs of k and j, the total completion time of the algorithm is at most

$$\max\{\alpha, 1 + \frac{1}{\alpha}\} + \max\{(1 + \frac{1}{\beta}) \cdot \alpha, 1 + \frac{1}{\alpha}, 1 + \beta\}.$$

When $\alpha = \beta = 1$ (which is the optimal selection), the competitive ratio is 4.

2.2 Our Observation

As stated by Albers and Eckl [1], $\alpha = \beta = 1$ is the optimal choice in their analysis framework. Therefore, it is not possible to find a better α and β to tighten the competitive ratio under the current framework. However, the framework can be improved via observations.

For example, given that $\alpha = \beta = 1$, consider two jobs k and j, where $(t_k, u_k, p_k) = (1 + \varepsilon, 1 + 3\varepsilon, 1 + 3\varepsilon)$ and $(t_j, u_j, p_j) = (1, 1 + 4\varepsilon, 1 + 2\varepsilon)$. By the (α, β)-SORT algorithm, both k and j are tested. The order of the tasks regarding these two jobs is t_j, t_k, p_j, and finally p_k. On the other hand, in the optimal schedule, $p_k^* = u_k = 1 + 3\varepsilon$ and $p_j^* = u_j = 1 + 4\varepsilon$. Since $k \leq_o j$, as shown in Fig. 1, both $c(k, j)$ and $c(j, k)$ are charged to $2p_k^*$, separately. Note that although $c(k, j) = t_k$ in this example, the worst-case nature of the analysis framework fails to capture the fact that the contribution from the tasks regarding k to the completion time of j is even smaller than p_k^*. This observation motivates us to establish a new analysis framework.

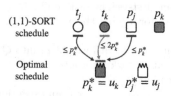

Fig. 1. An example where p_k^* is charged four times. The light blue and dark blue segments represent $c(k, j)$ and $c(j, k)$, respectively. The red segment represents p_k^*. (Color figure online)

3   Deterministic Algorithms

In this section, we first enhance the framework by equipping it with amortized analysis in Subsect. 3.1. Using amortized arguments, for any two jobs $k \leq_o j$, we manage to charge the sum of $c(k, j) + c(j, k)$ to p_k^*. The new framework not only improves the competitive ratio but also creates room for adjusting α and β.

Finally, in Subsect. 3.2, we improve the (α, β)-SORT algorithm based on our enhanced framework.

3.1   Amortization

We first bound $c(k, j) + c(j, k)$ for all pairs of jobs k and j with $k \leq_o j$ by a function $r(\alpha, \beta) \cdot c^*(k, j)$. Then, we can conclude that the algorithm is $r(\alpha, \beta)$-competitive by the following argument:

$$cost((\alpha, \beta)\text{-SORT}) = \sum_{j=1}^{n} \sum_{k=1}^{n} c(k, j) = \sum_{j=1}^{n} \left(\sum_{k <_o j} (c(k, j) + c(j, k)) + c(j, j) \right)$$

$$\leq \sum_{j=1}^{n} r(\alpha, \beta) \cdot \left(\sum_{k <_o j} c^*(k, j) + c^*(j, j) \right) = r(\alpha, \beta) \cdot cost(\text{OPT})$$

To bound $c(k, j) + c(j, k)$ by the cost of tasks k, we first observe that it is impossible that $c(k, j) = p_k^A$ and $c(j, k) = p_j^A$ at the same time. More specifically, depending on whether the jobs k and j are tested or not, the last task regarding these two jobs does not contribute to $c(k, j) + c(j, k)$. Furthermore, the order of these jobs' tasks in the priority queue provides a scheme to charge the cost of the tasks regarding j to the cost of tasks regarding k.

Figure 2 shows how the charging is done. Each row in the subfigures is a permutation of how the tasks regarding job j and k are executed. The gray objects are tasks regarding k, and the white objects are tasks regarding j. The circles, rectangles, and rectangles with the wavy top are testing tasks, execution tasks, and execution-untested tasks, respectively. The horizontal lines present the values of $c(k,j)$ (light blue) and $c(j,k)$ (dark blue). The red arrows indicate how the cost of a task regarding j is charged to that of a task regarding k according to the order of the tasks in the priority queue. The charging $c(k,j) + c(j,k)$ to the cost of tasks regarding k results in Lemmas 1 and 2. For the sake of space, the proof is provided in the full paper.

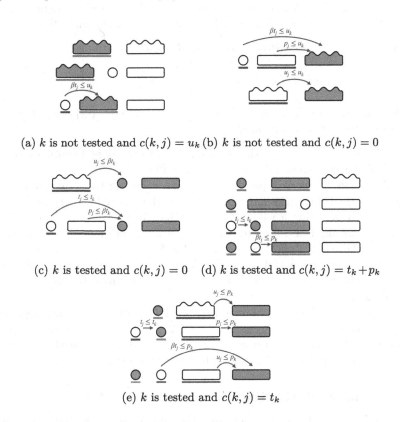

(a) k is not tested and $c(k,j) = u_k$ (b) k is not tested and $c(k,j) = 0$

(c) k is tested and $c(k,j) = 0$ (d) k is tested and $c(k,j) = t_k + p_k$

(e) k is tested and $c(k,j) = t_k$

Fig. 2. The red arrows illustrate how to charge $c(k,j) + c(j,k)$ to the cost of tasks regarding k. Each row in the sub-figures is a permutation of how the tasks are executed. The circles and rectangles are testing tasks and execution tasks after testing, respectively. The rectangles with curly tops are execution tasks without testing. The tasks in gray are from the job k, and the tasks in white are from the job j. The light blue and dark blue line segments under the tasks represent the contribution $c(k,j)$ and $c(j,k)$, respectively. (Color figure online)

Lemma 1. *If (α, β)-SORT does not test job k,*

$$c(k,j) + c(j,k) \leq (1 + \frac{1}{\beta})u_k.$$

Lemma 2. *If (α, β)-SORT tests job k,*

$$c(k,j) + c(j,k) \le \max\{2t_k + p_k, (1+\beta)t_k, t_k + (1 + \frac{1}{\beta})p_k\}.$$

Now, we can bound the competitive ratio of the (α, β)-SORT (Theorem 1). The idea is, depending on whether job k is tested or not by the optimal schedule, the expressions in Lemmas 1 and 2 can be written as a function of α, β, and p_k^*. By selecting the values of α and β carefully, we can balance the worst case ratio in the scenario where k is executed-untested by the algorithm (Lemma 1) and that in the scenario where k is tested by the algorithm (Lemma 2).

Theorem 1. *The competitive ratio of (α, β)-SORT is at most*

$$\max\{\alpha(1 + \frac{1}{\beta}), 1 + \frac{1}{\alpha} + \frac{1}{\beta}, 1 + \beta, 2, 1 + \frac{2}{\alpha}\} \tag{1}$$

By choosing $\alpha = \beta = \sqrt{2}$, (α, β)-SORT algorithm is $(1 + \sqrt{2})$-competitive. The choice is optimal for expression (1).

Note that by Theorem 1, the (α, β)-SORT algorithm is 3-competitive when $\alpha = \beta = 1$, which matches the observation in Fig. 1.

Our analysis framework provides room for adjusting the values of α and β. By selecting the values of α and β, we can tune the cost of tasks regarding k that is charged. By selecting a value of α other than 1, we can balance the penalty of making a wrong decision on testing a job or not. The capability of selecting a value of β other than 1 allows us to prioritize the testing tasks (which are scaled by β) and the execution tasks (which immediately decide a completion time of a job). Finally, the performance of the algorithm is tuned by finding the best values of α and β.

However, recall that the parameter α encodes the penalty for making a wrong guess on testing a job or not. When $\alpha = \sqrt{2}$, the penalty for testing a job we should not test is more expensive than that for executing-untested a job that we should test. It inspires us to improve the algorithm further.

3.2   An Improved Algorithm

Surprisingly, the introduction of amortization even sheds light on further improvement of the algorithm. We propose a new algorithm, *Prioritizing-Certain-Processing-time* ($\text{PCP}_{\alpha,\beta}$). The main difference between $\text{PCP}_{\alpha,\beta}$ and (α, β)-SORT is that in the $\text{PCP}_{\alpha,\beta}$ algorithm after a job j is tested, an item with weight $t_j + p_j$ is inserted into the queue instead of p_j (see Algorithm 3). Intuitively, we prioritize a job by its certain (total) processing time p_j^A, which can be $t_j + p_j$ or u_j. Then, we can charge the total cost of tasks regarding a wrong-ordered j to βt_k or p_k^A all at once.

The new algorithm $\text{PCP}_{\alpha,\beta}$ (Algorithm 1 combined with Algorithm 3) has an improved estimation of $c(k,j) + c(j,k)$ when $c(j,k) = t_j + p_j$. However, when there is only one task regarding j contributing to $c(j,k)$, the estimation of $c(k,j) + c(j,k)$ may increase. Formally, we have the following two lemas.

Algorithm 3. Procedure **Updated Queue-Execution** (Q)

---

procedure UPDATED QUEUE-EXECUTION(Q)
    while Q is not empty **do**
        $x \leftarrow$ Extract the smallest-weight task in Q
        if x is a testing task for a job j **then**
            Test job j ▷ It takes t_j time
            Insert an execution task with weight $t_j + p_j$ into Q
        else if x is an execution task for a job j **then**
            Execute (tested) job j ▷ It takes p_j time
        else   ▷ x is an execution-untested task for a job j
            Execute job j untested   ▷ It takes u_j time

Lemma 3. *Given two jobs* $k \leq_o j$, *if* $PCP_{\alpha,\beta}$ *does not test job* k,

$$c(k,j) + c(j,k) \leq (1 + \frac{1}{\beta})u_k.$$

Lemma 4. *Given two jobs* $k \leq_o j$, *if* $PCP_{\alpha,\beta}$ *tests job* k,

$$c(k,j) + c(j,k) \leq \max\{2t_k + p_k, \beta t_k, (1 + \frac{1}{\beta})(t_k + p_k)\}.$$

Similar to the proof of Theorem 1, we have the following competitiveness results of the $PCP_{\alpha,\beta}$ algorithm.

Theorem 2. *The competitive ratio of* $PCP_{\alpha,\beta}$ *is at most*

$$\max\{\alpha(1 + \frac{1}{\beta}), 1 + \frac{1}{\alpha} + \frac{1}{\beta} + \frac{1}{\alpha\beta}, \beta, 2, 1 + \frac{2}{\alpha}\} . \tag{2}$$

By choosing $\alpha = \frac{1+\sqrt{5}}{2}$ *and* $\beta = \frac{1+\sqrt{5}+\sqrt{2(7+5\sqrt{5})}}{4}$, *the competitive ratio of* $PCP_{\alpha,\beta}$ *is* $\frac{1+\sqrt{5}+\sqrt{2(7+5\sqrt{5})}}{4} \leq 2.316513$. *The choice is optimal for expression (2).*

The selection of golden ratio α balances the penalty of making a wrong guess for testing a job or not.

Note that using the analysis proposed in the work of Albers and Eckl [1] on the new algorithm that put $t_j + p_j$ back to the priority list after testing job j, the competitive ratio is $\max\{\alpha, 1 + \frac{1}{\alpha}\} + \max\{\alpha, 1 + \frac{1}{\alpha}, \beta\}$. The best choice of the values is $\alpha = \phi$ and $\beta \in [1, \phi]$, and the competitive ratio is at most 2ϕ.

3.3 Preemption

We show that preempting the tasks does not improve the competitive ratio. Intuitively, we show that given an algorithm A that generates a preemptive schedule, we can find another algorithm B that is capable of simulating A and performs the necessary merging of preempted parts. The simulation may make

---

Algorithm 4. Rand-PCP$_\beta$ algorithm

Initialize a priority queue Q
for $j = 1, 2, 3, \cdots, n$ do
    Let $r_j \leftarrow \frac{u_j}{t_j}$
    if $r_j < 1$ then
        $\mathbb{P}_j \leftarrow 0$
    else if $r_j > 3$ then
        $\mathbb{P}_j \leftarrow 1$
    else
        $\mathbb{P}_j = \frac{3r_j^2 - 3r_j}{3r_j^2 - 4r_j + 3}$

    Choose one of βt_j and u_j randomly with probability $\mathbb{P}_j$ for βt_j and $1 - \mathbb{P}_j$ for u_j
    Insert a testing task with weight βt_j into Q if βt_j is chosen, and insert an execution-untested task with weight u_j into Q otherwise
    Updated Queue-Execution(Q) $\triangleright$ See Algorithm 3

---

the timing of A's schedule gain extra information about the real processing times earlier due to the advance of a testing task. However, a non-trivial A can only perform better by receiving the information earlier. Thus, B's non-preemptive schedule has a total completion time at most that of A's schedule.

Lemma 5. *In the SUP problem, if there is an algorithm that generates a pre-emptive schedule, then we can always find another algorithm that generates a non-preemptive schedule and performs as well as the previous algorithm in terms of competitive ratios.*

4   Randomized Algorithm

The amortization also helps improve the performance of randomized algorithms. We combine the PCP$_{\alpha,\beta}$ algorithm with the framework in the work of Albers and Eckl [1], where instead of using a fixed threshold α, a job j is tested with probability $\mathbb{P}_j$, which is a function of u_j, t_j, and β.

Our Randomized Algorithm. For any job j with $\frac{u_j}{t_j} < 1$ or $\frac{u_j}{t_j} > 3$, we insert u_j or βt_j into the queue, respectively. For any job j with $1 \le \frac{u_j}{t_j} \le 3$, we insert βt_j into the queue with probability $\mathbb{P}_j$ and insert u_j with probability $1 - \mathbb{P}_j$. Once a testing task t_j is executed, we insert $t_j + p_j$ into the queue. (See Algorithms 4 and 3.)

Analysis. The following lema can be proven using Lemma 3 and Lemma 4.

Lemma 6. *The expected total completion time of the n jobs is at most*

$$\sum_j \sum_{k \le_o j} (1 + \frac{1}{\beta}) u_k (1 - \mathbb{P}_k) + \max\{2t_k + p_k, \beta t_k, (1 + \frac{1}{\beta})(t_k + p_k)\} \mathbb{P}_k,$$

where $\mathbb{P}_k$ is the probability that job k is tested.

Depending on whether the jobs are tested or not in the optimal schedule, the expected total completion time can be expressed by functions with the variables: the probability, the parameters of the jobs, and β. We design the probability $\mathbb{P}_k$ by balancing the costs between the worst cases where $p_k^* = u_k$ or $p_k^* = t_k + p_k$. Note that there are cases where the "ideal" value of $\mathbb{P}_k$ is outside the range $[0, 1]$. We take care of these special cases by setting $\mathbb{P}_k$ as 0 or 1 if the ideal value is smaller than 0 or larger than 1, respectively.

Theorem 3. *Let r_k denote $\frac{u_k}{t_k}$. The expected competitive ratio of Rand-PCP$_\beta$ is at most*

$$\max_k \frac{(1 + \frac{1}{\beta})u_k(1 - \mathbb{P}_k) + \max\{2t_k + p_k, \beta t_k, (1 + \frac{1}{\beta})(t_k + p_k)\}\mathbb{P}_k}{p_k^*}, \quad where$$

$$\mathbb{P}_k = \frac{(\beta + 1)(r_k - 1)}{\beta(\max\{\frac{2}{r_k} + 1, \frac{\beta}{r_k}, (1 + \frac{1}{\beta})(1 + \frac{1}{r_k})\} - \max\{2, \beta, 1 + \frac{1}{\beta}\} + r_k - 1) + r_k - 1}$$

if $r_k \in [1, 3]$, $\mathbb{P}_k = 0$ if $r_k < 1$, and $\mathbb{P}_k = 1$ if $r_k > 3$. By choosing $\beta = 2$, the ratio is $\frac{3(7 + 3\sqrt{6})}{20} \leq 2.152271$. The choice of β is optimal.

5   Conclusion

In this work, we study a scheduling problem with explorable uncertainty. We enhance the analysis framework proposed in the work [1] by introducing amortized perspectives. Using the enhanced analysis framework, we are able to balance the penalty incurred by different wrong decisions of the online algorithm. In the end, we improve the competitive ratio significantly from 4 to 2.316513 (deterministic) and from 3.3794 to 2.152271 (randomized). An immediate open problem is if one can further improve the competitive ratio by a deeper level of amortization.

Additionally, we show that preemption does not improve the competitive ratio in the current problem setting, where all jobs are available at first. It may not be true in the fully online setting, where jobs can arrive at any time. Thus, another open problem is to study the problem in the fully online model.

References

1. Albers, S., Eckl, A.: Explorable uncertainty in scheduling with non-uniform testing times. In: Kaklamanis, C., Levin, A. (eds.) WAOA 2020. LNCS, vol. 12806, pp. 127–142. Springer, Cham (2021). https://doi.org/10.1007/978-3-030-80879-2_9
2. Cardoso, J.M.P., Diniz, P.C., Coutinho, J.G.F.: Embedded Computing for High Performance: Efficient Mapping of Computations Using Customization, Code Transformations and Compilation. Morgan Kaufmann Publishers, Burlington (2017)

3. Caruso, S., Galatà, G., Maratea, M., Mochi, M., Porro, I.: Scheduling pre-operative assessment clinic via answer set programming. In: Benedictis, R.D., et al. (eds.) Proceedings of the 9th Italian workshop on Planning and Scheduling (IPS'21) and the 28th International Workshop on "Experimental Evaluation of Algorithms for Solving Problems with Combinatorial Explosion" (RCRA 2021) with CEUR-WS co-located with 20th International Conference of the Italian Association for Artificial Intelligence (AIxIA 2021), Milan, Italy (virtual), 29th–30th November 2021. CEUR Workshop Proceedings, vol. 3065. CEUR-WS.org (2021). https://ceur-ws.org/Vol-3065/paper3_196.pdf

4. Ding, B., Feng, Y., Ho, C., Tang, W., Xu, H.: Competitive information design for pandora's box. In: Bansal, N., Nagarajan, V. (eds.) Proceedings of the 2023 ACM-SIAM Symposium on Discrete Algorithms, SODA 2023, Florence, Italy, 22–25 January 2023, pp. 353–381. SIAM (2023). https://doi.org/10.1137/1.9781611977554.ch15

5. Dürr, C., Erlebach, T., Megow, N., Meißner, J.: Scheduling with explorable uncertainty. In: Karlin, A.R. (ed.) 9th Innovations in Theoretical Computer Science Conference, ITCS 2018, 11–14 January 2018, Cambridge, MA, USA. LIPIcs, vol. 94, pp. 30:1–30:14. Schloss Dagstuhl - Leibniz-Zentrum für Informatik (2018). https://doi.org/10.4230/LIPIcs.ITCS.2018.30

6. Dürr, C., Erlebach, T., Megow, N., Meißner, J.: An adversarial model for scheduling with testing. Algorithmica 82(12), 3630–3675 (2020). https://doi.org/10.1007/s00453-020-00742-2

7. Erlebach, T., Hoffmann, M.: Query-competitive algorithms for computing with uncertainty. Bull. EATCS 116 (2015). http://eatcs.org/beatcs/index.php/beatcs/article/view/335

8. Erlebach, T., Hoffmann, M., Kammer, F.: Query-competitive algorithms for cheapest set problems under uncertainty. Theor. Comput. Sci. 613, 51–64 (2016). https://doi.org/10.1016/j.tcs.2015.11.025

9. Esfandiari, H., Hajiaghayi, M.T., Lucier, B., Mitzenmacher, M.: Online pandora's boxes and bandits. In: The Thirty-Third AAAI Conference on Artificial Intelligence, AAAI 2019, The Thirty-First Innovative Applications of Artificial Intelligence Conference, IAAI 2019, The Ninth AAAI Symposium on Educational Advances in Artificial Intelligence, EAAI 2019, Honolulu, Hawaii, USA, 27 January–1 February 2019, pp. 1885–1892. AAAI Press (2019). https://doi.org/10.1609/aaai.v33i01.33011885

10. Feder, T., Motwani, R., O'Callaghan, L., Olston, C., Panigrahy, R.: Computing shortest paths with uncertainty. J. Algorithms 62(1), 1–18 (2007). https://doi.org/10.1016/j.jalgor.2004.07.005

11. Feder, T., Motwani, R., Panigrahy, R., Olston, C., Widom, J.: Computing the median with uncertainty. In: Yao, F.F., Luks, E.M. (eds.) Proceedings of the Thirty-Second Annual ACM Symposium on Theory of Computing, 21–23 May 2000, Portland, OR, USA, pp. 602–607. ACM (2000). https://doi.org/10.1145/335305.335386

12. Gupta, M., Sabharwal, Y., Sen, S.: The update complexity of selection and related problems. Theory Comput. Syst. 59(1), 112–132 (2016). https://doi.org/10.1007/s00224-015-9664-y

13. Halldórsson, M.M., de Lima, M.S.: Query-competitive sorting with uncertainty. Theor. Comput. Sci. 867, 50–67 (2021). https://doi.org/10.1016/j.tcs.2021.03.021

14. Hoffmann, M., Erlebach, T., Krizanc, D., Mihalák, M., Raman, R.: Computing minimum spanning trees with uncertainty. In: Albers, S., Weil, P. (eds.) STACS

2008, 25th Annual Symposium on Theoretical Aspects of Computer Science, Bordeaux, France, 21–23 February 2008, Proceedings. LIPIcs, vol. 1, pp. 277–288. Schloss Dagstuhl - Leibniz-Zentrum für Informatik, Germany (2008). https://doi.org/10.4230/LIPIcs.STACS.2008.1358

15. Kahan, S.: A model for data in motion. In: Koutsougeras, C., Vitter, J.S. (eds.) Proceedings of the 23rd Annual ACM Symposium on Theory of Computing, 5–8 May 1991, New Orleans, Louisiana, USA, pp. 267–277. ACM (1991). https://doi.org/10.1145/103418.103449

16. Lopes, J., Vieira, G., Veloso, R., Ferreira, S., Salazar, M., Santos, M.F.: Optimization of surgery scheduling problems based on prescriptive analytics. In: Gusikhin, O., Hammoudi, S., Cuzzocrea, A. (eds.) Proceedings of the 12th International Conference on Data Science, Technology and Applications, DATA 2023, Rome, Italy, 11–13 July 2023, pp. 474–479. SCITEPRESS (2023). https://doi.org/10.5220/0012131700003541

17. Nicolai, R.P., Dekker, R.: Optimal Maintenance of Multi-component Systems: A Review, pp. 263–286 (2008)

18. Olston, C., Widom, J.: Offering a precision-performance tradeoff for aggregation queries over replicated data. In: Abbadi, A.E., et al. (eds.) VLDB 2000, Proceedings of 26th International Conference on Very Large Data Bases, 10–14 September 2000, Cairo, Egypt, pp. 144–155. Morgan Kaufmann (2000). http://www.vldb.org/conf/2000/P144.pdf

19. Weitzman, M.L.: Optimal search for the best alternative. Econometrica **47**(3), 641–654 (1979). http://www.jstor.org/stable/1910412

20. Wiseman, Y., Schwan, K., Widener, P.M.: Efficient end to end data exchange using configurable compression. ACM SIGOPS Oper. Syst. Rev. **39**(3), 4–23 (2005). https://doi.org/10.1145/1075395.1075396

Total Completion Time Scheduling Under Scenarios

Thomas Bosman[1], Martijn van Ee[2], Ekin Ergen[3(✉)], Csanád Imreh[4],
Alberto Marchetti-Spaccamela[5,6], Martin Skutella[3], and Leen Stougie[6,7,8]

[1] Amsterdam, The Netherlands
[2] Netherlands Defence Academy, Den Helder, The Netherlands
M.v.Ee.01@mindef.nl
[3] Technische Universität Berlin, Berlin, Germany
ergen@math.tu-berlin.de, martin.skutella@tu-berlin.de
[4] Szeged, Hungary
[5] Universitá di Roma "La Sapienza", Rome, Italy
alberto@diag.uniroma.it
[6] Erable, INRIA, Paris, France
[7] Centrum voor Wiskunde en Informatica (CWI), Amsterdam, The Netherlands
leen.stougie@cwi.nl
[8] Vrije Universiteit Amsterdam, Amsterdam, The Netherlands

Abstract. Scheduling jobs with given processing times on identical parallel machines so as to minimize their total completion time is one of the most basic scheduling problems. We study interesting generalizations of this classical problem involving scenarios. In our model, a scenario is defined as a subset of a predefined and fully specified set of jobs. The aim is to find an assignment of the whole set of jobs to identical parallel machines such that the schedule, obtained for the given scenarios by simply skipping the jobs not in the scenario, optimizes a function of the total completion times over all scenarios.

While the underlying scheduling problem without scenarios can be solved efficiently by a simple greedy procedure (SPT rule), scenarios, in general, make the problem NP-hard. We paint an almost complete picture of the evolving complexity landscape, drawing the line between easy and hard. One of our main algorithmic contributions relies on a deep structural result on the maximum imbalance of an optimal schedule, based on a subtle connection to Hilbert bases of a related convex cone.

Keywords: machine scheduling · total completion time · scenarios · complexity

1 Introduction

For a set J of n jobs with given processing times p_j, $j \in J$, one of the oldest results in scheduling theory states that scheduling the jobs in order of non-

C. Imreh—Our co-author Csanád Imreh tragically passed away on January 5th, 2017.

© The Author(s), under exclusive license to Springer Nature Switzerland AG 2023
J. Byrka and A. Wiese (Eds.): WAOA 2023, LNCS 14297, pp. 104–118, 2023.
https://doi.org/10.1007/978-3-031-49815-2_8

decreasing processing times on identical parallel machines in a round robin procedure minimizes the total completion time, that is, the sum of completion times of all jobs [11].

We study an interesting generalization of this classical scheduling problem where the input in addition specifies a set of K scenarios $\mathcal{S} = \{S_1, \ldots, S_K\}$, with $S_k \subseteq J$, $k = 1, \ldots, K$. The task is then to find, for the entire set of jobs J, a parallel machine schedule, which is an assignment of all jobs to machines and on each machine an order of the jobs assigned to it. This naturally induces a schedule for each scenario by simply skipping the jobs not in the scenario. In particular, jobs not contained in a particular scenario do *not* contribute to the total completion time of that scenario and, in particular, do *not* delay later jobs assigned to the same machine.

We aim to find a schedule for the entire set of jobs that optimizes a function of the total completion times of the jobs over all scenarios. More specifically, we focus on two functions on the scheduling objectives: in the MinMax version, we minimize the maximum total completion time over all scenarios, and in the MinAvg version we minimize the average of the total completion times over all scenarios. In the remainder of the paper we refer to the MinMax version as MinMaxSTC (MinMax Scenario scheduling with Total Completion time objective) and to the MinAvg version as MinAvgSTC.

Optimization Under Scenarios. Scenarios are commonly used in optimization to model uncertainty in the input or different situations that need to be taken into account. A variety of approaches has been proposed that appear in the literature under different names. In fact, scenarios have been introduced to model discrete distributions over parameter values in Stochastic Programming [8], or as samples in Sampling Average Approximation algorithms for stochastic problems with continuous distributions over parameter values [20]. In Robust Optimization [6], scenarios describe different situations that should be taken into account and are often specified as ranges for parameter values that may occur. Moreover, in data-driven optimization, scenarios are often obtained as observations. The problems we consider also fit in the general framework of A Priori Optimization [7]: the schedule for the entire set of jobs can be seen as an a priori solution which is updated in a simple way to a solution for each scenario. In the scheduling literature, different approaches to modeling scenarios have been introduced, for which we refer to a very recent overview by Shabtay and Gilenson [25]. Another related and popular framework is that of Min-Max Regret, aiming at obtaining a solution minimizing the maximum deviation, over all possible scenarios, between the value of the solution and the optimal value of the corresponding scenario [21].

Not surprisingly, for many problems, multiple scenario versions are fundamentally harder than their single scenario counterparts. Examples are the shortest path problem with a scenario specified by the destination [16,21], and the metric minimum spanning tree problem with a scenario defining the subset of vertices to be in the tree [7]. Scenario versions of NP-hard combinatorial optimization problems were also considered in the literature such as, for example, set cover [1] and the traveling salesperson problem [27].

Related Work. As we have already discussed above, a variety of approaches to optimization under scenarios appear in the literature under different names. Here we mention work in the field of scheduling that is more closely related to the model considered in this paper. We refer to the survey [25] and references therein for an overview of scheduling under scenarios.

Closest to the problems considered in this paper is the work of Feuerstein et al. [13] who also consider scenarios given by subsets of jobs and develop approximation algorithms as well as non-approximability results for minimizing the makespan on identical parallel machines, both for the MinMax and the MinAvg version. In fact, the hardness results for our problem given in Proposition 1 below follow directly from their work.

In *multi-scenario models*, a discrete set of scenarios is given, and certain parameters (e.g., processing times) of jobs can have different values in different scenarios. Several papers follow this model, mainly focusing on single machine scheduling problems. Various functions of scheduling objectives over the scenarios are considered, that have the MinMax and the MinAvg versions as special cases. Yang and Yu [28], for example, study a multi-scenario model and show that the MinMax version of minimizing total completion time is NP-hard even on a single machine and with only 2 scenarios, whereas in our model 2-scenario versions are generally easy. (Notice that our model is different from simply assigning a processing time of 0 to a job in a scenario if the job is not present in that scenario.) Aloulou and Della Croce [3] present algorithmic and computational complexity results for several single machine scheduling problems. Mastrolilli, Mutsanas, and Svensson [22] consider the MinMax version of minimizing the weighted total completion time on a single machine and prove interesting approximability and inapproximability results. Kasperski and Zieliński [18] consider a more general single machine scheduling problem in which precedence constraints between jobs are present and propose a general framework for solving such problems with various objectives.

Kasperski, Kurpisz, and Zieliński [17] study multi-scenario scheduling with parallel machines and the makespan objective function, where the processing time of each job depends on the scenario; they give approximability results for an unbounded number of machines and a constant number of scenarios. Albers and Janke [2] as well as Bougeret, Jansen, Poss, and Rohwedder [9] study a budgetary model with uncertain job processing times; in this model, each job has a regular processing time while in each scenario up to Γ jobs fail and require additional processing time. The considered objective function is to minimize the makespan: [9] proposes approximate algorithms for identical and unrelated parallel machines while [2] analyses online algorithms in this setting.

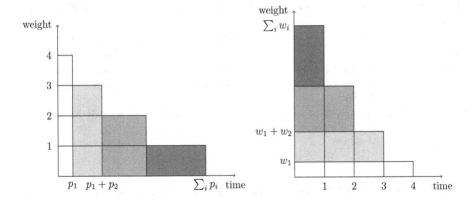

Fig. 1. Left: original schedule. Right: equivalent "weight-schedule". In both cases the objective value is equal to the total area of the rectangles

2   Preliminaries

We start by defining the problem formally, denoting the set of integers $\{1, \ldots, \ell\}$ simply by $[\ell]$. We are given a set of jobs $J = [n]$, machines $M = [m]$, non-negative job processing times p_j, $j \in [n]$, and scenarios $\mathcal{S} = \{S_1, \ldots, S_K\}$, where $S_k \subseteq J$, $k \in [K]$. We assume that the jobs are ordered by *non-increasing* processing times $p_1 \geq p_2 \geq \cdots \geq p_n$.[1] The task is to find a machine assignment, that is, a map $\varphi \colon [n] \to [m]$, or equivalently, a partitioning of the jobs $J_1, \ldots, J_m$ with the understanding that jobs in J_i shall be optimally scheduled on machine $i \in [M]$, that is, according to the *Shortest Processing Time first* (SPT) rule (i.e., in reverse order of their indices). Thus, the completion time of a particular job $j' \in J_i$ in scenario k is the sum over all processing times of jobs $j \in J_i \cap S_k$ with $j \geq j'$, and the contribution of jobs in J_i to the total completion in scenario $k \in [K]$ is thus:

$$\sum_{j' \in J_i \cap S_k} \sum_{j \in J_i \cap S_k : j' \leq j} p_j = \sum_{j \in J_i \cap S_k} p_j \cdot |\{j' \in J_i \cap S_k : j' \leq j\}| \qquad (1)$$

Observation 1 ([12]). *The above total unweighted completion time problem is equivalent to the total weighted completion time with jobs of unit length of weight $w_j := p_j$ that are processed in reverse order, i.e., in order of* non-increasing *weights.*

Indeed, this equivalence between total completion time for unweighted jobs scheduled in SPT order and total *weighted* completion time for unit length jobs

---

[1] In view of the SPT rule, this ordering might seem counterintuitive. But it turns out to be convenient as we argue below in Observation 1.

Table 1. The complexity landscape of MinMaxSTC on m machines with K scenarios

	$m = 2$	$m \in O(1)$	m part of input
$K = 2$	poly	poly	poly
$3 \leq K \in O(1)$	weakly NP-hard, pseudo-poly, FPTAS	weakly NP-hard, pseudo-poly, FPTAS	weakly NP-hard, poly if $w_j \in O(1)$
K part of input	strongly NP-hard [13], no $(2 - \varepsilon)$-approx [13], 2-approx	strongly NP-hard [13]	strongly NP-hard [13]

scheduled in order of non-decreasing weight was first observed by Eastman, Even, and Issacs [12]. The idea behind the equivalence (1) is best seen from a so-called 2-dimensional Gantt-chart; see Fig. 1 and the work of Goemans and Williamson [14], Megow and Verschae [23], or Cho, Shmoys, and Henderson [10].

In the introduction of this paper we prefer to present our results in terms of the more commonly known unweighted version, but for the remainder of this paper it is somewhat easier to argue about the weighted unit-processing time version; hence the ordering of jobs introduced above. The objective of (Min-MaxSTC) is then to minimize

$$\max_{k \in [K]} \sum_{i=1}^{m} \sum_{j \in J_i \cap S_k} w_j \cdot |\{j' \in J_i \cap S_k : j' \leq j\}|,$$

whereas MinAvgSTC aims to minimize

$$\frac{1}{K} \sum_{k=1}^{K} \sum_{i=1}^{m} \sum_{j \in J_i \cap S_k} w_j \cdot |\{j' \in J_i \cap S_k : j' \leq j\}|.$$

In the sequel in the MinAvgSTC we neglect the constant $1/K$-term and minimize the sum over all scenarios.

3   Our Contribution

We give a nearly complete overview of the complexity landscape of total completion time scheduling under scenarios. Tables 1 and 2 summarize our observations for MinMaxSTC and MinAvgSTC, respectively. The rows of both tables correspond to different assumptions on the number of scenarios K, whereas the columns specify assumptions on the given number of machines m.

First of all, it is not difficult to observe that both MinMaxSTC and MinAvgSTC are strongly NP-hard if K can be arbitrarily large; see last row of Tables 1 and 2. This even holds for the special case of unit length jobs and only two jobs per scenario. Moreover, for the case of MinMaxSTC on two machines, we get a tight non-approximability result and corresponding approximation algorithm, while for MinAvgSTC we can prove that the problem is APX-hard, i.e.,

Table 2. The complexity landscape of MinAvgSTC on m machines with K scenarios.

	$m = 2$	$m \in O(1)$	m part of input
$K = 2$	poly	poly	poly
$3 \leq K \in O(1)$	poly	poly	poly if $w_j \in O(1)$
K part of input	strongly NP-hard [13], no PTAS [15,19], 5/4-approx [24,26]	strongly NP-hard [13], no PTAS [15,19], (3/2-1/2m)-approx [24,26]	strongly NP-hard [13], no PTAS [15,19], (3/2-1/2m)-approx [24,26]

there is no PTAS, unless P = NP; see Sect. 4. For only $K = 2$ scenarios, however, both problems can be solved to optimality in polynomial time; see first row of Tables 1 and 2. Even better, in Sect. 4 we present a simple algorithm that constructs an 'ideal' schedule for the entire set of jobs simultaneously minimizing the total completion time in both scenarios. These results develop a clear complexity gap between the case of two and arbitrarily many scenarios.

A finer distinction between easy and hard can thus be achieved by considering the case of constantly many scenarios $K \geq 3$; see middle row in Tables 1 and 2. These results constitute the main contribution of this paper.

Our results on MinMaxSTC for constantly many scenarios $K \geq 3$ are presented in Sect. 5. Here it turns out that MinMaxSTC is weakly NP-hard already for $K = 3$ scenarios and $m = 2$ machines, but can be solved in pseudo-polynomial time for any constant number of scenarios and machines via dynamic programming. Moreover, the dynamic program together with standard rounding techniques immediately implies the existence of an FPTAS for this case. If the number of machines m is part of the input, however, our previous dynamic programming approach fails. But then again, we present a more sophisticated dynamic program that solves the problem efficiently if all job processing times are bounded by a constant.

MinAvgSTC with constantly many scenarios $K \geq 3$ is studied in Sect. 6. Somewhat surprisingly, and in contrast to MinMaxSTC, it turns out that MinAvgSTC remains easy as long as the number of machines m is bounded by a constant. This observation is again based on a dynamic programming algorithm. Moreover, we conjecture that the problem even remains easy if m is part of the input. More precisely, we conjecture that there always exists an optimal solution such that the imbalance between machine loads always remains bounded by $g(K)$ for some (exponential) function g that only depends on the number of scenarios K, but not on m or n. Using a subtle connection to the cardinality of Hilbert bases for convex cones, we prove that our conjecture is true for unit job processing times. In this case we obtain an efficient algorithm with running time $m^{h(K)} \cdot poly(n)$ for some function h.

Several of our results, in particular for MinMaxSTC, can be generalized to the Min-Max Regret framework; see the full version of the paper for details. Moreover, due to space restrictions, most proofs are omitted in this version of the paper and can be found in its full version.

4   Scheduling Under Arbitrary K is Hard, but $K = 2$ is Easy

4.1   NP-Hardness for Unbounded Number of Scenarios

For an unbounded number of scenarios, there is a straightforward proof that both MinMaxSTC and MinAvgSTC are NP-hard on $m \geq 3$ machines which relies on a simple reduction of the graph coloring problem: Given a graph, we interpret its nodes as unweighted unit length jobs and every edge as one scenario consisting of the two jobs that correspond to the end nodes of the edge. Obviously, the graph has an m-coloring without monochromatic edges if and only if there is a schedule such that the total completion time of each scenario is 2 (i.e., both jobs complete at time 1 on a machine of their own).

Since it is easy to decide whether the nodes of a given graph can be colored with two colors, the above reduction does not imply NP-hardness for $m = 2$ machines. For this case, however, the inapproximability results for the multi scenario makespan problem in [13] can be adapted to similar inapproximability results for our problems. Also these results already hold for unweighted unit length jobs. The proof is deferred to the full version of the paper.

Proposition 1. *For two machines and all jobs having unit lengths and weights, it is NP-hard to approximate MinMaxSTC within a factor $2 - \varepsilon$ and MinAvgSTC within ratio 1.011. The latter even holds if all scenarios contain only two jobs.*

Proof. (Sketches). We refrain from rigorous details of the proof because of similarly proven inapproximability results in [13].

For MinMaxSTC we can use essentially the same reduction as in the proof of Theorem 1 in [13]. Consider a set of scenarios with $2\ell + 1$ jobs each, where in each scenario the jobs can be partitioned in a perfectly balanced way. By the hardness of hypergraph balancing [5], it is NP-hard to find a solution where none of the scenarios puts all jobs to one of the machines. In a balanced solution the cost is $\sum_{i=1}^{\ell} i + \sum_{i=1}^{\ell+1} i = (\ell + 1)^2$. If we put all jobs on the same machine then the cost is $\sum_{i=1}^{2\ell+1} i = (2\ell + 1)(\ell + 1)$.

For the MinAvg version we adapt the reduction in the proof of Theorem 6 in [13] to the total completion time objective. Consider a MAX CUT instance and assign to each vertex a job with weight 1 and for each edge a scenario. Then a cut gives a partition of the jobs. If the scenario is in the cut then the cost is 2, if it is not in the cut then the jobs are on the same machine and the cost is 3. Thus the objective value is 3 times the number of edges minus the size of the cut. By this observation it follows that a $(1 + \alpha)$-approximation for MinAvgSTC yields a $(1 - 5\alpha)$-approximation for MAX CUT and our results follow from known inapproximability of MAX CUT [15, 19].   □

We notice that, for $m = 2$ machines, any algorithm for MinMaxSTC that assigns all jobs to the same machine (in SPT order) gives a 2-approximation. The approximability of MinMaxSTC for more than two machines is left as an interesting open question. The following approximation result for MinAvgSTC follows from a corresponding result for classical machine scheduling without scenarios.

Proposition 2. *For MinAvgSTC there is a $(3/2 - 1/2m)$-approximation algorithm for arbitrarily many machines, even if m is part of the input.*

Proof. The result is an immediate consequence of the following well known approximation result for the classical machine scheduling problem without scenarios: If all jobs are assigned to machines independently and uniformly at random, the expected total weighted completion time of the resulting schedule is at most a factor $3/2 - 1/2m$ away from the optimum; see, e.g., [24,26]. In our scenario scheduling model, this upper bound holds, in particular, for each single scenario which yields a randomized $(3/2 - 1/2m)$-approximation algorithm by linearity of expectation. Furthermore, this algorithm can be derandomized using standard techniques.   □

4.2   Computing an Ideal Schedule for Two Scenarios

For $K = 2$ scenarios, both MinMaxSTC and MinAvgSTC are polynomial time solvable on any number of machines. Actually, we prove an even stronger result: one can find, in polynomial time, a schedule that has optimal objective function value in each of the two scenarios simultaneously.

Theorem 2. *For the MinMaxSTC and the MinAvgSTC problem with two scenarios, after sorting the jobs in order of non-increasing weight, one can find in time linear in n a schedule that is simultaneously optimal for both scenarios.*

Proof. We show how to assign the jobs in *non-increasing* order of their weights in order to be optimal for both scenarios. As mentioned above, we want to assign the next job to a machine that, in each scenario, belongs to the least loaded machines in terms of the number of jobs already assigned to it. For this purpose we define two m-dimensional vectors s_1 and s_2 for scenarios 1 and 2, respectively, containing the relative loads on the machines. The relative load of a machine in a scenario is 0 if this machine belongs to the least loaded machines in this scenario; it is 1 if it has one job more than a least loaded machine, etc. In our assignment process, jobs will always be assigned to machines with relative load 0 in each of the scenarios they belong to. This ensures that we will end up with a schedule that is optimal for both scenarios simultaneously.

Initially, both vectors s_1 and s_2 are zero vectors, since no jobs have been assigned yet. When assigning job j, let μ_k be the lowest entry (lowest numbered machine) equal to zero in vector s_k. Moreover, let ν_k be the highest entry equal to zero in s_k. So initially, $\mu_1 = \mu_2 = 1$ and $\nu_1 = \nu_2 = m$. We apply the following assignment procedure, where we use $\mathbb{1}$ to denote the all-1 vector. For $j = 1$ to n, we apply the following case distinction:

- If $j \in S_1 \setminus S_2 \to$ assign j to machine μ_1 and increase μ_1 by 1.
- If $j \in S_2 \setminus S_1 \to$ assign j to machine μ_2 and increase μ_2 by 1.

- If $j \in S_1 \cap S_2 \rightarrow$ assign j to machine $\nu_1 = \nu_2$ and decrease both ν_1 and ν_2 by 1.
- If $s_1 = \mathbb{1} \rightarrow$ reset s_1 to become the all-0 vector. Reindex the machines such that s_2 becomes of the form $(1, \ldots, 1, 0, \ldots, 0)$. Reset $\mu_1 = 1$, $\nu_1 = m$, $\mu_2 = \mu_2 + m - \nu_2$, and $\nu_2 = m$. Do analogously if $s_2 = \mathbb{1}$.

We prove that for each job there is always a machine with relative load 0 in each scenario in which the job appears, thus implying the theorem. This is obviously true if job j appears in only one scenario, since after assigning $j-1$ jobs, $s_k \neq \mathbb{1}$, $k = 1, 2$. Hence there is always a machine with relative load 0. For job j appearing in both scenarios, we have to show that we maintain $\nu_1 = \nu_2$, in which case the same machine has relative load 0 to accommodate job j. The only way it can happen that $\nu_1 \neq \nu_2$ is if ever machine ν_1 was used for a job j' that only appeared in scenario 1. But that can only have happened if $\mu_1 = \nu_1$, in which case s_1 becomes an all-1 vector, and by resetting it to 0 and the renumbering of the machines, the relation $\nu_1 = \nu_2$ had been restored. □

5   The MinMax Version

5.1   Constant Number of Machines

In view of Theorem 2, the next theorem establishes a complexity gap for Min-MaxSTC when going from two to three scenarios.

Theorem 3. *On any fixed number $m \geq 2$ of machines and with $K = 3$ scenarios, MinMaxSTC is (weakly) NP-hard.*

The proof can be found in the full version of the paper. It reduces a variant of Partition, in which one is to partition a set of numbers into three sets of equal sum instead of two. This reduction establishes that MinMaxSTC is weakly NP-hard. For the case $m \in O(1)$, the weak NP-hardness cannot be strengthened to strong NP-hardness (unless P = NP) since on a fixed number of machines and scenarios an optimal solution can be found in pseudopolynomial time by dynamic programming. Moreover, via standard rounding techniques, one can obtain an FPTAS for MinMaxSTC. Details are given in the full version.

Theorem 4. *There exists a pseudopolynomial algorithm as well as a fully polynomial time approximation scheme for MinMaxSTC on a constant number of machines with a constant number of scenarios.*

The dynamic program runs in time $O(m(m^2 n^3 W)^{mK})$, where W denotes the largest (integer) job weight. The rounding for FPTAS redefines this weight to be $1 + mn^2/\varepsilon$, i.e., the runtime of the FPTAS is in $O(m(m^2 n^3 (mn^2/\varepsilon))^{mK})$, indeed polynomial in n and $1/\varepsilon$ assuming that m and K are constant.

An immediate consequence of the reduction that yields Theorem 3 is the NP-hardness of the robust version of scheduling with regret, as mentioned in [25], even when it is restricted to the parallel machine case. Similarly, the FPTAS given in Theorem 4 can easily be adapted to this model. A more elaborate discussion on the relations between our model and the scheduling with regret model can be found in the full version of the paper.

5.2   Any Number of Machines

If job weights are bounded by a constant, MinMaxSTC (and also MinAvgSTC) can be solved efficiently on any number of machines by dynamic programming. For simplicity, we only discuss the case of unit job weights here, but our approach can be easily generalized to the case of weights bounded by some constant. The DP leading to the following theorem is based on enumeration of machine configurations.

Theorem 5. *If the number of scenarios K is constant, and all jobs have unit weights, then MinMaxSTC and MinAvgSTC can be solved to optimality in polynomial time on any number of machines.*

The proof, which is omitted here, suggests an algorithm with runtime $O(mn^{2(2^K+K)})$.

6   The MinAvg Version

By Theorem 2, MinAvgSTC is solvable in polynomial time in the case of two scenarios and by Theorem 5 in case of a constant number of scenarios and bounded job weights. For general job weights, however, we need to design a different dynamic program (DP) that solves MinAvgSTC in polynomial time for any constant number of scenarios if there is also a constant number of machines; see Sect. 6.1.

In Sect. 6.2, we present a conjecture that, if true, leads to a polynomial time dynamic programming algorithm for *any* number of machines. We prove the conjecture for the special case of unit job weights which results in an efficient algorithm for MinAvgSTC in this case that is faster than the one given in the previous section as a function of the number of jobs, but slower as a function of the number of machines. Moreover, it is not clear yet if the techniques carry over to the more general case of job weights bounded by a constant.

6.1   Constant Number of Machines

We first describe the case of a constant number of machines. Recall that the objective function for MinAvgSTC is defined as

$$\sum_{k=1}^{K}\sum_{i=1}^{m}\sum_{j\in J_i\cap S_k} w_j \cdot |\{j' \in J_i \cap S_k : j' \le j\}|,$$

with J_i the set of jobs assigned to machine i.

It is clear from the objective function that the contribution of some job j to the cost of a solution depends only on the assignment of that job and any jobs with higher weight, i.e., jobs $1, \ldots, j-1$, to the various machines. In particular, if we want to compute the contribution of job j to the cost of a solution in some schedule, it is sufficient to know the following quantities for each $i \in [m], k \in [K]$

$$x_{ik}(j-1) := |\{j' \in J_i \cap S_k : j' < j\}|,$$

which together form a *state* of the scheduling process. If job j gets assigned to machine i in that schedule, its contribution to the overall cost would then be

$$\sum_{k \in [K]: j \in S_k} w_j(1 + x_{ik}(j-1)).$$

In light of these observations, one can derive a dynamic program, leading to the following theorem.

Theorem 6. *The MinAvgSTC problem with constant number of machines and constant number of scenarios can be solved in polynomial time.*

Proof. We define a state in a dynamic programming decision process as a partial schedule at the moment the first j jobs have been assigned and encode this by a $m \times K$ matrix $X(j)$, with $x_{ik}(j) = |\{j' \in J_i \cap S_k : j' \le j\}|$.

This leads to a simple DP, using the following recursion, where we use $f_j(X(j))$ to denote the minimum cost associated with the first j jobs in any schedule that can be represented by $X(j)$:

$$f_j(X(j)) = \min_{\ell \in [m]} \left\{ f_{j-1}(X(j-1, \ell)) + \sum_{k \in [K]: j \in S_k} w_j(1 + x_{\ell k}(j-1)) \right\},$$

where $X(j-1, \ell)$ is the matrix $X(j-1)$ from which $X(j)$ is obtained by assigning job j to machine ℓ. Equivalently, $X(j-1, \ell)$ is the matrix obtained from $X(j)$ by diminishing all positive entries in row ℓ of $X(j)$ by 1. It follows that $X(j-1, \ell)$ has entries $x_{ik}(j-1)$ that satisfy $x_{\ell k}(j-1) = x_{\ell k}(j) - \mathbf{1}_{j \in S_k}$ for all k, and $x_{ik}(j-1) = x_{ik}(j)$ for all $i \ne \ell$, and all k. Therefore, the computation of each state can be done in time $O(mK)$.

We initialize $f_0(\mathbf{0}) = 0$ (where $\mathbf{0}$ denotes the all-zero matrix) and set $f_0(X) = \infty$ for any other possible X. Thus in each of the n phases (partial job assignments) the number of possible states is bounded by $(n+1)^{m \times K}$, which, because m and K are constants, implies that the DP runs in polynomial time. $\square$

The proof implies that the problem can be solved in time $O(mKn^{mK+1})$, which is polynomial given that m and K are constants in this particular case.

6.2 Any Number of Machines

In this section we develop another efficient algorithm for MinAvgSTC on an arbitrary number of machines with a constant number of scenarios and unit job weights. In contrast to the dynamic program presented in the proof of Theorem 5, the running time is linear in n, the number of jobs, but polynomial in m with the power a function of K, that is, the running time is of the form $m^{h(K)} \cdot n$ for some function h.

More importantly, the technique that we use here is new and we believe it can be generalised to arbitrary job weights. For the time being it remains a fascinating open question whether MinAvgSTC can even be solved efficiently for arbitrary job weights. As we will explain, this is true under the assumption that the following conjecture holds, which we do believe but can only prove for the special case of unit job weights.

Conjecture 1. MinAvgSTC has an optimal solution such that for every scenario $k \in [K]$ and each $j \in [n]$, the j largest jobs are assigned to the machines in such a way that the difference in number of jobs assigned to each pair of machines is bounded by a function $g(K)$ of K only, or more formally

$$\max_{j \in [n], k \in [K]} \left\{ \max_{i \in [m]} |\{j' \in J_i \cap S_k : j' \leq j\}| - \min_{i \in [m]} |\{j' \in J_i \cap S_k : j' \leq j\}| \right\} \leq g(K).$$

We call the term on the left-hand side *full disbalance* of the schedule. To adjust the DP to run in polynomial time for any number of machines under the conjecture, we first observe that to compute a DP recursion step, the order of the rows in each matrix X representing a partial schedule is irrelevant: we only need to know how many machines have a certain number of jobs assigned to them under each scenario, not exactly which machines.

A first step to encode partial schedules is using vectors $\ell \in \mathcal{C} \subseteq \{0, \ldots, n\}^K$, where we call $\mathcal{C}$ the set of machine configurations. If a machine has configuration ℓ, it means that in the partial schedule ℓ_k jobs have been scheduled on this machine in scenario k. We then simply represent a partial schedule by storing the number of machines that have configuration ℓ. We consider the space of *states* that say how many machines have a certain configuration. We can further compress this space by storing the smallest number of jobs on any machine in a given scenario separately. That is

$$z_k = \min_{i \in [m]} x_{ik}, \ k \in [K]$$

The crucial observation now is that under Conjecture 1, there is an optimal solution such that for any partial schedule corresponding to that solution, $0 \leq x_{ik} - z_k \leq g(K)$. We may therefore take $\mathcal{C}$ to be $\{0, \ldots, g(K)\}^K$, i.e., the excess over z_k, so that $\mathcal{C}$ has constant size for constant K. Furthermore, we define

$$y_\ell = \sum_{i|x_{ik} - z_k = \ell_k, \ k \in [K]} 1, \ \ell \in \mathcal{C}.$$

Since the entries of y are bounded by m and the entries of z are bounded by n, the possible number of values for a pair (y, z) in the encoding above is $(m+1)^{(g(K)+1)^K} (n+1)^K$, yielding a polynomial time algorithm. Since the DP-computation for each state in each phase (job assignment) is $O(m)$ and there are

n consecutive job assignments in SPT-order in the DP, this yields a polynomial time algorithm.

It is still open whether there exists an upper bound on the full disbalance of the schedule depending only on K. In the following, we affirm this statement for the case where all jobs have unit weights. Note that, in this case, the j largest jobs are not well-defined. We therefore prove the stronger statement that for any ordering of the jobs, the conjecture holds. To do so, we utilize the power of integer programming, combining the theory of Hilbert bases with techniques of an algorithmic nature.

Theorem 7. *Conjecture 1 holds for unit weights and unit processing times, where the jobs are given by an arbitrary (but fixed) order and the number m of machines is part of the input.*

The proof can be found in the full version.

One may wonder whether $g(K)$ can be strengthened to be polynomial in K. In the case of arbitrary weights, one can establish exponential lower bounds, for which we refer to the full version of the paper as well.

7   Conclusion and Open Problems

We hope that our results inspire interest in the intriguing field of scenario optimization problems. There are some obvious open questions that we left unanswered. For example, is it true that MinMax versions are always harder than MinAvg versions? For researchers interested in exact algorithms and fixed parameter tractability (FPT) results we have the following question. For the scheduling problems that we have studied so far within the scenario model, we have seen various exact polynomial time dynamic programming algorithms for a constant number of scenarios K, but always with K in the exponent of a function of the number of jobs or the number of machines. Can these results be strengthened to algorithms that are FPT in K? Or are these problems W[1]-hard? Similarly, researchers interested in approximation algorithms may wonder how approximability is affected by introducing scenarios into a problem. We have given some first results here, but it clearly is a research area that has so far remained virtually unexplored.

Another interesting variation of the MinAvg version is obtained by assigning probabilities to scenarios (i.e., a discrete distribution over scenarios) with the objective to minimize the expected total (weighted) completion time. The DPs underlying the results of Theorems 5 and 6 easily generalize to this version. However, Theorem 7 does not.

We see as a main challenge to derive structural insights why multiple (constant number of) scenario versions are sometimes as easy as their single scenario versions, like the MinAvg versions of linear programming or of the min-cut problem [4] or of the scheduling problem that we have studied here, and for other problems, such as MinMaxSTC, become harder or even NP-hard.

Acknowledgements. We would like to thank Bart Litjens, Sven Polak, Lluís Vena and Bart Sevenster for providing a counterexample for a preliminary version of Conjecture 1, in which $g(K)$ was a linear function. Moreover, we would like to thank the anonymous referees for their valuable feedback and suggestions. Alberto Marchetti-Spaccamela was supported by the ERC Advanced Grant 788893 AMDROMA "Algorithmic and Mechanism Design Research in Online Markets", and the MIUR PRIN project ALGADIMAR "Algorithms, Games, and Digital Markets". Ekin Ergen and Martin Skutella were supported by the Deutsche Forschungsgemeinschaft (DFG, German Research Foundation) under Germany's Excellence Strategy—The Berlin Mathematics Research Center MATH+ (EXC-2046/1, project ID: 390685689). Leen Stougie was supported by NWO Gravitation Programme Networks 024.002.003, and by the OPTIMAL project NWO OCENW.GROOT.2019.015.

References

1. Adamczyk, M., Grandoni, F., Leonardi, S., Wlodarczyk, M.: When the optimum is also blind: a new perspective on universal optimization. In: 44th International Colloquium on Automata, Languages and Programming, ICALP, volume 80 of LIPIcs, pp. 35:1–35:15. Schloss Dagstuhl - Leibniz-Zentrum für Informatik (2017)

2. Albers, S., Janke, M.: Online makespan minimization with budgeted uncertainty. In: Lubiw, A., Salavatipour, M. (eds.) WADS 2021. LNCS, vol. 12808, pp. 43–56. Springer, Cham (2021). https://doi.org/10.1007/978-3-030-83508-8_4

3. Mohamed Ali Aloulou and Federico Della Croce: Complexity of single machine scheduling problems under scenario-based uncertainty. Oper. Res. Lett. **36**(3), 338–342 (2008)

4. Armon, A., Zwick, U.: Multicriteria global minimum cuts. Algorithmica **46**(1), 15–26 (2006)

5. Austrin, P., Hastad, J., Guruswami, V.: $(2 + \epsilon)$-SAT is NP-hard. In: Proceedings of 55th Annual Symposium on Foundations of Computer Science, pp. 1–10. IEEE (2014)

6. Ben-Tal, A., El Ghaoui, L., Nemirovski, A.: Robust Optimization, vol. 28. Princeton University Press, Princeton (2009)

7. Bertsimas, D., Jaillet, P., Odoni, A.R.: A priori optimization. Oper. Res. **38**(6), 1019–1033 (1990)

8. Birge, J.R., Louveaux, F.: Introduction to Stochastic Programming. Springer, Cham (2011). https://doi.org/10.1007/978-1-4614-0237-4

9. Bougeret, M., Jansen, K., Poss, M., Rohwedder, L.: Approximation results for makespan minimization with budgeted uncertainty. Theory Comput. Syst. **65**(6), 903–915 (2021)

10. Cho, W.-H., Shmoys, D.B., Henderson, S.G.: SPT optimality (mostly) via linear programming. Oper. Res. Lett. **51**(1), 99–104 (2023)

11. Conway, R.W., Maxwell, W.L., Miller, L.W.: Theory of Scheduling. Addison-Wesley Publishing Company, Boston (1967)

12. Eastman, W.L., Even, S., Isaacs, I.M.: Bounds for the optimal scheduling of n jobs on m processors. Manage. Sci. **11**(2), 268–279 (1964)

13. Feuerstein, E., et al.: Minimizing worst-case and average-case makespan over scenarios. J. Sched. **20**, 1–11 (2016)

14. Goemans, M.X., Williamson, D.P.: Two-dimensional Gantt charts and a scheduling algorithm of Lawler. SIAM J. Discret. Math. **13**(3), 281–294 (2000)

15. Håstad, J.: Some optimal inapproximability results. J. ACM **48**(4), 798–859 (2001)
16. Immorlica, N., Karger, D., Minkoff, M., Mirrokni, V.S.: On the costs and benefits of procrastination: approximation algorithms for stochastic combinatorial optimization problems. In: Proceedings of the Fifteenth Annual ACM-SIAM Symposium on Discrete Algorithms, pp. 691–700. Society for Industrial and Applied Mathematics (2004)
17. Kasperski, A., Kurpisz, A., Zieliński, P.: Parallel machine scheduling under uncertainty. In: Greco, S., Bouchon-Meunier, B., Coletti, G., Fedrizzi, M., Matarazzo, B., Yager, R.R. (eds.) IPMU 2012. CCIS, vol. 300, pp. 74–83. Springer, Heidelberg (2012). https://doi.org/10.1007/978-3-642-31724-8_9
18. Kasperski, A., Zieliński, P.: Single machine scheduling problems with uncertain parameters and the OWA criterion. J. Sched. **19**, 177–190 (2016)
19. Khot, S., Kindler, G., Mossel, E., O'Donnell, R.: Optimal inapproximability results for MAX-CUT and other 2-variable CSPs? SIAM J. Comput. **37**(1), 319–357 (2007)
20. Kleywegt, A.J., Shapiro, A., Homem-de-Mello, T.: The sample average approximation method for stochastic discrete optimization. SIAM J. Optim. **12**(2), 479–502 (2002)
21. Kouvelis, P., Yu, G.: Robust Discrete Optimization and Its Applications. Kluwer Academic Publishers, Boston (1997)
22. Mastrolilli, M., Mutsanas, N., Svensson, O.: Single machine scheduling with scenarios. Theoret. Comput. Sci. **477**, 57–66 (2013)
23. Megow, N., Verschae, J.: Dual techniques for scheduling on a machine with varying speed. SIAM J. Discret. Math. **32**(3), 1541–1571 (2018)
24. Schulz, A.S., Skutella, M.: Scheduling unrelated machines by randomized rounding. SIAM J. Discret. Math. **15**(4), 450–469 (2002)
25. Shabtay, D., Gilenson, M.: A state-of-the-art survey on multi-scenario scheduling. Eur. J. Oper. Res. **310**, 3–23 (2022)
26. Skutella, M.: Approximation and Randomization in Scheduling. PhD thesis, Technische Universität Berlin (1998)
27. van Ee, M., van Iersel, L., Janssen, T., Sitters, R.: A priori tsp in the scenario model. Discret. Appl. Math. **250**, 331–341 (2018)
28. Yang, J., Gang, Yu.: On the robust single machine scheduling problem. J. Comb. Optim. **6**(1), 17–33 (2002)

Approximating Fair k-Min-Sum-Radii in Euclidean Space

Lukas Drexler$^{(\boxtimes)}$ ⬤, Annika Hennes ⬤, Abhiruk Lahiri ⬤, Melanie Schmidt ⬤,
and Julian Wargalla ⬤

Heinrich-Heine-Universität, Düsseldorf, Germany
{lukas.drexler,annika.hennes,abhiruk.lahiri,mschmidt,
julian.wargalla}@hhu.de

Abstract. The k-center problem is a classical clustering problem in
which one is asked to find a partitioning of a point set P into k clusters
such that the maximum radius of any cluster is minimized. It is well-
studied. But what if we add up the radii of the clusters instead of only
considering the cluster with maximum radius? This natural variant is
called the k-min-sum-radii problem. It has become the subject of more
and more interest in recent years, inspiring the development of approximation algorithms for the k-min-sum-radii problem in its plain version
as well as in constrained settings.

We study the problem for Euclidean spaces $\mathbb{R}^d$ of arbitrary dimension
but assume the number k of clusters to be constant. In this case, a PTAS
for the problem is known (see Bandyapadhyay, Lochet and Saurabh [4]).
Our aim is to extend the knowledge base for k-min-sum-radii to the
domain of *fair* clustering. We study several group fairness constraints,
such as the one introduced by Chierichetti et al. [15]. In this model, input
points have an additional attribute (e.g., colors such as red and blue),
and clusters have to preserve the ratio between different attribute values
(e.g., have the same fraction of red and blue points as the ground set).
Different variants of this general idea have been studied in the literature.
To the best of our knowledge, no approximative results for the fair k-
min-sum-radii problem are known, despite the immense amount of work
on the related fair k-center problem.

We propose a PTAS for the fair k-min-sum-radii problem in $\mathbb{R}^d$ of
arbitrary dimension d for the case of constant k. To the best of our
knowledge, this is the first PTAS for the problem. It works for different notions of group fairness and for a more broad class of mergeable
constraints.

1 Introduction

The *k-min-sum-radii* problem (*k*-MSR for short) is a clustering problem that
resembles two well-known problems, namely the *k-center* problem and the *k-
median* problem. Given a set of points P and a number k, these problems ask for

Funded by *Deutsche Forschungsgemeinschaft* (DFG, German Research Foundation) -
Project 456558332.

© The Author(s), under exclusive license to Springer Nature Switzerland AG 2023
J. Byrka and A. Wiese (Eds.): WAOA 2023, LNCS 14297, pp. 119–133, 2023.
https://doi.org/10.1007/978-3-031-49815-2_9

a set C of k cluster centers and evaluate it according to the distances $d_{\min}(x) = \min\{d(c,x) \mid c \in C\}$ between points x and their closest center in C. The k-center objective $\max\{d_{\min}(x) \mid x \in P\}$ focuses on the radii of the resulting clusters, while the k-median objective $\sum_{x \in P} d_{\min}(x)$ sums up all individual point's costs. In the latter case, large individual costs can average out and so in scenarios where we really want to restrict the maximum cost an individual point can induce, k-center is the better choice. However, looking only for a single maximum distance completely ignores the fine-tuning of up to $k-1$ smaller clusters.

The k-min-sum-radii problem goes an intermediate way: It looks for k centers $c_1, \ldots, c_k$ and corresponding clusters $C_1, \ldots, C_k$ and sums up the radii, i.e., the objective is to minimize $\sum_{i=1}^{k} \max_{x \in C_i} d(x, c_i)$ (or in other words: to minimize the *average* radius). This objective allows for the fine-tuning of all clusters while still maintaining that the maximum cost of individual points is reasonably bounded (although it may be higher than for k-center by a factor of k). Another variation, known as the k-min-sum-diameter problem, aims to minimize the sum of the diameters of the clusters. The k-min-sum-radii problem has a close connection with the base station placement problem arising in wireless network design [22], where the objective is to minimize the energy required for wireless transmission which is proportional to the sum of the radii of coverage of the base stations. The mathematical model of this problem translates to the *minimum sum radii cover problem* where we have a set of client locations and a set of server locations. The objective is to cover the set of clients with a set of balls whose centers are located at a subset of server locations such that the sum of the radii of the balls is minimized.

There has been great interest in designing good approximation algorithms for the k-min-sum-radii problem. Charikar and Panigrahy [13] give an $O(1)$-approximation for the metric k-min-sum-radii problem (and the k-min-sum-diameter problem) based on the primal-dual framework by Jain and Vazirani for k-median. It was recently refined by Friggstad and Jamshidian [16] to obtain a 3.389-approximation for k-min-sum-radii which is currently the best-known approximation factor for the general case. For constrained k-min-sum-radii, lower bounds, outliers and capacities have been studied. Ahmadian and Swamy [1] built upon [13] to obtain a 3.83-approximation for the non-uniformly lower bounded k-min-sum-radii problem. They also give a $(12.365 + O(\varepsilon))$-approximation for k-MSR with outliers that runs in time $n^{O(1/\varepsilon)}$. Inamdar and Varadarajan [21] derive a 28-approximation for the uniformly capacitated k-min-sum-radii problem, but this algorithm is an FPT approximation algorithm with running time $O(2^{O(k^2)} \cdot n^{O(1)})$. Bandyapadhyay, Lochet, and Saurabh [4] also give an FPT-approximation: They develop a $(4 + \varepsilon)$-approximation algorithm with $2^{O(k \log(k/\varepsilon))} \cdot n^3$ running time for k-MSR with uniform capacities and a $(15 + \varepsilon)$-approximation algorithm for k-MSR with non-uniform capacities that runs in time $2^{O(k^2 \log k)} \cdot n^3$.

In the Euclidean case, it is possible to obtain better results. In the plane, every cluster in the optimal min-sum radii clustering lies inside some convex polygon drawn from the solution centers that partition the plane into k disjoint convex regions. The dual of that partition is an internally triangulated planar graph. Capolyleas et al. [11] use this fact to enumerate over $O(n^{6k})$ possible solutions to solve the problem exactly for $d = 2$. Gibson et al. [18] also give an exact algorithm for the Euclidean k-min-sum-radii problem in the plane. Their algorithm is based on an involved dynamic programming approach and has a running time of $O(n^{881})$ for $d = 2$. Bandyapadhyay, Lochet and Saurabh [4] give a randomized algorithm with running time $2^{O((k/\varepsilon^2)\log k)} \cdot dn^3$ which outputs a $(1 + \varepsilon)$-approximation with high probability. Their algorithm can handle capacitated k-min-sum-radii but allows the capacities to be violated by at most an ε-fraction. They also present a PTAS for k-MSR with strict capacities for both constant k and constant d with running time $2^{O(kd\log(\frac{k}{\varepsilon}))}n^3$.

In this paper, we advance the active research on fairness in clustering (see [12, 14] for surveys on the topic) and tackle the problem of k-min-sum-radii under a variety of different group fairness notions. These notions assume that the data points belong to different protected groups, represented by different colors. We will denote the set of colors by $\mathcal{H}$. For $X \subseteq P$ and $h \in \mathcal{H}$, let $col_h(X) \subseteq X$ denote the subset of points within X that carry color h. Now the notion of *exact fairness* requires that in every cluster the proportion of points of a certain color is the same as their proportion within the complete point set, i.e., a clustering $\mathscr{C}$ fulfills *exact fairness* if $\frac{|col_h(C)|}{|C|} = \frac{|col_h(P)|}{|P|}$ for every color $h \in \mathcal{H}$ and cluster $C \in \mathscr{C}$. This notion is for example defined in [24]. Our method can handle exact fairness but also other notions as it is indeed capable to handle the more general class of *mergeable constraints*. A clustering constraint is *mergeable* if the union $C \cup C'$ of any possible pair of clusters C, C' satisfying the constraint does itself satisfy the constraint (cf. [2]). In other words, merging clusters does not destroy the property of satisfying the constraint.

Important examples of mergeable constraints are (a) several *fairness constraints* (see the long version for a list), (b) *lower bound constraints* that require every cluster to contain at least a certain fixed number of points, and (c) *outliers* (see the end of the paper) in which a fixed number z of points can be ignored by any clustering (one can model this as a kind-of-mergeable constraint by viewing it as a $(k + z)$-clustering with the constraint that at most k clusters contain more than one point and the rest is singleton clusters). On the other hand, upper bounds on the cardinality of clusters (capacities) are not mergeable because merging clusters may violate the capacity constraint.

To the best of our knowledge, no results for k-min-sum-radii with fairness constraints are known, neither in the Euclidean setting (PTAS) nor in the metric setting (constant factor approximation) despite the huge amount of work on fair k-center and fair k-median (cf. [6,7,9,10,15,19,24])[1]. The reason for this may be that the k-min-sum-radii problem can behave quite counter-intuitively and has properties unlike both the k-center and k-median problem. One such property is that a k-min-sum-radii solution may actually cost *more* if we open more centers

---

[1] The result by [11] might be extendable to the setting of fair k-min-sum-radii to obtain an exact algorithm for constant k and $d = 2$, but an in-depth analysis would be required to verify this idea.

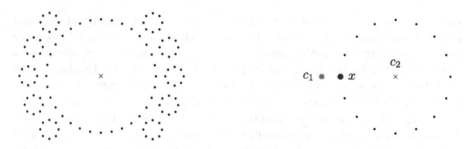

Fig. 1. Left side: An example where k-min-sum-radii rather opens one cluster than eleven. Right side: An example where the cheapest $k = 2$-clustering keeps c_1 as a singleton rather than combining it with x, despite the fact that c_1 is closer to x than the center c_2 of the big cluster. It is cheaper to *not* assign the blue point to the orange point even though that would be a closer center.

(see Fig. 1, left side) which cannot happen for (plain) k-center or k-median. This is a problem for the design of fair clustering algorithms because for k-center and k-median, these are built by computing fair micro-clusters (also called fairlets) first and then assembling the final fair clustering from the micro-clusters (cf. [15]). Another uncommon property of k-min-sum-radii is that even without any constraints, assigning points to centers does not have the locality property: It may be beneficial to assign to a further away center (see Fig. 1, right side). This has been observed for other clustering objectives when side constraints enter the picture, but for k-min-sum-radii, it already happens without any constraints.

Our Result. We present a simple PTAS for the Euclidean k-MSR problem with mergeable constraints that works for constant k and arbitrary dimension d. In particular, to the best of our knowledge, we provide the first approximate results for fair k-MSR.

Theorem 1. *For every $\varepsilon \in (0, \frac{1}{2})$, there exists an algorithm that computes a $(1 + \varepsilon)$-approximation for k-min-sum-radii with mergeable constraints in time $d \cdot \mathrm{poly}(n) \cdot f(k, \varepsilon)$, if the corresponding constrained k-center problem has a constant-factor polynomial time approximation algorithm. If no such k-center approximation exists, the running time increases to $d \cdot n^{\mathrm{poly}(k,1/\varepsilon)} \cdot f(k, \varepsilon)$ (Theorem 2).*

How We Obtain the PTAS. Our algorithm is based on an idea by Bădoiu, Har-Peled, and Indyk [3] who obtain a PTAS for k-center. Contrary to other k-center algorithms, the main idea of this algorithm – to iteratively construct minimum enclosing balls around subsets of optimum clusters until all points are covered – *does* carry over to k-MSR. However, we need to resolve significant obstacles that are due to the more complex structure of k-MSR, as illustrated in Fig. 1. In an optimal clustering, points do not necessarily get assigned to their closest center, so we cannot derive a lower bound for the initial size of the growing balls in the same manner as Bădoiu et al. This, however, is necessary to upper bound the running time of the algorithm.

Our approach to repair the analysis is mainly based on proving that there always exists a close-to-optimum k-MSR solution with a nice structure, as described in Sect. 2.3: (1) The Minimum Enclosing Balls (MEBs) around all clusters do not intersect, even if we enlarge all MEBs by some factor γ that depends on ε. We call such a solution γ-separated. (2) The ratio between the smallest and the largest radius in the solution is bounded by ε/k. We call a solution with this property ε-*balanced*. Achieving (2) is straightforward, but establishing (1) and (2) simultaneously requires a bit more work. Since we establish (1) mainly by merging close clusters, this technique still works under mergeable constraints.

After proving the existence of an approximately optimal solution that is sufficiently separated and balanced, we reconstruct this solution by adjusting the approach of Bădoiu et al. [3] appropriately. To ensure an upper bound on the running time, we have to extend their guessing oracle (that answers membership queries) to also provide approximate radii for all clusters. How this is done is outlined in Sect. 2.2. With the oracle in place, the structure of our algorithm is as follows:

- Initialize $S_i = \emptyset$ for $i = 1, \ldots, k$ and $P' = P$
- Ask the oracle for radii $\widetilde{r}_1, \widetilde{r}_2, \ldots, \widetilde{r}_k$
- Repeat until $P' = \emptyset$:
  1. Select an arbitrary point p_i from P'
  2. Query the oracle for an index j and add p_i to S_j
  3. If $|S_j| = 1$: Remove all points from P' that are within distance $\approx \varepsilon \widetilde{r}_j$ of p_i
  4. If $|S_j| > 1$: Compute the minimum enclosing ball of S_j, enlarge it by an appropriate factor and remove all points in the resulting ball from P'

We show that the algorithm will stop after $f(k, \varepsilon)$ iterations, resulting in a PTAS for constant k. Figure 2 shows an example run of the algorithm and also gives some details on what '$\approx \varepsilon \widetilde{r}_j$' and 'by an appropriate factor' mean. The respective constants are a result of the analysis and are discussed later.

Further Related Work. When k is part of the input, the metric k-min-sum-radii problem is known to be NP-hard, as shown in [23]. The same paper gives an exact algorithm with running time $O(n^{2k}/k!)$. Gibson et al. [17] provide a randomized algorithm for the metric k-min-sum-radii problem that runs in time $n^{O(\log n \log \Delta)}$ where Δ is the ratio between the largest and the smallest pairwise distance in the input and returns an optimal solution with high probability. They also show NP-hardness even for shortest path metrics in weighted planar graphs and for metrics of (large enough) constant doubling dimension. Bilò et al. [8] give a polynomial time algorithm for the problem when the input points are on a line. Behsaz and Salavatipour [5] show a polynomial time exact algorithm for the k-min-sum-radii problem when the metric is induced by an unweighted graph and no cluster contains only one point. Based on the constant-factor approximation

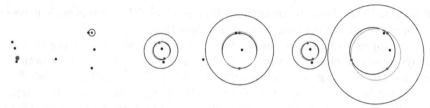

(a) The first two points are chosen and equipped with a small ball to start with.

(b) Two iterations later, each cluster has two points and the starting balls have been replaced.

(c) After a third point was added to the right cluster, all points in the point set are covered, the run ends.

Fig. 2. Example run of the algorithm for $\varepsilon = 0.2$. In every iteration, the purple points depict the points that were already chosen by the algorithm. For reference, the black circles represent the true minimum enclosing balls of the subsets S_i. These are not computed in the algorithm, but only the $(1 + \varepsilon)$-approximations of these, depicted in orange. The blue circles enclose the areas which we ignore when sampling new points. That is, for singleton clusters, it is $B(s_j, \frac{\varepsilon}{1+\varepsilon}\widetilde{r}_j)$, and once the algorithm found two points from a cluster, it computes an approximate MEB and enlarges it by a factor of $\gamma = 1 + \varepsilon + 2\sqrt{\varepsilon}$ to obtain the new blue ball. (Color figure online)

for k-min-sum-radii by Charikar and Panigrahy [13] mentioned in the introduction, Henzinger et al. [20] develop a data structure to efficiently maintain an $O(1)$-approximate solution under changes in the input.

For the k-center problem with exact fairness constraints as described earlier, Bercea et al. [7] give a 5-approximation. Further, several *balance* notions have been proposed. The simplest case with only two colors was proposed by Chierichetti et al. [15]. It requires that the minimum ratio between different colors within any cluster meets a given lower bound. For its most general formulation, there exists a 14-approximation for the k-center variant [24]. The definition by Böhm et al. [10] allows more colors but is stricter in that it demands the portions of colors in a cluster to be of equal size. The authors show how the k-center problem under this fairness notion can be reduced to the unconstrained case while increasing the approximation factor by 2, leading to a polynomial-time $O(1)$-approximation. They also give an $O(n^{\mathrm{poly}(k/\varepsilon)})$-time $(1+\varepsilon)$-approximation. A more general notion by Bera et al. [6] allows the number of cluster members of a certain color to lie in some color-dependent range. Harb and Shan [19] give a 5-approximation for the k-center problem under this constraint.

Preliminaries. For a given center $c \in \mathbb{R}^d$ and radius $r \in \mathbb{R}_{\geq 0}$, define the *ball of radius r around c* to be $B(c, r) = \{x \in \mathbb{R}^d \mid \|x - c\| \leq r\}$. We set $\mathrm{cost}\big(B(c, r)\big) = r$. Let $X \subset \mathbb{R}^d$ be a set of points. We say that a ball B *encloses* X if $X \subset B$. The ball with the smallest radius that encloses X is called the *minimum enclosing ball* (MEB) of X and we denote it as $\mathrm{MB}(X)$. The *cost* of X is defined as the cost of its minimum enclosing ball, $\mathrm{cost}(X) = \mathrm{cost}(\mathrm{MB}(X))$. A *$k$-clustering* $\mathscr{C} = \{C_1, \ldots, C_k\}$ of a given finite set of points P is a partitioning of P into k disjoint (possibly empty) sets. Its cost is the sum of the costs of all its individual clusters, i.e. $\text{cost}(\mathscr{C}) = \sum_i \text{cost}(C_i)$. Now we can define the *Euclidean k-min-sum-radii problem*: Given a finite set of points P in the d-dimensional Euclidean space $\mathbb{R}^d$ and a number $k \in \mathbb{N}$, find a k-clustering of P with minimal cost. We can also formulate the problem in the following form: Find at most k centers $c_1, \ldots, c_k \in \mathbb{R}^d$ and radii $r_1, \ldots, r_k \geq 0$ such that the union of balls $B(c_1, r_1) \cup \ldots \cup B(c_k, r_k)$ covers P and the sum of the radii $\sum_{i=1}^{k} r_i$ is minimized.

2 k-Min-Sum-Radii with Mergeable Constraints

Algorithm 1 gives a detailed description of our method. Instead of an oracle, this pseudo code assumes that it is given a string $u \in \{1, \ldots, k\}^*$ as answers to membership queries and it is also given estimates for the radii $\tilde{r}_1, \ldots, \tilde{r}_k$. Despite looking a bit more technical, the algorithm follows the plan outlined above: Iteratively construct balls to cover the point set. A ball shall always cover exactly one optimum cluster (or approximately optimal cluster). It starts out when the first point from that cluster is discovered. This point will be the center of a small starting ball. The starting radius is related to the true radius of the cluster (for which a good estimate has been provided). Whenever a point from a cluster is discovered, the MEB around all its discovered points is computed and the ball is increased to an enlarged version of that MEB. We are done when all points are covered by the balls.

To set up the analysis, we introduce a few more definitions. In the preliminaries, we have defined a clustering to be a partitioning of the underlying space. In the following section, however, it will be helpful to occasionally conceive clusterings as collections of balls that cover P. To avoid confusion, we term the latter coverings.

Definition 1. *We say that a set of balls $\{B(c_1, r_1), \ldots B(c_k, r_k)\}$ forms a covering of P, if $P \subseteq \bigcup_i B(c_i, r_i)$. It is a* disjoint *covering, if $B(c_i, r_i) \cap B(c_j, r_j) = \emptyset$ for all $i \neq j$.*

The relation between coverings and clusterings, as it concerns this paper, is straightforward. Every disjoint covering $B_1, \ldots, B_k$ of P yields a unique corresponding clustering $\mathscr{C} = \{B_1 \cap P, \ldots, B_k \cap P\}$ of P. Conversely, every k-clustering $\mathscr{C} = \{C_1, \ldots, C_k\}$ yields a corresponding covering $\{\text{MB}(C_1), \ldots, \text{MB}(C_k)\}$.

We increase the size of the balls computed during Algorithm 1 by some multiplicative factor to ensure that they grow reasonably fast, so we want them to be not only disjoint but actually separated by some positive amount.

Definition 2. *Let $\gamma \geq 1$. Two balls $B(c_1, r_1)$ and $B(c_2, r_2)$ are said to be γ-separated if $B(c_1, \gamma r_1) \cap B(c_2, \gamma r_2) = \emptyset$. A covering is γ-separated if all its balls are pairwise γ-separated.*

Algorithm 1: SELECTION

Input   : An ordered set $P \subseteq \mathbb{R}^d$, a number $k \in \mathbb{N}$, a string
$u \in \{1, \ldots, k\}^*$, a set $\{\tilde{r}_1, \ldots, \tilde{r}_k\}$ of k radii, a value $0 < \varepsilon < 1$
Output: Balls $B_1, \ldots, B_k \subseteq \mathbb{R}^d$, such that $P \subseteq \bigcup_j B_j$

1   $S_i \leftarrow \emptyset$ for $i = 1, \ldots, k$;
2   $\gamma \leftarrow 1 + \varepsilon + 2\sqrt{\varepsilon}$;
3   $X \leftarrow \emptyset$;   /* points that have been covered so far */
4   **for** $i = 1, \ldots, |u|$ **do**
5 | $I = \{j \mid S_j = \{s_j\}$ is a singleton$\}$;
6 | $R \leftarrow \bigcup_{j \in I} \mathrm{B}\left(s_j, \frac{\varepsilon}{1+\varepsilon}\tilde{r}_j\right)$;   /* put small balls around singletons
     | */
7 | Let p_i be the point from $P \setminus (X \cup R)$ that is first in the order induced
     | by P;
8 | $S_{u_i} \leftarrow S_{u_i} \cup \{p_i\}$;
9 | **if** $|S_{u_i}| \geq 2$ **then**
10 | | $\mathrm{B}(c, r) \leftarrow (1 + \varepsilon)$-approximation of $\mathrm{MB}(S_{u_i})$;
11 | | $B_{u_i} \leftarrow \mathrm{B}(c, \gamma r)$;
12 | | $X \leftarrow X \cup (B_{u_i} \cap P)$;
13 | **end**
14 **end**
15 **forall** $S_i = \{s_i\}$ **do**
16 | $B_i \leftarrow \mathrm{B}(s_i, 0)$;
17 **end**
18 **return** $B_1, \ldots, B_k$;

2.1   The Main Algorithm and the Main Lemma

The following constitutes the main technical lemma. Claim 1(a) is mainly an observation: For some sequence of oracle guesses (i.e., for some $u \in \{1, \ldots, k\}^*$), we always guess the cluster correctly and thus S_i is always a subset of some optimum cluster. In fact, in the end each S_i can be viewed as a compact approximation in the following sense: it is a small set whose MEB covers almost all of its corresponding optimum cluster. Claims 1(b) and 1(c) state that during the algorithm, the balls that are placed around the sampled points are always disjoint. This is important for the rest of the analysis. Then Claim (2) and (3) are the core part of the original analysis by [3]: Whenever we add a point, the ball for the cluster grows by an appropriate factor, and once a certain threshold of points has been reached, the ball covers the true cluster we are looking for.

Lemma 1. *Let* $\mathscr{B} = \{B(c_1^*, r_1^*), \ldots, B(c_k^*, r_k^*)\}$ *be an arbitrary covering of* P *and* $\widetilde{r}_1, \ldots, \widetilde{r}_k$ *a set of radii such that* $r_i^* \leq \widetilde{r}_i \leq (1+\varepsilon)r_i^*$ *for all* $i \in \{1, \ldots, k\}$. *If* $\mathscr{B}$ *is* $(1+\varepsilon)\gamma$-separated, *with* $\gamma \geq 1 + \varepsilon + 2\sqrt{\varepsilon}$, *then there is an element* $u \in \{1, \ldots, k\}^*$, *such that*

1. *At each stage of* SELECTION$(P, k, u, \{\widetilde{r}_1, \ldots, \widetilde{r}_k\}, \varepsilon)$, *the following holds for all* i:

   (a) $S_i \subseteq B(c_i^*, r_i^*)$,

   (b) $B(c_j^*, (1+\varepsilon)\gamma r_j^*) \cap B(s_i, \frac{\varepsilon}{1+\varepsilon}\widetilde{r}_i) = \emptyset$ *for all* $j \neq i$ *whenever* $S_i = \{s_i\}$ *is a singleton*,

   (c) $B(c_j^*, (1+\varepsilon)\gamma r_j^*) \cap B_i = \emptyset$ *for all* $j \neq i$.

2. *With every addition of a new point,* $MB(S_i)$ *grows by a factor of at least* $1 + \frac{\varepsilon^2}{16}$.

3. *For any index* i *it holds that* $B(c_i^*, r_i^*) \subset B_i$, *at the latest when* $|S_i| \geq \frac{32(1+\varepsilon)}{\varepsilon^3}$.

Proof. 1. We construct u by recording the proper assignments in SELECTION. This is possible because the algorithm is deterministic and because assignments do not have to be specified before points have been selected. During the first iteration, if $p_1 \in B(c_{i_1}^*, r_{i_1}^*)$, set $u_1 = i_1$, and so on. Note that the covering is disjoint, so this assignment is unambiguous. Obviously, $S_i \subseteq B(c_i^*, r_i^*)$ has to hold for all i necessarily and (a) follows. To prove (b), assume that there exists a singleton $S_i = \{s_i\}$ in the current iteration. From the previous point, we know that $s_i \in B(c_i^*, r_i^*)$ and so $B(s_i, \frac{\varepsilon}{1+\varepsilon}\widetilde{r}_i) \subset B(s_i, \varepsilon r_i^*) \subseteq B(c_i^*, (1+\varepsilon)r_i^*)$. Since $\mathscr{B}$ is at least $(1+\varepsilon)$-separated, this proves the second point. To prove (c), we have to look at how each B_i is constructed. We start with a $(1+\varepsilon)$-approximation $B(c_i, r_i)$ of $MB(S_i)$. By definition, $B(c_i, r_i) \subseteq B(c_i^*, (1+\varepsilon)r_i^*)$ and setting $B_i = B(c_i, \gamma r_i)$ ensures that $B_i \subseteq B(c_i^*, (1+\varepsilon)\gamma r_i^*)$. The claim thus again follows from the assumption that $\mathscr{B}$ is $(1+\varepsilon)\gamma$-separated.

2. This part of the proof does not deviate significantly from Bădoiu et al. [3]. The main difference is that we are working with $(1+\varepsilon)$-approximations of radii, which adds another layer of complexity. It can be found in the full version of the paper.

3. Each ball B_i is a $(1+\varepsilon)\gamma$-approximation of $MB(S_i)$. Since $S_i \subset B(c_i^*, r_i^*)$ in each iteration, it follows that $B_i \subset B(c_i^*, (1+\varepsilon)\gamma r_i^*)$ also holds throughout. No other ball (neither from (b) nor (c)) can intersect $B(c_i^*, r_i^*)$ and so, by continually selecting new points, at some point $B(c_i^*, r_i^*) \subset B_i$ must hold. We now want to show that this happens relatively quickly and that it is necessary to add at most $\frac{32(1+\varepsilon)}{\varepsilon^3}$ points to S_i until this state is reached. Of course, it may happen that $B(c_i^*, r_i^*)$ is covered at an earlier point in time and that fewer points have to be added to S_i. Assume that we get at least to an iteration, where S_i contains two points. Since we ignored all points that were at a distance of at most $\frac{\varepsilon}{(1+\varepsilon)}\widetilde{r}_i \leq \varepsilon r_i^*$ from the first selected point, $MB(S_i)$ has to have an initial radius of at least $\frac{\varepsilon}{2(1+\varepsilon)}\widetilde{r}_i \geq \frac{\varepsilon r_i^*}{2(1+\varepsilon)}$. As we saw in point (2), any subsequent additions of new points to S_i further increase the

radius by a multiplicative factor of at least $(1 + \frac{\varepsilon^2}{16})$. Combining both of these observations gives us an upper bound on the number of iterations. First, note that the initial radius of $\frac{\varepsilon r_i^*}{2(1+\varepsilon)}$ grows by at least $\frac{\varepsilon^2}{16} \cdot \frac{\varepsilon r_i^*}{2(1+\varepsilon)} = \frac{\varepsilon^3 r_i^*}{32(1+\varepsilon)}$ when the next point is added. Since the radii only grow larger, each subsequent update also increases the radius by at least this amount. At the same time, r_i^* is clearly an upper bound for the radius of $\mathrm{MB}(S_i)$, so we can add at most $\frac{32(1+\varepsilon)}{\varepsilon^3}$ many points to S_i. $\square$

This lemma shows that we can reconstruct well-separated coverings (or rather, the corresponding clusterings) using a reasonably small oracle for the assignments, given that we know the radii up to an ε-factor.

2.2   Guessing the Radii

Let us now consider how we can compute such approximate radii. We split this problem into two parts. First, we guess the largest radius of the covering and in the next step, we guess the remaining radii, assuming that they cannot be too small compared to this largest radius.

There are two different approaches to guessing the largest radius. The first one makes use of a relation between the largest radius in an optimal k-MSR solution and the value of an optimal k-center solution. If we have access to a constant-factor approximation algorithm for k-center under the given constraint, we can use it to compute a candidate set of small size for the largest radius. The second approach uses results from the theory of ε-coresets and works for arbitrary mergeable constraints with the trade-off that the set of candidates is larger and, in the end, depends exponentially on k. We focus on the first approach but refer to the long version for the details and for the second approach. The following lemma establishes a useful connection between k-center and k-MSR.

Lemma 2. *Let r_α denote the value of an α-approximate k-center solution and r_1^* the largest radius of a β-approximative k-MSR solution for the same instance. Then it holds that $r_1^* \in \left[\frac{r_\alpha}{\alpha}, \beta \cdot k^2 \cdot r_\alpha\right]$, even if we impose the same clustering constraints on both problems.*

This means that, by running a (polynomial time) constant-factor approximation algorithm for k-center, we can obtain an interval I which necessarily contains the radius r_1^* of the *largest* cluster in an optimal min-sum-radii solution. By utilizing standard discretization techniques, we are then able to obtain a finite candidate set such that (a) its size only depends on ε, k, α and β, and (b) it contains a $(1 + \varepsilon)$-approximation for each value in I. The details can be found in the long version.

Once we have a guess for r_1^*, we can apply a similar technique to obtain a candidate set for the remaining radii. However, this requires that the other radii are not too small in comparison. More precisely, we assume that the covering we are interested in is ε-balanced. Later on, we will show that this requirement can easily be met.

Definition 3. *Let $\varepsilon > 0$. A covering $\{B(c_1, r_1), \ldots, B(c_k, r_k)\}$ of P is ε-balanced, if $r_i \geq \frac{\varepsilon}{k} \max\limits_j r_j$ for all $i \in \{1, \ldots, k\}$.*

Given such an ε-balanced covering, we can conclude this part with the following statement, whose proof can also be found in the long version.

Lemma 3. *Let $\varepsilon > 0$ and let $\mathscr{B}^*$ be an ε-balanced covering with radii $r_1^*, \ldots, r_k^*$. Then we can compute a set of size $O(\log_{(1+\varepsilon)} k)$ that contains a number r_1 with $r_1^* \leq r_1 \leq (1 + \varepsilon)r_1^*$, and a set of size $O(\log_{(1+\varepsilon)} \frac{k}{\varepsilon})$ that contains for each r_i^*, $i \geq 2$, a number r_i with $r_i^* \leq r_i \leq (1 + \varepsilon)r_i^*$.*

2.3 Cheap, Separable and Balanced Coverings

In the main technical lemma (Lemma 1), we have proven that SELECTION is able to reconstruct well-separated coverings (clusterings), given approximate values for the radii. How those latter approximations could be computed was then outlined in Sect. 2.2. What now remains to be shown is that there actually exist cheap, well-separated and balanced coverings, so that these results can be applied.

Lemma 4. *Let $\mathscr{C} = \{C_1, \ldots, C_k\}$ be a min-sum-radii solution for P. Then for all $\varepsilon > 0$ and $\gamma \geq 1$, there exists an ε-balanced and γ-separated covering $\mathscr{B} = \{B_1, \ldots, B_{k'}\}$ of P with $k' \leq k$ and $\mathrm{cost}(\mathscr{B}) \leq (1 + \varepsilon)^k \gamma^{k-1} \mathrm{cost}(\mathscr{C})$. Additionally, if $\mathscr{C}$ satisfies a given mergeable constraint, then so does the corresponding clustering $\{B_1 \cap P, \ldots, B_{k'} \cap P\}$.*

Proof. Starting with $\{\mathrm{MB}(C_1), \ldots, \mathrm{MB}(C_k)\}$, we construct $\mathscr{B}$ from $\mathscr{C}$ in phases consisting of two steps: (1) merge balls that are currently too close to each other and thus not γ-separated, (2) ensure that the current covering is ε-balanced by increasing the radii of balls that are too small. The first step increases the cost by a multiplicative factor of at most γ and the second by a multiplicative factor of at most $(1 + \varepsilon)$. Both steps are alternatively applied in phases until the resulting clustering is both ε-balanced and γ-separated. Now, even though step (1) might yield a covering that is neither ε-balanced nor γ-separated and step (2) might yield a clustering that is not γ-separated, since the number of balls reduces with every phase, except maybe the first, there can only be k phases altogether. At that point, only one ball would remain and the clustering necessarily has to satisfy both properties. The resulting covering will cost at most $(1 + \varepsilon)^k \gamma^{k-1} \mathrm{cost}(\mathscr{C})$.

Let $B_1 = \mathrm{B}(c_1, r_1), \ldots, B_{k'} = \mathrm{B}(c_{k'}, r_{k'})$ denote the covering that has been constructed up to this point. For step (1), construct a graph G on top of $\mathscr{B}$, where two balls B_i and B_j are connected, iff $\|c_i - c_j\| \leq \gamma(r_i + r_j)$. In other words, two balls are connected by an edge, if and only if they are not γ-separated. We try to construct a γ-separated covering by merging all balls that belong to the same connected component. This just means that we replace all balls of the connected component with the minimal-enclosing-ball of the connected component. Take any connected component Z of G and consider two arbitrary points $x, y \in \bigcup_{B_\lambda \in Z} B_\lambda$, say $x \in B_i$ and $y \in B_j$. We can upper-bound the distance between them as follows: For any path, $B_i = B_{i_0}, \ldots B_{i_\ell} = B_j$ in G that connects B_i and B_j we have

$$\|x - y\| \leq \|x - c_i\| + \sum_{\lambda=0}^{\ell-1} \|c_{i_\lambda} - c_{i_{\lambda+1}}\| + \|y - c_j\|$$

$$\leq r_i + r_j + \sum_{\lambda=0}^{\ell-1} \gamma(r_{i_\lambda} + r_{i_{\lambda+1}}) \leq \gamma \sum_{B_\lambda \in Z} 2r_\lambda$$

In other words, the radius of the resulting ball is larger than the sum of the previous radii by a factor of at most γ. At this point, we might end up in a situation similar to the one with which we started; there might again be balls that are too close together and thus not γ-separate. However, we have reduced the number of balls by at least one and so this step can be performed at most $k - 1$ times.

For step (2) let $r_{i_1}, \ldots, r_{i_\ell}$ denote all radii with $r_{i_j} < \frac{\varepsilon}{k} \max_i r_i$. If we just set $r_{i_j} = \frac{\varepsilon}{k} \max_i r_i$ for all $j \in \{1, \ldots, \ell\}$, this increases the cost of the covering by at most $\ell \frac{\varepsilon}{k} \max_i r_i \leq \varepsilon \max_i r_i \leq \varepsilon \operatorname{cost}(\mathscr{C}')$. The resulting covering is necessarily ε-balanced. If it is not γ-separated we add another phase, starting with step (1).

$\square$

Algorithm 2: CLUSTERING

Input   : An ordered set $P \subseteq \mathbb{R}^d$, a number $k \in \mathbb{N}$, a value $0 < \varepsilon < 1$, a $(1 + \varepsilon)$-approximation $r_{\max}$ for largest radius

Output: A $(1 + \varepsilon)$-approximative k-clustering $\mathscr{C}$ of P

1 $\mathscr{C} \leftarrow \{P, \emptyset, \ldots, \emptyset\}$;   /* A feasible clustering to start with */

2 **forall** $(r_2, \ldots, r_k) \in \{(1 + \varepsilon)^i \frac{\varepsilon}{k} r_{max} \mid i \in \{0, \ldots, \lceil \log_{1+\varepsilon}(\frac{k}{\varepsilon}) \rceil\}\}^{k-1}$ **do**

3 | **forall** $u \in \{1, \ldots, k\}^{\frac{32k(1+\varepsilon)}{\varepsilon^3}}$ **do**

4 | | $B_1, \ldots, B_k \leftarrow$ SELECTION$(P, k, u, \{r_{\max}, r_2, \ldots, r_k\}, \varepsilon)$;

5 | | $\mathscr{C}' \leftarrow C_1, \ldots, C_k$, where $C_i = B_i \cap P$;

6 | | **if** $\mathscr{C}'$ *is a valid clustering and* $\operatorname{cost}(\mathscr{C}') < \operatorname{cost}(\mathscr{C})$ **then**

7 | | | $\mathscr{C} \leftarrow \mathscr{C}'$;

8 | | **end**

9 | **end**

10 **end**

11 **return** $\mathscr{C}$;

2.4 The Main Result

Now we are ready to prove the main theorem of this paper.

Theorem 1. *For every $\varepsilon \in (0, \frac{1}{2})$, there exists an algorithm that computes a $(1 + \varepsilon)$-approximation for k-min-sum-radii with mergeable constraints in time $d \cdot \mathrm{poly}(n) \cdot f(k, \varepsilon)$, if the corresponding constrained k-center problem has a constant-factor polynomial time approximation algorithm. If no such k-center approximation exists, the running time increases to $d \cdot n^{\mathrm{poly}(k,1/\varepsilon)} \cdot f(k, \varepsilon)$ (Theorem 2).*

Proof. Set $\varepsilon' = \left(\frac{\varepsilon}{12k}\right)^2$, $\gamma = (1 + \varepsilon' + 2\sqrt{\varepsilon'})$ and let $\mathscr{C}^{\mathrm{opt}}$ be an optimal min-sum-radii solution that satisfies the mergeable constraint. Lemma 4 shows that there is a $(1+\varepsilon')\gamma$-separated and ε'-balanced covering $\mathscr{B}^* = \{\mathrm{B}(c_1^*, r_1^*), \ldots \mathrm{B}(c_k^*, r_k^*)\}$ with $\mathrm{cost}(\mathscr{B}^*) \leq (1 + \varepsilon')^k \gamma^{k-1} \mathrm{cost}(\mathscr{C}^{\mathrm{opt}})$. Denote the corresponding clustering by $\mathscr{C}^* = \{\mathrm{B}(c_1^*, r_1^*) \cap P, \ldots, \mathrm{B}(c_1^*, r_1^*) \cap P\}$ and assume that the balls are ordered such that r_1^* is the largest radius. Using Lemma 3, we can compute approximate radii $\widetilde{r}_1, \ldots, \widetilde{r}_k$, such that $\mathrm{cost}(C_i^*) \leq \widetilde{r}_i \leq (1 + \varepsilon') \mathrm{cost}(C_i^*)$ for all i. Consider now, for u^* as in Lemma 1, the variables at the end of $\mathrm{SELECTION}(P, k, u^*, \{\widetilde{r}_1, \ldots, \widetilde{r}_k\}, \varepsilon')$. Since none of the B_i overlap and $C_i^* \subseteq B_i$ for all i, Algorithm 1 is able to fully reconstruct $\mathscr{C}^*$. Additionally, since the length of u^* does not exceed $k\frac{32(1+\varepsilon')}{\varepsilon'^3}$, this necessarily happens in one of the iterations of $\mathrm{CLUSTERING}(P, k, \varepsilon', \widetilde{r}_1)$. As such, running Algorithm 2 for all possible guesses of maximal radii provided by Lemma 3 guarantees an approximation ratio of $(1 + \varepsilon')^k \gamma^{k-1} \leq (1 + 3\sqrt{\varepsilon'})^{2k}$. Substituting $\varepsilon' = \left(\frac{\varepsilon}{12k}\right)^2$, we get an approximation ratio of $(1 + 3\sqrt{\varepsilon'})^{2k} \leq (1 + \frac{\varepsilon}{4k})^{2k} \leq e^{\varepsilon/2} \leq 1 + \varepsilon$ for $\varepsilon \leq 1/2$.

For the running time, we start by analyzing the time needed for one call to $\mathrm{SELECTION}$. Initializing the S_i takes $O(k)$ and so does the final loop in line 14. The main for-loop iterates over u, which has length $32k(1 + \varepsilon')/\varepsilon'^3$. The computation of the $(1 + \varepsilon')$-approximation of the minimum enclosing balls in line 9 can be done in $O(|S_{u_i}| \cdot d/\varepsilon') = O(d \cdot \mathrm{poly}(k, 1/\varepsilon'))$ with the algorithm from [25]. Thus, a single call to $\mathrm{SELECTION}$ takes $O(d \cdot \mathrm{poly}(k, 1/\varepsilon'))$.

In $\mathrm{CLUSTERING}$, we have two nested for-loops, that go through $k^{\frac{32k(1+\varepsilon')}{\varepsilon'^3}}$ and $O((\log_{1+\varepsilon'}(k/\varepsilon'))^{k-1})$ iterations respectively, in each of which $\mathrm{SELECTION}$ is invoked. Line 6, which checks whether the clustering covers the whole set and satisfies the constraint, takes at most $O(\mathrm{poly}(n))$ time (depending on the constraint, this time might even be linear in n). Thus, one call to $\mathrm{CLUSTERING}$ takes $d \cdot \mathrm{poly}(n) \cdot k^{O(\mathrm{poly}(k,1/\varepsilon))} \cdot (\log_{1+\varepsilon'} \mathrm{poly}(k, 1/\varepsilon))^{k-1}$ time.

Finally, $\mathrm{CLUSTERING}$ has to be called for every candidate for $r_{\max}$. There are at most $O(k + \log_{1+\varepsilon} \gamma^{k-1} k)$ such candidates and so the overall running time is $d \cdot \mathrm{poly}(n) \cdot f(k, \varepsilon)$. $\qquad \square$

Using the other method of guessing the largest radius extends the result to *all* mergeable clustering constraints. The trade-off is a worse running time.

Theorem 2. *For every $0 < \varepsilon < 1/2$, there exists an algorithm that computes a $(1 + \varepsilon)$-approximation for min-sum-radii with mergeable constraints in time $d \cdot n^{\mathrm{poly}(k,1/\varepsilon)} \cdot f(k, \varepsilon)$.*

Proof. We follow the arguments in Theorem 1, with the only difference being the computation of the candidate set R for the largest radius. We show in the long version how to compute a candidate set of size $n^{O(1/\varepsilon')}$ that contains a $(1 + \varepsilon')$-approximation for the largest radius in time $\frac{d}{\varepsilon'^2}n^{O(1/\varepsilon')}$. Substituting $\varepsilon' = \left(\frac{\varepsilon}{12k}\right)^2$, we get the purported runtime. $\square$

A Word on Outliers. We have stated the allowance of outliers as a mergeable constraint in the introduction. Note, however, that there is one issue: When we want to achieve ε-separation, we may not start to merge outliers into clusters with more than one point. So clustering with outliers is not strictly mergeable. However, the algorithm can be suitably adapted: Only make sure that the non-outlier clusters are separated, and during the oracle calls, only let the oracle decide whether a point is an outlier or not, and if not, to which cluster it belongs. We do not derive the details of such an algorithm in this paper.

References

1. Ahmadian, S., Swamy, C.: Approximation algorithms for clustering problems with lower bounds and outliers. In: Proceedings of the 43rd International Colloquium on Automata, Languages, and Programming (ICALP), vol. 55, pp. 69:1–69:15 (2016). https://doi.org/10.4230/LIPIcs.ICALP.2016.69
2. Arutyunova, A., Schmidt, M.: Achieving anonymity via weak lower bound constraints for k-median and k-means. In: Proceedings of the 38th International Symposium on Theoretical Aspects of Computer Science (STACS), vol. 187, pp. 7:1–7:17 (2021). https://doi.org/10.4230/LIPIcs.STACS.2021.7
3. Badoiu, M., Har-Peled, S., Indyk, P.: Approximate clustering via core-sets. In: Proceedings on 34th Annual ACM Symposium on Theory of Computing (STOC), pp. 250–257. ACM (2002). https://doi.org/10.1145/509907.509947
4. Bandyapadhyay, S., Lochet, W., Saurabh, S.: FPT constant-approximations for capacitated clustering to minimize the sum of cluster radii. In: 39th International Symposium on Computational Geometry (SoCG) (2023, to appear). https://doi.org/10.48550/arXiv.2303.07923
5. Behsaz, B., Salavatipour, M.R.: On minimum sum of radii and diameters clustering. Algorithmica **73**(1), 143–165 (2015). https://doi.org/10.1007/s00453-014-9907-3
6. Bera, S.K., Chakrabarty, D., Flores, N., Negahbani, M.: Fair algorithms for clustering. In: Proceedings of the Annual Conference on Neural Information Processing Systems (NeurIPS 2019), pp. 4955–4966 (2019). https://proceedings.neurips.cc/paper/2019/hash/fc192b0c0d270dbf41870a63a8c76c2f-Abstract.html
7. Bercea, I.O., et al.: On the cost of essentially fair clusterings. In: Proceedings of APPROX/RANDOM 2019, vol. 145, pp. 18:1–18:22 (2019). https://doi.org/10.4230/LIPIcs.APPROX-RANDOM.2019.18
8. Bilò, V., Caragiannis, I., Kaklamanis, C., Kanellopoulos, P.: Geometric clustering to minimize the sum of cluster sizes. In: Brodal, G.S., Leonardi, S. (eds.) ESA 2005. LNCS, vol. 3669, pp. 460–471. Springer, Heidelberg (2005). https://doi.org/10.1007/11561071_42

9. Böhm, M., Fazzone, A., Leonardi, S., Menghini, C., Schwiegelshohn, C.: Algorithms for fair k-clustering with multiple protected attributes. Oper. Res. Lett. **49**(5), 787–789 (2021)

10. Böhm, M., Fazzone, A., Leonardi, S., Schwiegelshohn, C.: Fair clustering with multiple colors. arXiv preprint arXiv:2002.07892 (2020)

11. Capoyleas, V., Rote, G., Woeginger, G.: Geometric clusterings. J. Algorithms **12**(2), 341–356 (1991). https://www.sciencedirect.com/science/article/pii/019667749190007L

12. Caton, S., Haas, C.: Fairness in machine learning: a survey. arXiv preprint arXiv:2010.04053 (2020)

13. Charikar, M., Panigrahy, R.: Clustering to minimize the sum of cluster diameters. J. Comput. Syst. Sci. **68**(2), 417–441 (2004). https://doi.org/10.1016/j.jcss.2003.07.014

14. Chhabra, A., Masalkovaitė, K., Mohapatra, P.: An overview of fairness in clustering. IEEE Access **9**, 130698–130720 (2021)

15. Chierichetti, F., Kumar, R., Lattanzi, S., Vassilvitskii, S.: Fair clustering through fairlets. In: Advances in Neural Information Processing Systems, vol. 30 (2017)

16. Friggstad, Z., Jamshidian, M.: Improved polynomial-time approximations for clustering with minimum sum of radii or diameters. In: 30th Annual European Symposium on Algorithms (ESA), vol. 244, pp. 56:1–56:14 (2022). https://doi.org/10.4230/LIPIcs.ESA.2022.56

17. Gibson, M., Kanade, G., Krohn, E., Pirwani, I.A., Varadarajan, K.R.: On metric clustering to minimize the sum of radii. Algorithmica **57**(3), 484–498 (2010). https://doi.org/10.1007/s00453-009-9282-7

18. Gibson, M., Kanade, G., Krohn, E., Pirwani, I.A., Varadarajan, K.R.: On clustering to minimize the sum of radii. SIAM J. Comput. **41**(1), 47–60 (2012). https://doi.org/10.1137/100798144

19. Harb, E., Lam, H.S.: KFC: a scalable approximation algorithm for k-center fair clustering. Adv. Neural. Inf. Process. Syst. **33**, 14509–14519 (2020)

20. Henzinger, M., Leniowski, D., Mathieu, C.: Dynamic clustering to minimize the sum of radii. In: Proceedings of the 25th Annual European Symposium on Algorithms (ESA), vol. 87, pp. 48:1–48:10 (2017). https://doi.org/10.4230/LIPIcs.ESA.2017.48

21. Inamdar, T., Varadarajan, K.R.: Capacitated sum-of-radii clustering: an FPT approximation. In: Proceedings of the 28th Annual European Symposium on Algorithms (ESA), vol. 173, pp. 62:1–62:17 (2020). https://doi.org/10.4230/LIPIcs.ESA.2020.62

22. Lev-Tov, N., Peleg, D.: Polynomial time approximation schemes for base station coverage with minimum total radii. Comput. Netw. **47**(4), 489–501 (2005). https://doi.org/10.1016/j.comnet.2004.08.012

23. Proietti, G., Widmayer, P.: Partitioning the nodes of a graph to minimize the sum of subgraph radii. In: Asano, T. (ed.) ISAAC 2006. LNCS, vol. 4288, pp. 578–587. Springer, Heidelberg (2006). https://doi.org/10.1007/11940128_58

24. Rösner, C., Schmidt, M.: Privacy preserving clustering with constraints. In: 45th International Colloquium on Automata, Languages, and Programming, (ICALP) 2018, vol. 107, pp. 96:1–96:14 (2018). https://doi.org/10.4230/LIPIcs.ICALP.2018.96. arXiv:1802.02497

25. Yildirim, E.A.: Two algorithms for the minimum enclosing ball problem. SIAM J. Optim. **19**(3), 1368–1391 (2008). https://doi.org/10.1137/070690419

Online Hitting Set of d-Dimensional Fat Objects

Shanli Alefkhani[1], Nima Khodaveisi[1], and Mathieu Mari[1,2(✉)]

[1] IDEAS-NCBR, Warsaw, Poland
[2] University of Warsaw, Warsaw, Poland
mathieu.mari@lirmm.fr

Abstract. We consider an online version of the geometric minimum hitting set problem that can be described as a game between an adversary and an algorithm. For some integers d and N, let P be the set of points in $(0, N)^d$ with integral coordinates, and let $\mathcal{O}$ be a family of subsets of P, called objects. Both P and $\mathcal{O}$ are known in advance by the algorithm and by the adversary. Then, the adversary gives some objects one by one, and the algorithm has to maintain a valid hitting set for these objects using points from P, with an immediate and irrevocable decision. We measure the performance of the algorithm by its competitive ratio, that is the ratio between the number of points used by the algorithm and the offline minimum hitting set for the sub-sequence of objects chosen by the adversary.

We present a simple deterministic online algorithm with competitive ratio $((4\alpha + 1)^{2d} \log N)$ when objects correspond to a family of α-fat objects. Informally, α-fatness measures how cube-like is an object. We show that no algorithm can achieve a better ratio when α and d are fixed constants. In particular, our algorithm works for two-dimensional disks and d-cubes which answers two open questions from related previous papers in the special case where the set of points corresponds to all the points of integral coordinates with a fixed d-cube.

Keywords: Online algorithms · Minimum hitting set · Euclidean Plane

1 Introduction

The hitting set problem is one of the fundamental problems in combinatorial optimization. Let $(X, \mathcal{R})$ be a range space where X is a set of elements and $\mathcal{R}$ is a family of subsets of X, $|X| = n, |\mathcal{R}| = m$. A subset $H \subseteq X$ is a *hitting set* for $\mathcal{R}$ if and only if, for every range $R \in \mathcal{R}$ the intersection of H and R is non-empty. In the offline setting, the goal is to find a hitting set of minimum size. Note that by interchanging the roles of subsets and elements, the hitting set problems is equivalent to the set cover problem. The hitting set problem is a classic NP-hard problem [10], and the best approximation factor achievable in polynomial time (assuming $P \neq NP$) is $\Theta(\log n)$ [3,7,9,12].

© The Author(s), under exclusive license to Springer Nature Switzerland AG 2023
J. Byrka and A. Wiese (Eds.): WAOA 2023, LNCS 14297, pp. 134–144, 2023.
https://doi.org/10.1007/978-3-031-49815-2_10

There is a line of work that considered the hitting set problem in a geometrical setting. The set of elements X is a subset of *points* of the d-dimensional plane $\mathbb{R}^d$ and $\mathcal{R}$ corresponds to a family of geometrical objects[1], e.g., disks, squares, rectangles, etc., for $d = 2$. The hitting set problem remains NP-hard even for simple geometric objects like unit disks or unit squares in $\mathbb{R}^2$ [8]. For some families of geometric objects, there are better approximation ratios than for the general case, e.g., a PTAS for axis-parallel squares and disks [13], and more generally for fat objects in a fixed dimension [2].

In this paper, we consider an *online* version of the problem. It is convenient to define this problem as a game between an adversary and an algorithm. Initially, a range space $(X, \mathcal{R})$ is known in advance by both the algorithm and the adversary. The game consists of a series of turns until the adversary decides to stop the game. In each turn, the adversary gives a subset $R \in \mathcal{R}$, and the algorithm has to choose a point $p \in X$, such that $p \in R$, if none of the points previously chosen by the algorithm are contained in R. The algorithm is allowed to select several points during the same turn and may decide to select new points even though the current subset S is already hit by one of its previous points. The goal of the algorithm is to minimize the total number of points selected at the end of the game. See Fig. 1 for an illustration of this game. We measure the performance of an algorithm by its *competitive ratio*, which corresponds to the ratio between the number of points selected by the algorithm and the minimum size of a hitting set of the sub-family of $\mathcal{R}$ given by the adversary during the game.

In the online setting, Alon et al. [1] introduced an (essentially) tight $O(\log n \log m)$-competitive algorithm for the general case. There are also a few works that considered the special case of geometrical objects. Even et al. [5] presented an $O(\log n)$-competitive algorithm for intervals ($d = 1$), for unit-disks ($d = 2$), and later for half-planes in dimension two [6]. Khan et al. [11] presented $O(\log N)$-competitive algorithm for axis-parallel squares of arbitrary sizes, assuming that all points have integral coordinates in $[0, N)^2$. De et al. [4] looked at the problem in dimension d, when the algorithm is allowed to use any point of integral coordinates. They showed an $O(d^2)$-competitive algorithm for unit hypercubes and an $O(d^4)$-competitive algorithm for unit balls in dimension d. They also showed that any deterministic online algorithm for hypercubes has a competitive ratio of at least $d + 1$. Even and Smorodinsky also showed a lower bound of $\Omega(\log n)$ for intervals and arbitrary points [5].

In this paper, we are interested in two open questions mentioned in these papers:

1. Can one obtain an $o(\log^2 n)$-competitive algorithm for disks (dimension two)? [6]
2. Can one obtain an $o(\log n \log m)$-competitive algorithm for cubes (dimension three)? [11]

---

[1] to simplify, we consider that a subset of points $D \subseteq X \subseteq \mathbb{R}^2$ is a disk (or a square, or another type of geometric object) if there exists a disk $D' \subset \mathbb{R}^2$ such that $D' \cap X = D$. This allows us to consider a subset of X as a geometrical object.

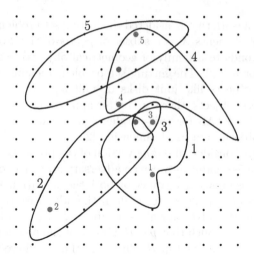

Fig. 1. Here, the set of points X corresponds to the points of integral coordinates of the Euclidean plane (small black dots). During the game, the adversary has given 5 subsets. In each turn, the algorithm has chosen one point (green) that is contained in the object given by the adversary. In total, the algorithm has used 5 points while the offline minimum hitting set consists of only two points (purple). (Color figure online)

1.1 Our Contribution

In this paper, we are interested in d-dimensional fat objects, that generalize disks, squares, hypercubes, etc. An object $O \in \mathbb{R}^d$ is α-fat, for some $\alpha \geq 1$ if the ratio of the sizes of the smallest hypercube containing O and the biggest hypercube contained in O is at most α.

We answer the two open questions mentioned above in the case where X corresponds to the set of points with integral coordinates that are contained in a fixed hypercube. More precisely, let d and N be some integers. Let $P = (0, N)^d \cap \mathbb{Z}^d$ be the set of points of integral coordinates that are contained in $(0, N)^d$. In particular $n = |P| = (N - 1)^d$. Let $\mathcal{O}$ be a family of subsets of P.

Theorem 1. *There is an $((4\alpha+1)^{2d} \log N)$-competitive algorithm for minimum hitting set on $(P, \mathcal{O})$ when $\mathcal{O}$ corresponds to a family of d-dimensional α-fat objects in $(0, N)^d$.*

Notice that disks are $\sqrt{2}$-fat and cubes are 1-fat. Thus, Theorem 1 settles both questions from the introduction in the affirmative.

This algorithm is $O(\log n)$-competitive ratio for disks of arbitrary sizes in the 2-dimensional plane, and 3-dimensional cubes.

Our algorithm works as follows. It associates to each point in P a color in $\{0, \ldots, \lfloor \log N \rfloor - 1\}$. Then, when an object arrives, if it is already hit we do nothing, otherwise we pick all the points with the maximum color inside the object. Our coloring guarantees that we add at most $(4\alpha + 1)^d$ points in each step. We show that for each color l, and each point p in the offline solution, the adversary cannot give more than $(4\alpha + 1)^d$ objects that are not already hit at their arrival time, that contain p and are of level[2] l. This will help us to prove our competitive ratio.

We also show a lower bound of $\Omega(\frac{\log N}{1+\log \alpha})$ on the competitiveness of any algorithm for the problem. This implies that no algorithm can achieve a better ratio when α and d are fixed constants.

2   The Algorithm

In this section, we present our online algorithm for hitting set of d-dimensional α-fat objects. We start with some useful definitions. Let d and N be two integers and P be the set of points with integral coordinates in $(0, N)^d$. A d-cube is an axis-parallel d-dimensional hypercube. The *width* of a d-cube is the length of any of its sides. For simplicity, we assume that all geometrical objects $O \subset \mathbb{R}^d$ considered in this paper are open sets.

Definition 1 (α-fat). *Let $O \subset \mathbb{R}^d$. The* in-width *of O is the length of the largest d-cube contained in O. The* out-width *of O is the length of the smallest d-cube containing O. We say that O is α-fat, for some $\alpha \geq 1$ if the ratio of its out-width over its in-width is at most α.*

For instance d-disks are $O(\sqrt{d})$-fat. Also, notice that d-cubes are 1-fat.

Definition 2 (level of a point). *Let $i \in \mathbb{N}$, we denote $\ell(i)$ the maximum number k such that i is a multiple of 2^k. For a point $x = (x_1, \ldots, x_d) \in P$, we define the level of a point to be $\ell(x) = \min_{i=1}^{d} \ell(x_i)$.*

See Fig. 2 for an illustration of the levels of the points of P. We denote $\mathcal{L} = \{0, \ldots, \lfloor \log N \rfloor - 1\}$. It is clear that for each point $x \in P$, we have $\ell(x) \in \mathcal{L}$.

Definition 3 (level of an object). *For a geometric object $O \subseteq (0, N)^d$, we define its level $\ell(O)$ as the maximum level over all the points in $O \cap P$. For each $l \in \mathcal{L}$, we denote $n_l(O)$ as the number of points of level l in $O \cap P$ and $n_{\geq l}(O)$ the number of points of level at least l in $O \cap P$.*

We now describe the algorithm. We maintain a hitting set P' that is initially empty. In each round, we are given an α-fat object $O \subseteq (0, N)^d$.

– If O is already hit by a point in P', then we do nothing.
– Otherwise, we add all the points in $P \cap O$ that are of level $\ell(O)$ to P'.

It is clear that at the end of each turn, P' is a hitting set of the objects given so far. Notice that in each step, the action made by the algorithm only depends on P' and the current object, but not on the previous objects given by the adversary.

Now we prove that the competitive ratio of this algorithm is $((4\alpha + 1)^{2d} \log N)$.

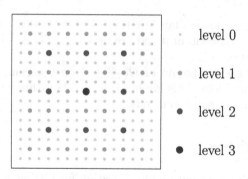

Fig. 2. Levels of points in P for $N = 16$.

Lemma 1. *The in-width of an object $O \subseteq (0, N)^d$ is less than $2^{\ell(O)+1}$.*

See Fig. 3 for an example.

Proof (Proof of Lemma 1.) Let $l = \ell(O)$. We define S_O as the largest d-cube contained in O. We assume for the sake of a contradiction that the width of S_O is at least 2^{l+1} and we show that there exists a point $q \in S_O$ such that $\ell(q) \geq l+1$. Since the width of S_O is at least 2^{l+1}, for each i, $1 \leq i \leq d$, there exists an integer k_i, such that the hyperplane with i-th coordinate $k_i 2^{l+1}$ intersects S_O. Then, the $q = (k_1 2^{l+1}, \ldots, k_d 2^{l+1})$ is contained in S_O, and its level is $\ell(q) \geq l+1$. This proves the lemma.

Corollary 1. *Let $O \subseteq (0, N)^d$ be an α-fat object, for some $\alpha \geq 1$. Then, the out-width of O is at most $2^{\ell(O)+1}\alpha$.*

Lemma 2. *Let $l \in \mathcal{L}$ and $\alpha \geq 1$. A d-cube of width at most $\alpha 2^{l+2}$ contains at most $(4\alpha + 1)^d$ points of level l.*

See Fig. 4 for an example.

Proof (Proof of Lemma 2). Let $C \subseteq (0, N)^d$ be a d-cube of width $w \leq \alpha 2^{l+2}$. For each i, $1 \leq i \leq d$, let Λ_i be the set of integers λ such that (i) λ is a multiple of 2^l and, (ii) the hyperplane whose i-th coordinate is λ intersects C. For each i, $1 \leq i \leq d$, we have $|\Lambda_i| \leq \frac{w}{2^l} + 1 \leq 4\alpha + 1$.

It is easy to see that $\prod_{i=1}^{d} \Lambda_i$ is the set of points of C that are of level at least l. Thus,

$$n_l(C) \leq n_{\geq l}(C) \leq |\prod_{i=1}^{d} \Lambda_i| = \prod_{i=1}^{d} |\Lambda_i| \leq (4\alpha + 1)^d.$$

This finishes the proof.

---

[2] The level of an object is the maximum color of the points contained inside it.

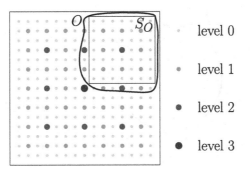

Fig. 3. This figure shows an object O of level $l = 2$ in dimension two. $N = 16$ and the in-width of O is $7 \leq 2^{l+1}$.

Corollary 2. *Let* $O \subseteq (0, N)^d$ *be an* α-*fat object for some* $\alpha \geq 1$. *Then,* $n_{\ell(O)}(O) \leq (4\alpha + 1)^d$.

We now analyze the competitive ratio of our algorithm and show the following bound.

Lemma 3. *Our algorithm is* $((4\alpha + 1)^{2d} \log N)$-*competitive.*

Proof. Let $S \subseteq O$ denote the sequence of objects given by the adversary and $S' \subseteq S$ denotes the sub-sequence of objects that are not already hit at their arrival. For each $O \in S$, our algorithm picks the points of level $\ell(O)$ inside O if O is not already hit, i.e., if $O \in S'$. By Corollary 2, we know that O contains at most $(4\alpha + 1)^d$ points of level l, and thus, our algorithm returns a hitting set of size

$$|P'| \leq |S'|(4\alpha + 1)^d \tag{1}$$

We now establish an upper bound on the size of the minimum hitting set OPT $\subseteq$ P of S.

For each $l \in \mathcal{L}$ and each $p \in P$, we denote $S'_{l,p} \subseteq S'$ the set of objects of level l in S' that contain p.

Claim. $|S'_{l,p}| \leq (4\alpha + 1)^d$.

Assume for the sake of a contradiction, that $|S'_{l,p}| > (4\alpha + 1)^d$. We denote B_p as the d-cube of width $2^{l+2}\alpha$ centered in p. By Lemma 2, we know that there are at most $(4\alpha + 1)^d$ points of level l in B_p. Also, it is clear by Corollary 1 that any object of level l containing p is inside B_p. Then, by the pigeonhole principle, there are two objects $O, O' \in S'_{l,p}$ such that they both contain the same point q of level l. See Fig. 5. Without loss of generality, let us assume that O arrived before O' in the sequence of objects given by the adversary. Then, our algorithm picks q when O arrives since both O and q have level l. On the other side, we know that $q \in O'$, meaning that O' is already hit when it is given by

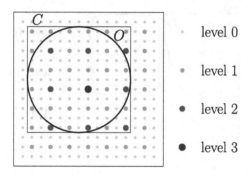

Fig. 4. In this figure, O is a two-dimensional disk. In particular, O is α-fat for $\alpha = \sqrt{2}$. Let $l = 1$. The smallest 2-cube containing O is C, which has width less than $11 \leq \alpha 2^{l+2} \approx 11.31$. The number of points of level l contained in C is $27 \leq (4\sqrt{2} + 1)^2 \approx 44.31$.

the adversary, which is a contradiction with the fact that O' is in $\mathcal{S}'$. Therefore, for each $l \in \mathcal{L}$ and each $p \in P$, we have $|\mathcal{S}'_{l,p}| \leq (4\alpha + 1)^d$.

Since OPT is a hitting set for $\mathcal{S}$, it is also an hitting set for $\mathcal{S}'$, which implies that

$$\mathcal{S}' \subseteq \bigcup_{p \in \text{OPT}} \bigcup_{l \in \mathcal{L}} \mathcal{S}'_{l,p}.$$

With the previous upper bound on the size of $\mathcal{S}'_{l,p}$, we obtain that

$$|\mathcal{S}'| \leq |\text{OPT}| \cdot |\mathcal{L}| \cdot (4\alpha + 1)^d \leq |\text{OPT}| \cdot (\lfloor \log N \rfloor) \cdot (4\alpha + 1)^d,$$

which together with the bound of Eq. (1) implies that our algorithm is $((4\alpha + 1)^{2d} \log N)$-competitive.

3   Lower Bound

In this section, we prove the lower bound for the problem (Theorem 2). For any object $O \subset \mathbb{R}^d$, we say that object $O' \subset \mathbb{R}^d$ is a *dilation* of O if it is the result of a translation and a homothety of O with positive scale factor. More formally, O' is a dilation of O if there exists $\beta > 0$ and translation vector v such that $O' = \beta O + v$. Let $\mathcal{D}(O)$ denote the set of dilations of O. Notice that if O is α-fat, for some $\alpha \geq 1$, then a dilation of O is also α-fat.

Recall that the online hitting set problem can be formalized as a game between an adversary and an algorithm.

Theorem 2. *Consider the range space $(P, \mathcal{D}(O))$, for any α-fat object $O \subset \mathbb{R}^d$. The adversary has a strategy that forces the algorithm to place at least $\frac{\log N}{1 + \log \alpha}$ points, whereas the optimum offline solution only requires one point.*

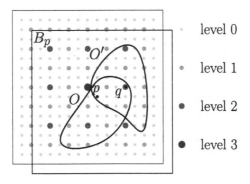

Fig. 5. Here, $l = 2$. We show in the proof of Lemma 3 that the adversary gave two objects O and O' of level $l = 2$, containing a common point p, that share a common point q of level two. This cannot happen if O and O' are both not already hit at their arrival time.

Proof. The adversary produces a sequence $O_1, O_2, \ldots, O_s \in \mathcal{D}(O)$ as follows. In the first step, the adversary chooses $O_1 \in \mathcal{D}(O)$ such that the smallest d-cube enclosing O_1 is $(0, N)^d$. Then, for each $j \geq 1$, the adversary does the following. At step j, with $j \geq 2$, let O_j be the largest dilation of O that is contained in O_{j-1} and that does not contain any point from the algorithm. If O_j does not contain any point in P, the game ends. Otherwise, the adversary gives O_j to the algorithm; See Fig. 6.

Let $O_1, O_2, \ldots, O_s$ be the sequence of objects obtained at the end of the game. Since $O_1 \supseteq O_2 \supseteq \cdots \supseteq O_s$, and there exists a point $p \in P \cap O_s$, the set $\{p\}$ is a hitting a hitting set of $\{O_1, O_2, \ldots, O_s\}$.

Now we give a lower bound on the number of points used by the algorithm. For each j, $1 \leq j \leq s$, let k_j be the number of points added by the algorithm during step j. It is clear that the algorithm uses in total $\sum_{j=1}^{s} k_j$ points. We show that $\sum_{j=1}^{s} k_j \geq \frac{\log N}{1 + \log \alpha}$.

For each j, $1 \leq j \leq s$, let C_j (C'_j) be the smallest (largest, resp.) d-cube containing (contained in, resp.) O_j and let w_j (w'_j, resp.) denote its width. Notice that w_j (w'_j) is the out-width (in-width, resp.) of O_j. See Fig. 7. We claim that $w_{j+1} \geq \frac{w_j}{\alpha(k_j + 1)}$.

Let P_j be the set of points from the algorithm that are contained in C'_j at the end of step j. We have $|P_j| \leq k_j$ since O_j is not already hit at the beginning of step j, and the algorithm adds k_j new points during that step. We prove that there exists a d-cube of width at least $\lfloor \frac{w'_j}{k_j + 1} \rfloor$ inside C'_j which does not contain any points of P_j. Let $p = (x_1, \ldots, x_d)$ be a corner of C'_j such that for any other point $p' = (x'_1, \ldots, x'_d) \in C'_j$ and any $1 \leq i \leq d$, we have $x_i \leq x'_i$. For each i, $1 \leq i \leq d$, consider the set of $k_j + 1$ intervals $\{[x_i + h\frac{w'_j}{k_j + 1}, x_i + (h+1)\frac{w'_j}{k_j + 1}] \mid 0 \leq h \leq k_j, h \in \mathbb{Z}\}$. Since $|P_j| \leq k_j$, there exists an integer $0 \leq h_i \leq k_j$ such that the i-th coordinate of every point in P_j is not in $(x_i + h_i\frac{w'_j}{k_j + 1}, x_i + (h_i + 1)\frac{w'_j}{k_j + 1})$.

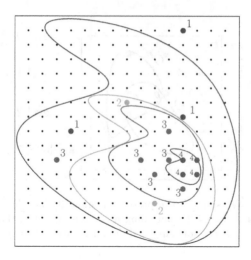

Fig. 6. This figure illustrates the strategy of the adversary. In each turn, the adversary finds the biggest dilation of O that is not hit by any point inside the previous object. The points with the same number are the set of points chosen by the algorithms in each step.

Therefore, if we consider the d-cube $S_j = \prod_{i=1}^{d}[x_i + h_i\frac{w'_j}{k_j+1}, x_i + (h_i+1)\frac{w'_j}{k_j+1}]$, S_j does not contain any points of P_j, and the width of S_j is $\frac{w'_j}{k_j+1}$ which implies that w_{j+1} would be at least $\frac{w'_j}{k_j+1}$ (See Fig. 7). Also, note that S_j is inside C'_j. Moreover, since O_i is α-fat, we have $w_{j+1} \geq \frac{w'_j}{k_j+1} \geq \frac{w_j}{\alpha(k_j+1)}$.

We conclude that $w_{s+1} \geq \frac{w_1}{\alpha^s \prod_{j=1}^{s}(k_j+1)} = \frac{N}{\alpha^s \prod_{j=1}^{s}(k_j+1)}$. Also, it is obvious that the adversary stops when $w_{s+1} \leq 1$. Hence $\frac{N}{\alpha^s \prod_{j=1}^{s}(k_j+1)} \leq w_{s+1} \leq 1$ and since $(1+x) \leq e^x$ for all $x \in \mathbb{R}$, it holds that

$$N \leq \alpha^s \prod_{j=1}^{s} e^{k_j} = e^{s\log\alpha + \sum_{j=1}^{s} k_j}.$$

Now, we apply log to this equation, and we obtain $\log N \leq s\log\alpha + \sum_{j=1}^{s} k_j$. Also, recall that for each j, $1 \leq j \leq s$, O_j is not hit at the time of its arrival so we have $k_j \geq 1$, which implies $s \leq \sum_{j=1}^{s} k_j$. Therefore, we obtain $\frac{\log N}{1+\log\alpha} \leq \sum_{j=1}^{s} k_j$ which finishes the proof.

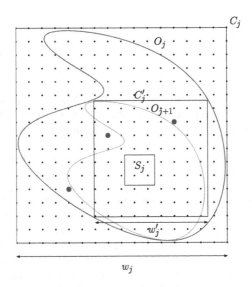

Fig. 7. This figure shows the existence of a d-cube of width at least $\frac{w_j}{\alpha(k_j+1)}$ at the end of j-th step of the game that does not contain any point of the algorithm. Red points are the points chosen by the algorithm in the j-th step.

4   Conclusion

We have presented a tight $O(\log n)$-competitive algorithm for the online hitting set problem of fat objects of fixed dimension and fixed aspect ratio, when the set of points corresponds to all the points of integral coordinates contained in a fixed hypercube of width $N = n^{1/d}$. We finish with some open questions.

- For any d, when the algorithm is allowed to use any point in $\mathbb{Z}^d \cap (0, N)^d$, can one either design an online algorithm for α-*fat* objects with a better competitive ratio than $((4\alpha+1)^{2d} \log N)$ or improve the lower bound of $\frac{\log N}{1+\log \alpha}$ to tighten the gap?
- When $d = 2$, and the algorithm is allowed to use any point in a fixed square of width N, can one design an online algorithm for *rectangles* with competitive ratio $O(\log N)$?
- When $d = 2$, and the set of points is a fixed subset (known in advance) $P \subseteq \mathbb{Z}^2 \cap (0, N)^2$, can one design an online algorithm for *disks* with competitive ratio $O(\log N)$?
- When $d = 3$, and the set of points is a fixed subset (known in advance) $P \subseteq \mathbb{Z}^3 \cap (0, N)^3$, can one design an online algorithm for 3-*cubes* with competitive ratio $O(\log N)$?
- When $d = 2$, and the set of points is a fixed subset (known in advance) $P \subseteq \mathbb{R}^2$ of size n, can one design an online algorithm for *squares* with competitive ratio $O(\log n)$?

References

1. Alon, N., Awerbuch, B., Azar, Y., Buchbinder, N., Naor, J.: The online set cover problem. SIAM J. Comput. **39**(2), 361–370 (2009). https://doi.org/10.1137/060661946

2. Chan, T.M.: Polynomial-time approximation schemes for packing and piercing fat objects. J. Algorithms **46**(2), 178–189 (2003). https://doi.org/10.1016/S0196-6774(02)00294-8, https://www.sciencedirect.com/science/article/pii/S0196677402002948

3. Chvátal, V.: A greedy heuristic for the set-covering problem. Math. Oper. Res. **4**(3), 233–235 (1979). https://doi.org/10.1287/moor.4.3.233

4. De, M., Singh, S.: Hitting geometric objects online via points in $\mathbb{Z}^d$. In: Zhang, Y., Miao, D., Möhring, R.H. (eds.) COCOON 2022. LNCS, vol. 13595, pp. 537–548. Springer, Cham (2022). https://doi.org/10.1007/978-3-031-22105-7_48

5. Even, G., Smorodinsky, S.: Hitting sets online and vertex ranking. In: Demetrescu, C., Halldórsson, M.M. (eds.) ESA 2011. LNCS, vol. 6942, pp. 347–357. Springer, Heidelberg (2011). https://doi.org/10.1007/978-3-642-23719-5_30

6. Even, G., Smorodinsky, S.: Hitting sets online and unique-max coloring. Discret. Appl. Math. **178**, 71–82 (2014). https://doi.org/10.1016/j.dam.2014.06.019

7. Feige, U.: A threshold of ln n for approximating set cover. J. ACM **45**(4), 634–652 (1998). https://doi.org/10.1145/285055.285059

8. Fowler, R.J., Paterson, M., Tanimoto, S.L.: Optimal packing and covering in the plane are np-complete. Inf. Process. Lett. **12**(3), 133–137 (1981). https://doi.org/10.1016/0020-0190(81)90111-3

9. Johnson, D.S.: Approximation algorithms for combinatorial problems. J. Comput. Syst. Sci. **9**(3), 256–278 (1974). https://doi.org/10.1016/S0022-0000(74)80044-9

10. Karp, R.M.: Reducibility among combinatorial problems. In: Miller, R.E., Thatcher, J.W. (eds.) Proceedings of a symposium on the Complexity of Computer Computations, held 20–22 March 1972, at the IBM Thomas J. Watson Research Center, Yorktown Heights, New York, USA, pp. 85–103. The IBM Research Symposia Series, Plenum Press, New York (1972). https://doi.org/10.1007/978-1-4684-2001-2_9

11. Khan, A., Lonkar, A., Rahul, S., Subramanian, A., Wiese, A.: Online and dynamic algorithms for geometric set cover and hitting set. CoRR abs/2303.09524 (2023). https://doi.org/10.48550/arXiv.2303.09524

12. Lovász, L.: On the ratio of optimal integral and fractional covers. Discret. Math. **13**(4), 383–390 (1975). https://doi.org/10.1016/0012-365X(75)90058-8

13. Mustafa, N.H., Ray, S.: PTAS for geometric hitting set problems via local search. In: Hershberger, J., Fogel, E. (eds.) Proceedings of the 25th ACM Symposium on Computational Geometry, Aarhus, Denmark, 8–10 June 2009, pp. 17–22. ACM (2009). https://doi.org/10.1145/1542362.1542367

Approximation Schemes Under Resource Augmentation for Knapsack and Packing Problems of Hyperspheres and Other Shapes

Vítor Gomes Chagas[iD], Elisa Dell'Arriva[(✉)][iD], and Flávio Keidi Miyazawa[iD]

Institute of Computing, Universidade Estadual de Campinas, Campinas, Brazil
{vitor.chagas,elisa.arriva,fkm}@ic.unicamp.br

Abstract. The problems we investigate consist in packing hyperspheres in bins optimizing some resource, such as minimizing the number or the size of the bins, or maximizing the total profit associated with the packed items. We present an approximation scheme under resource augmentation for the circle knapsack problem, i.e., a polynomial-time algorithm that, for any constant $\varepsilon > 0$, obtains a solution whose value is within a factor of $1 - \varepsilon$ of the optimal value, using augmented bins of height increased by a factor of ε. To the best of our knowledge, this is the first approximation scheme for this problem. Additionally, our technique can be extended to accomplish PTASs for other packing problems, like the multiple strip packing problem and the problem of minimizing the size of the bins. Our technique is not restricted to circles and hyperspheres, working for items, bins and strip bases of different convex shapes, such as squares, regular polygons with bounded number of sides, ellipses, among others, and for their generalizations to the d-dimensional case, for constant d.

Keywords: Multiple knapsack · Hypersphere packing · Resource augmentation · PTAS

1 Introduction

Hypersphere packing problems consist in packing hyperspheres in bins optimizing some resource, such as minimizing the number or the size of the bins and maximizing the profit associated with the packed items. From the mathematical viewpoint, the problem of packing spheres in the Euclidean space has been investigated for centuries and poses a great challenge. For instance, in the 17th century, Kepler [23] conjectured a bound on the average density of any packing of spheres in the three-dimensional Euclidean space. This remained an open question for centuries until, only in 2006, Hales [16] presented a formal proof in the affirmative. More recently, Viazovska [31] gave an optimal packing of equal spheres in the 8-dimensional space, which was extended to 24 dimensions

© The Author(s), under exclusive license to Springer Nature Switzerland AG 2023
J. Byrka and A. Wiese (Eds.): WAOA 2023, LNCS 14297, pp. 145–159, 2023.
https://doi.org/10.1007/978-3-031-49815-2_11

by Cohn et al. [12]. From the computational angle, it is known that several geometric packing problems are NP-hard [5,13,14,24,25]. Nevertheless, there are many heuristics and exact algorithms for the problem of maximizing the packing density [2,7,15,19,30], as well as for the problem of minimizing the size of the container [1,6,8,9,32]. We refer the reader to the survey of Hifi and M'Hallah [18].

In the context of approximation algorithms, however, the literature is not so vast and most of the results regard rectangular shapes and d-dimensional boxes. For packing rectangles into rectangular bins, the best-known result is an asymptotic 1.405-approximation due to Bansal and Khan [4], while in the d-dimensional context, there is an APTAS for the hypercube bin packing problem, given by Bansal et al. [3]. For the rectangle strip packing, Keynon and Rémila [22] gave an APTAS. We refer the reader to the works of Christensen et al. [10] and Coffman et al. [11] for an extensive review. Regarding the knapsack variant, Jansen et al. [20] gave a PTAS for the problem of packing hypercubes in a multidimensional knapsack. Merino and Wiese [27] studied the packing of convex polygons and gave a quasi-polynomial-time constant approximation algorithm, as well as a quasi-polynomial-time algorithm that computes a solution of optimal value under resource augmentation in all dimensions. For the hypersphere bin and strip packing problems, Miyazawa et al. [28] gave an APTAS under resource augmentation in only one dimension. Their technique extends to items and containers of varied convex forms, such as spheres under the L_p-norm, hypercubes, and other regular polytopes. Lintzmayer et al. [26] derived a PTAS under resource augmentation for the particular case of the circle knapsack problem where the profits of the circles are their respective area (note that, in this case, the objective becomes to maximize the packing density). For a review of techniques for circle and hypersphere packing, we recommend the survey due to Miyazawa and Wakabayashi [29].

In this paper, we present a technique that leads to approximation schemes for several packing problems, comprising objective functions that range from maximization of profits to minimization of occupied volume. We now describe the problems covered by our technique. In the *hypersphere knapsack* problem, we have a collection of hyperspheres associated with profits and a hyperrectangular knapsack, and the objective is to pack a subset of the hyperspheres maximizing the sum of the profits of the packed items. For this problem, we give an approximation scheme under resource augmentation in only one dimension, i.e., given a constant $\varepsilon > 0$, our algorithm finds a packing of profit at least $1 - \mathcal{O}(\varepsilon)$ of the optimal value in a knapsack whose height is increased by a factor of ε of the original height. With some increments, our technique can be applied to other versions of packing problems. Namely, we derive an approximation scheme under resource augmentation for the *hypersphere multiple knapsack* problem, in which we have m hyperrectangular knapsacks, rather than just one; we also achieve PTASs for the *hypersphere multiple strip packing* problem, in which all the hyperspheres must be packed in at most m strips with a hyperrectangular base and minimum height, and the *hypersphere multiple minimum-sized bin* problem, in which we want to pack all the hyperspheres into at most m hypercubes of minimum side length. We make use of an algebraic system to check the feasibility of a packing of a set of items into one bin. This allows generalizations to items and bins of other shapes, such as regular polygons with bounded number of sides, ellipses, among others, as well as for their correspondents in the d-dimensional case, for constant d.

In the following, we summarize the main ideas behind our technique. We first show that there is an almost optimal packing that respects some convenient structural properties. In short, the process is as follows: The spheres are partitioned into groups according to their radii, then a low-profit subset of these groups is discarded (by guessing) so that the remaining spheres can be organized in sets $S_0, S_1, S_2...$, where the spheres of S_j are much smaller than the spheres of S_{j-1}, for $j \geq 1$. Then, we strategically consider bins of appropriate size for each S_j to guarantee that bins used to pack S_j are small compared to the circles of S_{j-1}. Finally, we show how to modify an optimal solution to obey such properties, obtaining what we call a structured packing, while losing little profit and using resource augmentation. We then proceed to give an algorithm that obtains a good structured packing.

In a structured packing, we can look at the spheres as if they were in levels, each S_j defining one level. If we needed to pack all the spheres minimizing the occupied volume, such as in the bin packing problem, a good strategy would be to independently pack each S_j in their corresponding bins, filling each bin as much as possible, and then distribute bins of further levels (with smaller items) in the space left free in bins of previous levels (with bigger items). For the knapsack problem, however, such greedy approach is not sufficient. Since the available volume is limited by the capacity of the knapsack and we need to select a subset of items in respect to their profits, the best local choices for each level may not be the best choices for the overall packing. If we have small items that are much more profitable than some big items, for instance, it may be a better choice to pack fewer of the bigger items, leaving extra free space to pack more of the smaller but more profitable items. This aspect makes the knapsack problem more complex, forcing us to design a more sophisticated strategy. Our technique makes use of a combination of configuration-based integer programs. A first IP is used to obtain an optimal global solution that gives us two crucial bounds: the amount of volume used for each level, and the amount of volume left free to subsequent levels. Then, once fixed a level, a second IP is used to obtain an optimal solution that respect the bounds obtained previously. This guarantee that, despite packing levels independently, each one uses the same amount of volume as in a globally optimal solution, while the objective function of the IP certifies that the most profitable items are chosen. The need for this approach in two phases arises from the fact that the fractional solution given by the linear relaxation of the first IP may have many non-null variables concentrated in just one level. This becomes a problem due to the rounding strategy we adopt. We simply round the value of each fractional variable up to the next integer, which results in extra bins that need to be accommodated in the knapsack (thus the resource augmentation). If the non-null variables are concentrated in just one level, specially if in a level of big items, then the accommodation of the extra bins may require a too large of an increase in the size of the knapsack. To the best of our knowledge, our approach with integer linear programs in two phases is new in the context of the packing problems investigated in this work.

The text is organized as follows. In Sect. 2, we introduce preliminaries concepts and definitions. In Sect. 3, we present our main results. Finally, in Sect. 4 we offer some thoughts on the problems addressed in this work.

2   Preliminaries

Given an integer n, we write $[n] = \{1, \ldots, n\}$, and given an n-dimensional vector $x = (x_1, \ldots, x_n)$, we define the *ceil* of x as $\lceil x \rceil = (\lceil x_1 \rceil, \ldots, \lceil x_n \rceil)$. We assume that all objects lie in the Euclidean space. If p and q are two points in the plane, their Euclidean distance is denoted by $\mathrm{dist}(p, q)$. Given a set $\mathcal{S} = \{s_1, \ldots, s_n\}$ of n circles, we denote the radius and the diameter of each circle $s_i \in \mathcal{S}$ by r_i and d_i, respectively. For a rectangle B of rational width w and height h, we write $B_{w \times h}$ and we call $w \times h$ the *size* of B. When the context is clear, we may omit the size from the notation. If D is a circle or a rectangle, we denote its area by $\mathrm{Area}(D)$. If D is a set of circles or rectangles, then $\mathrm{Area}(D) = \sum_{A \in D} \mathrm{Area}(A)$.

A packing of a set of circles into bins consists in an attribution of the center position of each circle to rational coordinates such that no two circles overlap and each circle is entirely contained in some bin. Circle packing problems raise an intrinsic issue: It is not known if, for every instance of the problem, there always exists an optimal solution where the center of every circle is given by rational coordinates. For the circle bin packing problem (CBP), Miyazawa et al. [28] handle this issue with an algorithm that produces rational solutions, but in augmented bins. Briefly, the idea is to render a packing into a system of polynomial equations where the variables correspond to the center position of the circles. When a packing is possible, the center positions are given by roots of polynomials, which may be irrational numbers. Adjusting them to rational coordinates may cause overlaps with the borders of the bin or among circles. Their algorithm then applies a shifting strategy to rearrange the circles within the bin until there is no overlap, resulting in an increase in the height of the bin by a small constant. We state this result in the next lemma. In the following, we denote an instance of the CBP by a tuple $(\mathcal{I}, w, h)$, where $\mathcal{I}$ is the set of circles and w, h are the dimensions of the bins, and we denote the optimal value of the instance by $\mathrm{OPT}^{\mathrm{BP}}_{w \times h}(\mathcal{I})$.

Lemma 1 (Miyazawa et al. [28]). *Let $(\mathcal{I}, w, h)$ be an instance of the circle bin packing problem, where $w, h \in \mathcal{O}(1)$ and $|\mathcal{I}| = n$, and such that $\min_{1 \leq i \leq n} r_i \geq \delta$ and $|\{r_1, \ldots, r_n\}| \leq K$, for constants δ and K. Given a number $\gamma > 0$, there exists an algorithm that produces a packing of $\mathcal{I}$ into at most $\mathrm{OPT}^{\mathrm{BP}}_{w \times h}(\mathcal{I})$ bins of size $w \times (1 + \gamma)h$, in polynomial time on n.*

3 The Circle Knapsack Problem Under Resource Augmentation

Formally, an instance of the *circle multiple knapsack problem* (CMKP) is defined as a tuple $(\mathcal{I}, w, h, p, m)$ where $w, h \in \mathbb{Q}_+$ are the dimensions of the knapsacks, with $w \leq h$, $\mathcal{I} = \{s_1,, \dots, s_n\}$ is a set of n circles, each circle $s_i \in \mathcal{I}$ with diameter $d_i \in \mathbb{Q}_+$ and $d_i \leq w$, $p \colon \mathcal{I} \to \mathbb{Q}_+$ is a function of profit on the circles, and $m \in \mathbb{Z}_+$ is the number of available knapsacks. We denote the profit of a circle s_i as p_i. If A is a set of circles, we say its profit is $p(A) = \sum_{s_i \in A} p_i$. The objective of the CMKP is to find a packing of a subset $I \subseteq \mathcal{I}$ of circles in at most m knapsacks of size $w \times h$, maximizing $p(I)$. We denote the optimal value of CMKP for instance $(\mathcal{I}, w, h, p, m)$ by $\mathrm{OPT}^{\mathrm{MKP}}_{w \times h}(\mathcal{I}, m)$. The *circle knapsack problem* (CKP) is the particular case of CMKP where $m = 1$. We denote an instance of the CKP by the tuple $(\mathcal{I}, w, h, p)$ and its optimal value by $\mathrm{OPT}^{\mathrm{KP}}_{w \times h}(\mathcal{I})$.

In this section, we first describe an approximation scheme under resource augmentation for the CKP, i.e., given an instance $(\mathcal{I}, w, h, p)$ and a constant $\varepsilon > 0$, we give a polynomial-time algorithm that finds a packing of a subset $I \subseteq \mathcal{I}$ into a knapsack of size $w \times (1 + \mathcal{O}(\varepsilon))h$ such that $p(I) \geq (1 - \mathcal{O}(\varepsilon))\mathrm{OPT}^{\mathrm{KP}}_{w \times h}(\mathcal{I})$. Moreover, we first assume that w and h are bounded by constants, and later, we extend the result for the CMKP and for knapsacks of unconstrained size.

3.1 Transforming an Optimal Solution

Let $(\mathcal{I}, w, h, p)$ be an instance of the CKP and let $\varepsilon > 0$ be a constant. We define $r = 1/\varepsilon$ and, without loss of generality, we assume $\varepsilon \leq 1/3$ and that r and hr/w are integers. Let $I^* \subseteq \mathcal{I}$ be the set of circles of an optimal solution. We partition $\mathcal{I}$ into groups $G_i = \{s_j \in \mathcal{I} : \varepsilon^{2i}w \geq d_j > \varepsilon^{2(i+1)}w\}$, for $i \geq 0$. Then we partition these groups into sets $H_\ell = \{G_i : i \equiv \ell \pmod{r}\}$, for $0 \leq \ell < r$. For some $1 \leq t < r$, there must be a set H_t such that $p(H_t \cap I^*) \leq \frac{1}{(r-1)}p(I^*) \leq 2\varepsilon p(I^*)$. Now we remove the set H_t from the instance, causing only a small loss of the profit of an optimal solution, and then arrange the remaining groups into sets of groups such that there is a significant gap on the radii of circles of any two consecutive sets. For that purpose, we define sets $S_j = \bigcup_{i=t+(j-1)r+1}^{t+jr-1} G_i$, for $j \geq 0$. See Fig. 1 for an illustrative sketch. We denote $\mathcal{S}(\mathcal{I}) = \bigcup_{j \geq 0} S_j$ and say that $H_t, S_0, S_1, \dots$ is a *gap-structured partition* of $\mathcal{I}$. The minimum and maximum radii of S_j are denoted by $r^j_{\min}$ and $r^j_{\max}$, respectively. The strategy is to pack each S_j in bins of appropriate dimensions according to the size of the circles. We set $w_0 = w$, $h_0 = h$, representing the knapsack itself, and for $j \geq 1$, we set $w_j = h_j = \varepsilon^{2(t+(j-1)r)+1}w$. We say that the bins of size $w_j \times h_j$ for $j \geq 1$ *respect* $w \times h$. Additionally, we say a *grid* of *size* $w_j \times h_j$ over a bin B divides B into a set $\mathsf{G}_j(B)$ of square *cells* of size $w_j \times h_j$. To avoid verbosity, hereafter we refer to each j as *level j*. Many times throughout the text we refer to circles of S_j and bins of size $w_j \times h_j$ simply as circles and bins of level j.

					H_t
S_0:			G_0 ... G_{t-1}		G_t
S_1:	G_{t+1}	G_{t+2} ...	G_r ...	G_{t+r-1}	G_{t+r}
S_2:	G_{t+r+1}	G_{t+r+2} ...	G_{2r} ...	G_{t+2r-1}	G_{t+2r}
S_j:	$G_{t+(j-1)r+1}$	G_{t+jr+2} ...	G_{3r} ...	G_{t+jr-1}	G_{t+jr}

Fig. 1. Sketch to illustrate the partition of the original instance.

We highlight two important properties regarding the sets S_j: i) within the same level, circles are small compared to bins; and ii) between two consecutive levels j and $j+1$, $j \geq 0$, circles and bins of level $j+1$ are much smaller than circles and bins of level j. This indicates that after packing circles of a level only in bins of that same level, the area left unoccupied can accommodate a great number of circles (and bins) of the subsequent level. The idea is to recursively use grids to build a packing respecting a certain structure: For each level j, circles are packed in bins of their respective levels, over which it is drawn a grid of size $w_{j+1} \times h_{j+1}$; the cells of this grid are then used to pack circles of S_{j+1}. For the sake of clarity, from level 1 onward, we say subbins instead of just bins. In the following, we present a formal definition.

Definition 1. *Consider a set $\mathcal{I}$ of circles. We say that a packing of $S(\mathcal{I})$ in a bin $B_{w \times h}$ is a structured packing if the following holds:*

- *S_0 is packed in B;*
- *for every $j \geq 1$, S_j is packed in a subset $D_j \subseteq \mathsf{G}_j(B)$ of subbins of size $w_j \times h_j$; and*
- *for every subbin $D' \in D_j$, D' does not intersect any circle from S_ℓ, for $\ell < j$.*

Miyazawa et al. [28] showed that, given an optimal packing of the circle bin packing problem for instance $(\mathcal{I}, w, h)$, they can derive a structured packing of $S(\mathcal{I})$, using a small amount of extra area.

Lemma 2 (Miyazawa et al. [28]). *Let $(\mathcal{I}, w, h)$ be an instance of the circle bin packing problem and let $H, S_0, S_1, \ldots$ be a gap-structured partition of $\mathcal{I}$. There exists a structured packing of $\mathcal{I} \setminus H$ into a set of bins D that respect $w \times h$ such that $\mathrm{Area}(D) \leq (1 + 44\varepsilon)\mathrm{OPT}^{BP}_{w \times h}(\mathcal{I})wh$.*

Using this result, given an optimal packing for the CKP, it is possible to obtain a structured packing in an augmented knapsack with only a small loss of the total profit.

Corollary 1. *Let $(\mathcal{I}, w, h, p)$ be an instance of the CKP. There exists a structured packing of a subset $I \subseteq \mathcal{I}$ into a knapsack of size $w \times (1 + 44\varepsilon)h$ such that $p(I) \geq (1 - \varepsilon)\mathrm{OPT}_{w \times h}^{KP}(\mathcal{I})$.*

In the following subsection, we present an algorithm that gives an almost optimal structured packing into an augmented knapsack. We define $\widehat{h} = (1 + 44\varepsilon)h$ to acknowledge the increase in the knapsack.

3.2   A Structured Packing of the Original Instance

From Corollary 1, there is an almost optimal structured packing for the instance $(\mathcal{I}, w, h, p)$, if we allow some increase in the size of the knapsack. Thus, we can now focus on obtaining structured packings. We remove H_t from $\mathcal{I}$ and design an algorithm to find an optimal structured packing only of $\mathcal{S}(\mathcal{I})$. Thus, in this subsection, we are dealing with the instance $(\mathcal{S}(\mathcal{I}), w, \widehat{h}, p)$.

Now we need some more notation. For $j \geq 0$, let $\widehat{\mathcal{T}}_j = \{t_1, \ldots, t_{\widehat{T}_j}\}$ be the set of different radii among circles of S_j, where $\widehat{T}_j = |\widehat{\mathcal{T}}_j|$. Each set $\widehat{\mathcal{T}}_j$ is associated with a tuple $(\widehat{n}_j^1, \ldots, \widehat{n}_j^{\widehat{T}_j})$ of *demands*, where $\widehat{n}_j^k$ is the number of circles of radius t_j^k contained in S_j, for $k = 1, \ldots, \widehat{T}_j$. A *configuration* of S_j is a tuple $C = (c_1, \ldots, c_{\widehat{T}_j})$ where each c_k is the number of circles of radius t_j^k in C, for $k = 1, \ldots, \widehat{T}_j$. We define $|C| = \sum_{k=1}^{\widehat{T}_j} c_k$ and we say C has $|C|$ circles. The area of a configuration C, denoted by $\mathrm{Area}(C)$, is the sum of the area of every circle in C. We say a configuration C of S_j is *feasible* if its circles can be packed in exactly one bin of level j. We denote the set of all feasible configurations of S_j by $\widehat{\mathcal{C}}_j$. The next lemma states bounds on the number of circles that fit in a bin and the number of feasible configurations.

Lemma 3. *For any level $j \geq 0$ and configuration $C \in \widehat{\mathcal{C}}_j$, if $h/w \in \mathcal{O}(1)$ then $|C|$ is bounded by a constant and $|\widehat{\mathcal{C}}_j|$ is bounded by a polynomial in n.*

Hereafter, we refer to a feasible configuration simply as a configuration. We want to determine a subset of configurations (of all levels) that together lead to an optimal structured packing of $\mathcal{S}(\mathcal{I})$. For a configuration $C \in \widehat{\mathcal{C}}_j$, let $\widehat{f}_j(C)$ be the number of empty subbins of size $w_{j+1} \times \widehat{h}_{j+1}$ available for circles of level $j + 1$ onward. Consider the following decision variables:

- x_j^C: the number of times configuration $C \in \widehat{\mathcal{C}}_j$ is used in level j;
- b_j: the number of empty bins of size $w_j \times \widehat{h}_j$ available for circles of level j;
- z_i: binary variable that indicates if circle $s_i \in \mathcal{S}(\mathcal{I})$ is packed or not.

We present an integer program, named $\mathcal{F}_{\text{exact}}$, to find an optimal structured packing of $\mathcal{S}(\mathcal{I})$ into a knapsack of size $w \times h$.

$$(\mathcal{F}_{\text{exact}}) \quad \max \quad \sum_{s_i \in \mathcal{S}(\mathcal{I})} z_i p_i \tag{1a}$$

$$\text{s.t.} \quad \sum_{C \in \widehat{\mathcal{C}}_j} x_j^C c_k \leq \widehat{n}_j^k \qquad \forall j \geq 0, k \in [\widehat{T}_j], \tag{1b}$$

$$\sum_{s_i \in S_j : r_i = t_j^k} z_i = \sum_{C \in \widehat{\mathcal{C}}_j} x_j^C c_k \qquad \forall j \geq 0, k \in [\widehat{T}_j], \tag{1c}$$

$$\sum_{C \in \widehat{\mathcal{C}}_j} x_j^C = b_j \qquad \forall j \geq 0, \tag{1d}$$

$$b_j \leq \sum_{C \in \widehat{\mathcal{C}}_{j-1}} \widehat{f}_{j-1}(C) x_{j-1}^C \quad \forall j \geq 1, \tag{1e}$$

$$b_0 = 1, \tag{1f}$$

$$z_i \in \{0, 1\} \qquad \forall s_i \in \mathcal{S}(\mathcal{I}), \tag{1g}$$

$$x_j^C, b_j \in \mathbb{Z}_+ \qquad \forall j \geq 1, C \in \widehat{\mathcal{C}}_j. \tag{1h}$$

Constraints (1b) assure that the demand of each size is not surpassed. Constraints (1c) determine which circles are packed, based on the chosen configurations. Note that the objective function enforces that among circles of the same radius, the ones of highest profit are selected. Constraints (1d) define the number of bins used in each level, while constraints (1e) limit the number of empty bins available for the subsequent levels, based on the chosen configurations. Finally, constraint (1f) guarantees that only one knapsack is used and constraints (1g) and (1h) define the scope of the variables.

Note that the number of variables and constraints of $\mathcal{F}_{\text{exact}}$ is bounded by a polynomial in n, therefore it is possible to solve its linear relaxation in polynomial time. However, a fractional solution of $\mathcal{F}_{\text{exact}}$ may have too many fractional variables, which could prevent our rounding strategy to yield a solution that causes only a small increase in the knapsack. For this reason, we modify the instance and consider a similar integer program, as described in the following subsection.

3.3 A Structured Packing of a Modified Instance

We modify the original instance by rounding the radii of the circles so that we have a constant number of different radii in each level. For this purpose, let $R_j = \{r_{\min}^j (1 + \varepsilon)^k : k \geq 0, r_{\min}^j (1 + \varepsilon)^k < r_{\max}^j\} \cup \{r_{\max}^j\}$. The circles of S_0 remain the same. For $j \geq 1$, we round up the radius of the circles of S_j to the closest value in R_j. We denote the rounded radius of a circle s_i by $\bar{r}_i$. We define $\mathcal{T}_j = \{t_j^1, \ldots, t_j^{T_j}\}$ and $(n_j^1, \ldots, n_j^{T_j})$ for the rounded circles analogously as in

Sect. 3.2. The following lemma shows that the number of different radii in each level is now constant.

Lemma 4. *For any level $j \geq 1$, the number T_j of different rounded radii is at most $2r^2 \ln(r)$.*

Since the number of different radii is now constant, we have a better bound on the number of configurations from level 1 onward.

Lemma 5. *For any level $j \geq 1$, the number of different configurations of rounded circles of S_j is bounded by a constant.*

With these new bounds and Lemma 1, we can check the feasibility of a configuration and find its corresponding packing in constant time.

Lemma 6. *For a level j, given a configuration C of rounded circles of S_j, we can decide if C is feasible, and in the affirmative case, for any constant $\gamma > 0$, we obtain a packing of C in a bin of size $w_j \times (1 + \gamma)h_j$, in constant time.*

For $j \geq 1$, we use Lemma 6 to determine the sets $\mathcal{C}_j$ of all feasible configurations of rounded circles of S_j. Since the circles of S_0 are not modified, we set $\mathcal{C}_0 = \widehat{\mathcal{C}}_0$. To compensate the possible increase in the radius of the circles after the rounding, we use augmented bins of size $w'_j \times h'_j$, where $w'_j = (1+\varepsilon)w_j$ and $h'_j = (1+\varepsilon)(1+16\varepsilon)\widehat{h}_j$. We now use another IP, similar to $\mathcal{F}_{\text{exact}}$, to find an optimal structured packing of $\mathcal{S}(\mathcal{I})$ after the rounding. We make two adjustments for the new IP: We fix a configuration for level 0, and instead of computing $\widehat{f}_j(C)$, we estimate its value using the following result given by Miyazawa et al. [28].

Lemma 7 (Miyazawa et al. [28]). *Let $A \subseteq S_j$ be a set of circles packed in a bin $B_{w_j \times h_j}$ and $D \subseteq G_{j+1}(B)$ be the subset of grid cells of size $w_{j+1} \times h_{j+1}$ intersecting but not entirely contained in circles of A. Then $\text{Area}(D) \leq 16\varepsilon \text{Area}(A)$.*

Based on this lemma, we define $f_j(C) = \dfrac{w'_{j-1}h'_{j-1} - (1+16\varepsilon)\text{Area}(C)}{w'_j h'_j}$,

which is a lower bound on the number of empty subbins of size $w'_{j+1} \times h'_{j+1}$ after packing a configuration $C \in \mathcal{C}_j$ in a bin of size $w'_j \times h'_j$. Given a configuration $C_0 \in \mathcal{C}_0$, the IP $\mathcal{F}_{\text{rounded}}(C_0)$ finds an optimal structured packing of the rounded circles assuming that C_0 is used in level 0. Decision variables x, z and b have the same meaning as in $\mathcal{F}_{\text{exact}}$.

$(\mathcal{F}_{\text{rounded}}(C_0))$

$$\max \quad \sum_{s_i \in \mathcal{S}(\mathcal{I})} z_i p_i \tag{2a}$$

s.t.   (1b)–(1d) with regard to rounded circles,

$$b_j \leq \sum_{C \in \mathcal{C}_{j-1}} f_{j-1}(C) x_{j-1}^C \qquad \forall j \geq 1, \tag{2b}$$

$$x_0^{C_0} = 1, \ x_0^{C'} = 0 \qquad \forall C' \in \mathcal{C}_0 \setminus C_0, \tag{2c}$$

$$z_i \in \{0,1\}, x_j^C \in \mathbb{Z}_+, b_j \in \mathbb{Z}_+ \quad \forall s_i \in \mathcal{S}(\mathcal{I}), j \geq 1, C \in \mathcal{C}_j. \tag{2d}$$

Despite the increase of the circles and the error caused by the function $f_j(C)$, $\mathcal{F}_{\text{rounded}}$ still gives a good solution if we increase the size of the knapsack by a small factor. The next lemma states that if we use a knapsack of size $w' \times h'$, the optimum value of $\mathcal{F}_{\text{rounded}}$, when given an optimal configuration for level 0, is at least the optimum value given by $\mathcal{F}_{\text{exact}}$.

Lemma 8. *Let $(\mathcal{I}, w, h, p)$ be an instance of CKP and $C_0^* \in \mathcal{C}_0$ be the configuration of S_0 used in an optimal structured solution. Then $\text{OPT}(\mathcal{F}_{\text{rounded}}(C_0^*)) \geq \text{OPT}(\mathcal{F}_{\text{exact}})$ if $\mathcal{F}_{\text{rounded}}(C_0^*)$ considers an augmented knapsack of size $w' \times h'$.*

We use $\mathcal{F}_{\text{rounded}}$ to obtain an optimal fractional solution. However, we seek a solution where the number of non-null x variables in each level is bounded by a constant. We say that such a solution is *balanced*. This property is crucial to assure that the total increase in the height of the knapsack due to the rounding of the fractional solution is small. To obtain a balanced fractional solution, we define another integer program: $\mathcal{F}_{\text{level}}^j(A, B)$ finds an optimal solution for level j using A bins, and leaving B empty subbins available to level $j+1$. Again, the variables x, b and z have the same meaning as in $\mathcal{F}_{\text{exact}}$.

$(\mathcal{F}_{\text{level}}^j(A,B)) \quad \max \sum_{s_i \in S_j} z_i p_i \tag{3a}$

$$\text{s.t.} \quad \sum_{C \in \mathcal{C}_j} x_j^C c_k \leq n_j^k \qquad \forall k \in [T_j], \tag{3b}$$

$$\sum_{s_i \in S_j : \bar{r}_i = t_j^k} z_i = \sum_{C \in \mathcal{C}_j} x_j^C c_k \qquad \forall k \in [T_j], \tag{3c}$$

$$\sum_{C \in \mathcal{C}_j} x_j^C = A, \tag{3d}$$

$$B = \sum_{C \in \mathcal{C}_j} f_j(C) x_j^C, \tag{3e}$$

$$x_j^C \in \mathbb{Z}_+, z_i \in \{0,1\} \quad \forall C \in \mathcal{C}_j, s_i \in S_j. \tag{3f}$$

Given a feasible solution (x, b, z) to $\mathcal{F}_{\text{rounded}}$, we use $\mathcal{F}_{\text{level}}^j(A, B)$ independently for each level $j \geq 1$, with parameters $A = \sum_{C \in \mathcal{C}_j} x_j^C$ and $B = b_{j+1}$.

We call such procedure BALANCED-FRACTIONAL-SOLUTION. This way we can exploit the fact that, despite the number of levels being at most n, the number of constraints in each level is constant. Lemma 9 comes from the fact that $\mathcal{F}_{\text{level}}$ has a constant number of constraints.

Lemma 9. *Given an instance $(\mathcal{I}, w, h, p)$ of CKP, let $H, S_0, S_1, \ldots$ be a gap-structured partition of $\mathcal{I}$, and C_0 be a configuration of rounded circles of S_0. There is an optimal solution $(\tilde{x}, \tilde{b}, \tilde{z})$ to the linear relaxation of $\mathcal{F}_{\text{rounded}}(C_0)$ such that for each level $j \geq 1$, there are at most $2T_j + 2$ non-null variables $\tilde{x}_j$.*

Let (x^*, b^*, z^*) be an optimal balanced fractional solution of $\mathcal{F}_{\text{rounded}}(C_0)$ given by the BALANCED-FRACTIONAL-SOLUTION procedure. We round the variables x^* up to the next integer, yielding a collection of configurations represented by the vector $\lceil x^* \rceil$. The total extra area necessary to contemplate the extra bins created by the rounding is small.

Lemma 10. *Let (x^*, b^*, z^*) be an optimal balanced fractional solution of model $\mathcal{F}_{\text{rounded}}(C_0^*)$. The extra bins created after rounding the variables x^* to $\lceil x^* \rceil$ fit into a strip of size $w' \times \varepsilon h'$.*

Observe that a solution of the linear program $\mathcal{F}_{\text{rounded}}$ gives a set of configurations used in each level, where each configuration represents a bin. To build a packing, for each configuration, we obtain a packing in a bin of its respective level. Then we distribute these packings (bins) into the knapsack.

Lemma 11. *For each level j, let X_j be a collection (allowing duplication) of configurations of the rounded circles of S_j, considering bins of size $w'_j \times h'_j$. Given a constant $\gamma > 0$, there is an algorithm that finds a packing of maximum profit of the original circles, corresponding to the configurations of X_j, in bins of size $w'_j \times (1 + \gamma)h'_j$, in polynomial time.*

Finally, we give an algorithm that, for an instance $(\mathcal{I}, w, h, p)$ of CKP where $h/w \in \mathcal{O}(1)$, and a positive constant $\varepsilon \leq 1/3$, it produces an almost optimal solution under resource augmentation. See Algorithm 1.

Theorem 1. *Given an instance $(\mathcal{I}, w, h, p)$ of CKP with $h/w \in \mathcal{O}(1)$ and a constant $\varepsilon \leq 1/3$, Algorithm 1 obtains a packing of a subset $I \subseteq \mathcal{I}$ of circles in a knapsack of size $w \times (1 + 703\varepsilon)h$ such that $p(I) \geq (1 - 3\varepsilon)\text{OPT}_{w \times h}^{KP}(\mathcal{I})$, in polynomial time on the size of the instance.*

With few modifications, Algorithm 1 works for CMKP as well.

Lemma 12. *Let $(\mathcal{I}, w, h, p, m)$ be an instance of CMKP. If $h/w \in \mathcal{O}(1)$, then for any constant $\varepsilon > 0$ we can obtain, in polynomial time, a packing of $I \subseteq \mathcal{I}$ in up to m bins of size $w \times (1 + \mathcal{O}(\varepsilon))h$ such that $p(I) \geq (1 - \mathcal{O}(\varepsilon))\text{OPT}_{w \times h}^{MKP}(\mathcal{I}, m)$.*

Our algorithm runs in polynomial time only if $h/w \in \mathcal{O}(1)$. However, by splitting the knapsack into strips of bounded ratio and using Lemma 12, we show a PTAS for CKP even with unbounded ratio between w and h.

Algorithm 1: APPROXIMATION-SCHEME

Input: Instance $(\mathcal{I}, p, w, h)$ of CKP; and constant $\varepsilon \leq 1/3$.
Output: An almost optimal packing in an augmented knapsack.

1 Let $r = 1/\varepsilon$.
2 Define $G_i = \{s_j \in \mathcal{I} : \varepsilon^{2i}w \geq d_j > \varepsilon^{2(i+1)}w\}$, for $i \geq 0$.
3 Define $H_\ell = \{G_i : i \equiv \ell \pmod{r}\}$, for $0 \leq \ell < r$.
4 **for** *each t from 1 to $r-1$ such that $p(H_t) \leq 2\varepsilon p(\mathcal{I})$* **do**
5   Define $S_j = \bigcup_{i=t+(j-1)r+1}^{t+jr-1} G_i$, for every integer $j \geq 0$.
6   Define $w_0 = w$, $h_0 = h$, and $w_j = h_j = \varepsilon^{2(t+(j-1)r)+1}w$, for $j \geq 1$.
7   For $j \geq 1$, round up the radii of the circles of S_j.
8   **for** *each $C_0 \in \mathcal{C}_0$* **do**
9          $(x, b, z) \leftarrow$ BALANCED-FRACTIONAL-SOLUTION$(\mathcal{F}_{\text{rounded}}(C_0))$.
10   Let (x^*, b^*, z^*) be a balanced fractional solution of highest profit.
11   Given $\lceil x^* \rceil$, build a packing P_t into a knapsack of size $w \times (1 + \mathcal{O}(\varepsilon))h$.
12 **return** *packing P_t of maximum profit.*

Theorem 2. *Let $(\mathcal{I}, w, h, p)$ be an instance of CKP and $\varepsilon > 0$ be a constant. There is a polynomial-time algorithm that finds a packing of a subset $I \subseteq \mathcal{I}$ in a knapsack of size $w \times (1 + \mathcal{O}(\varepsilon))h$ such that $p(I) \geq (1 - \mathcal{O}(\varepsilon))\text{OPT}_{w \times h}^{KP}(\mathcal{I})$.*

Finally, by applying the ideas behind Theorem 2 and Lemma 12, we conclude that there is a PTAS also for CMKP with unbounded ratio between w and h.

Theorem 3. *Let $(\mathcal{I}, w, h, p, m)$ be an instance of CMKP and $\varepsilon > 0$ be a constant. There is a polynomial-time algorithm that finds a packing of a subset $I \subseteq \mathcal{I}$ in at most m knapsacks of size $w \times (1 + \mathcal{O}(\varepsilon))h$ such that $p(I) \geq (1 - \mathcal{O}(\varepsilon))\text{OPT}_{w \times h}^{MKP}(\mathcal{I}, m)$.*

4 Final Remarks

Although packing of circles and hyperspheres has a long research history, there are few approximation algorithms for such problems. One difficulty of packing spheres comes from the representation of optimal packings, since it is an open question if it is always possible to represent optimal packings with rational positions, even when the radii and bin size are given by rational numbers [28]. For this reason, the use of resource augmentation is justified. Another concern is to make a good use of the space of the bin. Ideally, we want to fill the bin completely. For hypercubes there are known algorithms that guarantee a good packing density of small items, such as NFDH [3,17]. That is not the case for hyperspheres, since the packing density becomes very small as the dimension increases. Kabatianskii and Levenshtein [21] presented an upper bound of $2^{-(0.5990...+o(1))d}$ on the packing density of congruent hyperspheres. This makes the packing of hyperspheres more challenging compared to the packing of hypercubes, for instance. To handle this difficulty, it was necessary to use a more particular and refined approach for classifying the items and partitioning the space.

In this work, we presented an approximation scheme under resource augmentation for the hypersphere multiple knapsack problem. To the best of our knowledge, this is the first approximation scheme for the problem. We point to the fact that although the resource augmentation factor in our algorithm is dependent on ε, this is not in detriment of the approximation ratio, i.e., both the approximation ratio and the resource augmentation can be arbitrarily small. In addition, our technique can be used to yield a PTAS for the multiple minimum-sized bin problem and the multiple strip packing problem. The strength of our technique lies on the ease of adapting integer programs. For instance, it can easily handle problems with demand on the items, such as in the cutting stock problem. Moreover, the flexibility on the shape of the items and bins allows us to extend our results to several variants of the same problems.

Acknowledgements. We thank the anonymous reviewers for the valuable comments and suggestions. This research was financially supported by CNPq (grants 161030/2021-1, 163645/2021-3, 313146/2022-5) and FAPESP (grant 2022/05803-3).

References

1. Akeb, H., Hifi, M., M'Hallah, R.: A beam search algorithm for the circular packing problem. Comput. Oper. Res. **36**(5), 1513–1528 (2009). https://doi.org/10.1016/j.cor.2008.02.003
2. Amore, P.: Circle packing in regular polygons. Phys. Fluids (2023). https://doi.org/10.1063/5.0140644
3. Bansal, N., Correa, J.R., Kenyon, C., Sviridenko, M.: Bin packing in multiple dimensions: inapproximability results and approximation schemes. Math. Oper. Res. **31**(1), 31–49 (2006). https://doi.org/10.1287/moor.1050.0168
4. Bansal, N., Khan, A.: Improved approximation algorithm for two-dimensional bin packing. In: Proceedings of the Twenty-Fifth Annual ACM-SIAM Symposium on Discrete Algorithms (SODA), SODA 2014, pp. 13–25. Society for Industrial and Applied Mathematics (2014). https://doi.org/10.1137/1.9781611973402.2
5. Berman, F., Leighton, F.T., Snyder, L.: Optimal tile salvage (1981)
6. Birgin, E.G., Bustamante, L.H., Callisaya, H.F., Martínez, J.M.: Packing circles within ellipses. Int. Trans. Oper. Res. **20**(3), 365–389 (2013). https://doi.org/10.1111/itor.12006
7. Birgin, E.G., Lobato, R.D.: A matheuristic approach with nonlinear subproblems for large-scale packing of ellipsoids. Eur. J. Oper. Res. **272**(2), 447–464 (2019). https://doi.org/10.1016/j.ejor.2018.07.006
8. Birgin, E.G., Lobato, R.D., Martínez, J.M.: Packing ellipsoids by nonlinear optimization. J. Glob. Optim. **65**(4), 709–743 (2015). https://doi.org/10.1007/s10898-015-0395-z
9. Birgin, E.G., Sobral, F.N.C.: Minimizing the object dimensions in circle and sphere packing problems. Comput. Oper. Res. **35**(7), 2357–2375 (2008). https://doi.org/10.1016/j.cor.2006.11.002

10. Christensen, H.I., Khan, A., Pokutta, S., Tetali, P.: Approximation and online algorithms for multidimensional bin packing: a survey. Comput. Sci. Rev. **24**, 63–79 (2017). https://doi.org/10.1016/j.cosrev.2016.12.001

11. Coffman, E.G., Csirik, J., Galambos, G., Martello, S., Vigo, D.: Bin packing approximation algorithms: survey and classification. In: Pardalos, P.M., Du, D.-Z., Graham, R.L. (eds.) Handbook of Combinatorial Optimization, pp. 455–531. Springer, New York (2013). https://doi.org/10.1007/978-1-4419-7997-1_35

12. Cohn, H., Kumar, A., Miller, S., Radchenko, D., Viazovska, M.: The sphere packing problem in dimension 24. Ann. Math. **185**(3) (2017). https://doi.org/10.4007/annals.2017.185.3.8

13. Demaine, E.D., Fekete, S.P., Lang, R.J.: Circle packing for origami design is hard. In: Origami[5]: Proceedings of the 5th International Conference on Origami in Science, Mathematics and Education (OSME 2010), pp. 609–626. A K Peters, Singapore (2010). https://doi.org/10.48550/arXiv.1008.1224

14. Fowler, R.J., Paterson, M.S., Tanimoto, S.L.: Optimal packing and covering in the plane are NP-complete. Inf. Process. Lett. **12**(3), 133–137 (1981). https://doi.org/10.1016/0020-0190(81)90111-3

15. Fu, L., Steinhardt, W., Zhao, H., Socolar, J.E.S., Charbonneau, P.: Hard sphere packings within cylinders. Soft Matter **12**(9), 2505–2514 (2016). https://doi.org/10.1039/c5sm02875b

16. Hales, T., Ferguson, S.: A formulation of the Kepler conjecture. Discret. Comput. Geom. **36**, 21–69 (2006). https://doi.org/10.1007/s00454-005-1211-1

17. Harren, R.: Approximation algorithms for orthogonal packing problems for hypercubes. Theor. Comput. Sci. **410**(44), 4504–4532 (2009). https://doi.org/10.1016/j.tcs.2009.07.030

18. Hifi, M., M'Hallah, R.: A literature review on circle and sphere packing problems: models and methodologies. Adv. Oper. Res. (2009). https://doi.org/10.1155/2009/150624

19. Hifi, M., Yousef, L.: A local search-based method for sphere packing problems. Eur. J. Oper. Res. **274**(2), 482–500 (2019). https://doi.org/10.1016/j.ejor.2018.10.016

20. Jansen, K., Khan, A., Lira, M., Sreenivas, K.V.N.: A PTAS for packing hypercubes into a knapsack. In: 49th International Colloquium on Automata, Languages, and Programming, ICALP 2022, Paris, France, 4–8 July 2022. LIPIcs, vol. 229, pp. 78:1–78:20. Schloss Dagstuhl - Leibniz-Zentrum für Informatik (2022). https://doi.org/10.4230/LIPIcs.ICALP.2022.78

21. Kabatyanskii, G.A., Levenshtein, V.I.: Bounds for packings on a sphere and in space (Russian). Problemy Peredači Informacii **14**, 3–25 (1978). English translation in Probl. Inf. Transm. **14**, 1–17 (1978)

22. Kenyon, C., Rémila, E.: A near-optimal solution to a two-dimensional cutting stock problem. Math. Oper. Res. **25**(4), 645–656 (2000). https://doi.org/10.1287/moor.25.4.645.12118

23. Kepler, J.: Strena seu de nive sexangula (the six-cornered snowflake) (1611)

24. Kim, H., Miltzow, T.: Packing segments in a simple polygon is APX-hard. In: European Conference on Computational Geometry (EuroCG 2015), pp. 24–27 (2015)

25. Leung, J.Y.T., Tam, T.W., Wong, C.S., Young, G.H., Chin, F.Y.L.: Packing squares into a square. J. Parallel Distrib. Comput. **10**(3), 271–275 (1990). https://doi.org/10.1016/0743-7315(90)90019-L

26. Lintzmayer, C.N., Miyazawa, F.K., Xavier, E.C.: Two-dimensional Knapsack for circles. In: Bender, M.A., Farach-Colton, M., Mosteiro, M.A. (eds.) LATIN 2018.

LNCS, vol. 10807, pp. 741–754. Springer, Cham (2018). https://doi.org/10.1007/978-3-319-77404-6_54

27. Merino, A., Wiese, A.: On the two-dimensional knapsack problem for convex polygons. In: Czumaj, A., Dawar, A., Merelli, E. (eds.) 47th International Colloquium on Automata, Languages, and Programming (ICALP 2020). Leibniz International Proceedings in Informatics (LIPIcs), vol. 168, pp. 84:1–84:16. Schloss Dagstuhl-Leibniz-Zentrum für Informatik, Dagstuhl, Germany (2020). https://doi.org/10.4230/LIPIcs.ICALP.2020.84

28. Miyazawa, F.K., Pedrosa, L.L.C., Schouery, R.C.S., Sviridenko, M., Wakabayashi, Y.: Polynomial-time approximation schemes for circle and other packing problems. Algorithmica **76**, 536–568 (2015). https://doi.org/10.1007/978-3-662-44777-2_59

29. Miyazawa, F.K., Wakabayashi, Y.: Techniques and results on approximation algorithms for packing circles. São Paulo J. Math. Sci. **16**(1), 585–615 (2022). https://doi.org/10.1007/s40863-022-00301-3

30. Romanova, T.E., Stetsyuk, P.I., Fischer, A., Yaskov, G.M.: Proportional packing of circles in a circular container. Cybern. Syst. Anal. **59**(1), 82–89 (2023). https://doi.org/10.1007/s10559-023-00544-8

31. Viazovska, M.: The sphere packing problem in dimension 8. Ann. Math. **185**(3) (2017). https://doi.org/10.4007/annals.2017.185.3.7

32. Zeng, Z., Yu, X., He, K., Huang, W., Fu, Z.: Iterated tabu search and variable neighborhood descent for packing unequal circles into a circular container. Eur. J. Oper. Res. **250**(2), 615–627 (2016). https://doi.org/10.1016/j.ejor.2015.09.001

Hitting Sets when the Shallow Cell Complexity is Small

Sander Aarts[✉][ID] and David B. Shmoys[ID]

Cornell University, Ithaca, NY 14580, USA
{sea78,david.shmoys}@cornell.edu

Abstract. The hitting set problem is a well-known NP-hard optimization problem in which, given a set of elements and a collection of subsets, the goal is to find the smallest selection of elements, such that each subset contains at least one element in the selection. Many geometric set systems enjoy improved approximation ratios, which have recently been shown to be tight with respect to the shallow cell complexity of the set system. The algorithms that exploit the cell complexity, however, tend to be involved and computationally intensive. This paper shows that a slightly improved asymptotic approximation ratio for the hitting set problem can be attained using a much simpler algorithm: solve the linear programming relaxation, take one initial random sample from the set of elements with probabilities proportional to the LP-solution, and, while there is an unhit set, take an additional sample from it proportional to the LP-solution. Our algorithm is a simple generalization of the elegant net-finder algorithm by Nabil Mustafa. To analyze this algorithm for the hitting set problem, we generalize the classic Packing Lemma, and the more recent Shallow Packing Lemma, to the setting of weighted epsilon-nets.

Keywords: Hitting set · Set cover · Approximation algorithms · Computational geometry · Shallow cell complexity · Wireless coverage

1 Introduction

The input to the hitting set problem is a finite *set system* – a ground set X of m elements, or *points*, and a collection $\mathcal{R}$ of n subsets, or *ranges*, of X. This can also be understood as a hypergraph, with vertices X and hyper-edges $\mathcal{R}$. A *hitting set* is a subset of elements $H \subseteq X$ such that every set $R \in \mathcal{R}$ is hit by H, i.e. $R \cap H \neq \emptyset$, for all $R \in \mathcal{R}$. This is a vertex cover under the hypergraph view. The set system can be encoded as a set-element incidence matrix $A \in \{0,1\}^{n \times m}$, in which the (i,j)th entry a_{ij} is 1 if range R_i contains point x_j, and 0 otherwise.

This material is based on work supported by the NSF under Grant CNS-1952063.

© The Author(s), under exclusive license to Springer Nature Switzerland AG 2023
J. Byrka and A. Wiese (Eds.): WAOA 2023, LNCS 14297, pp. 160–174, 2023.
https://doi.org/10.1007/978-3-031-49815-2_12

The IP of the minimum hitting set problem is

$$\min_{y} \sum_{j:x_j \in X} y_j$$

$$\text{s.t.} \sum_{j:x_j \in X} a_{ij}y_j \geq 1, \qquad \forall i : R_i \in \mathcal{R}; \qquad (1)$$

$$y_j \in \{0,1\}, \qquad \forall j : x_j \in X,$$

where variable $y_j \in \{0,1\}$ indicates whether element x_j is in the solution H.

Hitting sets and set covers are intimately connected; a hitting set for A is a set cover of A^T. Both problems' decision versions are NP-complete [9]. There exists an $\mathcal{O}(\log m)$-approximation algorithm, and this bound is tight unless P = NP [8,12]. However, there are algorithms that exploit additional structure in A to attain improved approximation ratios[1]. Indeed, our work is motivated by the problem of exploiting structure when covering large numbers of wireless LoRaWAN transmitters with wireless receivers. Transmitters can be viewed as points, which are considered to be covered if they are in the line of sight of a wireless receiver, which in turn drives transmission quality in LoRaWAN [20]. The area in the line of sight of a receiver roughly resembles a simple shape.

Many geometric set systems enjoy better approximation ratios via *epsilon-nets*, or ϵ-nets. A set system is said to be *geometric* whenever its elements can be encoded as points in Euclidean space, and sets are derived from containment of the points in geometric shapes, such as half-spaces, balls or rectangles[2]. The seminal work of Brönnimann and Goodrich [3], and Even *et al.* [7], connects the approximability of a hitting set instance to the size of weighted ϵ-nets. Given non-negative weights on the points, $\mu : X \to \mathbb{R}_{\geq 0}$, a *weighted ϵ-net* with respect to weights μ is a subset $H \subseteq X$ that hits all ϵ-heavy sets:

$$\forall R \in \mathcal{R} \text{ with } \mu(R) \geq \epsilon \cdot \mu(X) : \quad R \cap H \neq \emptyset, \qquad (2)$$

where the weight of any subset $S \subseteq X$ is defined as $\mu(S) = \sum_{x \in S} \mu(x)$. Even *et al.* [7] reduce the problem of finding a small hitting set to finding a small ϵ-net via a reformulation of the linear programming relaxation of the hitting set problem (1). The reformulated LP (3) is a program for finding the largest ϵ, and corresponding weights μ, subject to the constraint that an ϵ-net with respect to

---

[1] For example when A has bounded row or column sums [2,5].

[2] Some definition allow for uncountably many geometric shapes in $\mathcal{R}$, e.g. all squares. However, because the number of points X is finite, there are nevertheless a finite number of unique sets induced by these shapes.

weights μ is a hitting set.

$$\max_{\epsilon,\mu} \epsilon$$

$$\text{s.t.} \sum_{j:x_j \in X} a_{ij}\mu_j \geq \epsilon, \qquad \forall i : R_i \in \mathcal{R};$$

$$\sum_{j:x_j \in X} \mu_j = 1; \qquad (3)$$

$$\mu_j \geq 0, \qquad \forall j : x_j \in X.$$

The first constraint requires that each set R is ϵ-heavy; the second constraint normalizes the weights. Let (ϵ^*, μ^*) denote an optimal solution to LP (3), with $\mu^* = (\mu_1^*, \ldots, \mu_n^*)$. Let z^* be the optimal value to the LP relaxation of the original program (1). The first constraint ensures that an ϵ^*-net with respect to weights μ^* is a hitting set. Moreover, the reciprocal optimal value $1/\epsilon^*$ is equal to the optimal LP value z^* [7]. In particular, an ϵ^*-net of size $g(1/\epsilon^*)$ for some function $g(\cdot)$ is a hitting set of size of $g(z^*)$. Hence, to find a small hitting set it suffices to solve LP (3) and find a small ϵ^*-net with respect to weights μ^*.

Haussler and Welzl [11] show that set systems with bounded VC-dimension admit small ϵ-nets, and develop a simple algorithm to find them. The VC-dimension is a measure of the set system's complexity. Given a subset $S \subseteq X$, the *projection* of $\mathcal{R}$ to S is the set system formed by elements S and sets $\mathcal{R}|_S = \{R \cap S : R \in \mathcal{R}\}$. The VC-dimension of $\mathcal{R}$ is the size of the largest subset $S \subseteq X$ such that $\mathcal{R}|_S$ shatters S, i.e. the largest set S such that $\mathcal{R}|_S$ contains all subsets of S. In particular, Clarkson [6], and Haussler and Welzl [11], show that any set system with VC-dimension d has a weighted ϵ-net of size $\mathcal{O}\left(\frac{d}{\epsilon} \log \frac{1}{\epsilon}\right)$. This is remarkable, as the size is independent of both the size of X and $\mathcal{R}$. Moreover, the algorithm for finding such an ϵ-net is simple: Select a subset $H \subseteq X$ by sampling each element x in X independently.

Theorem 1 (ϵ-net Theorem [11,13]). *Let $(X, \mathcal{R})$ be a set system with VC-dimension d, and let $\mu : X \to \mathbb{R}_{\geq 0}$ be element weights with $\mu(X) = 1$. Then for any $\epsilon, \gamma \in (0,1)$:*

$$H \leftarrow \text{ pick each } x \in X \text{ with probability } \min\left\{1, \frac{2\mu(x)}{\epsilon} \cdot \max\left\{\log \frac{1}{\gamma}, d \log \frac{1}{\epsilon}\right\}\right\}$$

is a weighted ϵ-net with respect to weights μ with probability at least $1 - \gamma$.

Throughout, we define $\mu(S) = \sum_{x \in S} \mu(x)$ for all subsets $S \subseteq X$. For general set systems of VC-dimension d, this bound is tight in expectation [13]. However, there are alternative ways to parameterize the complexity of set systems.

1.1   Shallow Cell Complexity

The *shallow cell complexity* (SCC) is a finer parameterization of the complexity of set systems. [1,4,19]. Readers are referred to Mustafa and Varadarajan [18] for more background. A *cell* in a binary matrix A is a collection of identical rows. A cell has *depth* k if the number of 1's in any of its rows is exactly k, i.e., if each set in the cell contains k elements. For a non-decreasing function $\varphi(\cdot, \cdot)$ we say binary matrix A has *shallow cell complexity* (SCC) $\varphi(\cdot, \cdot)$ if, for all $1 \le k \le l \le m$, the number of cells of depth at most k in any submatrix A^* of A of at most l columns, is at most $\varphi(l, k)$. A set system $(X, \mathcal{R})$ is said to have SCC $\varphi(l, k)$ if its set-element incidence matrix A does. Often $\varphi(l, k) = \mathcal{O}(\varphi(l) k^c)$ for some constant $c > 0$ and single-variable function $\varphi(\cdot)$, in which case the dependence on k is can be dropped and the SCC denoted by $\varphi(l)$. Examples of geometric set systems with small shallow cell complexity are discs in the plane with $\varphi(l, k) = \mathcal{O}(k)$, and axis-parallel rectangles with $\varphi(l, k) = \mathcal{O}(lk^2)$ [16].

As is true for VC-dimension, there are algorithms that find hitting sets or ϵ-nets with sizes bounded in terms of the shallow cell complexity. A prominent example is the quasi-uniform sampling algorithm of Chan *et al.* [4]. Given non-negative weights $\mu : X \to \mathbb{R}_{\ge 0}$, and a value $\epsilon > 0$, the algorithm finds a hitting set while maintaining an upper bound on the probability of selecting any given element.

Theorem 2 (Quasi-uniform sampling [4]). *Suppose a set system defined by A has SCC $\varphi(l, k) = \varphi(l) k^c$ for some $c > 0$. Then there is a randomized poly-time algorithm that returns a hitting set of expected size $\mathcal{O}(\max\{1, \log(\varphi(m))\})$ times the LP optimum.*

The algorithm attains the optimal approximation ratio with respect to the SCC[3]. However, the sampling procedure is involved, and may require enumeration over all sets $\mathcal{R}$, of which there can be $n = \Omega(m^c)$ for some constant $c > 0$ [15].

Taking a different approach, Mustafa and colleagues [14,15,17] develop a net-finder for asymptotically optimal-sized *unweighted* ϵ-nets with respect to the SCC. The algorithm is remarkably simple: Take an initial sample from X, and while there are unhit sets, choose an unhit set arbitrarily, and add $\mathcal{O}(1)$ randomly chosen elements from this set to the original sample. The algorithm assumes access to an oracle that returns an unhit set. This oracle is called at most $\mathcal{O}(1/\epsilon)$ times in expectation. While the size of the returned ϵ-net is asymptotically on par with the quasi-uniform sampling algorithm, there are large constants in the upper bound [15].

This algorithm is not directly applicable to the hitting set problem via the LP-reduction above, although it can be used via a standard reduction. The analysis of the algorithm applies to only uniform weights, and the optimal weights μ^* of the LP-formulation (3) are not generally uniform. Nevertheless, it is possible to reduce the problem of finding a weighted ϵ-net to that of finding a uniform ϵ'-net following a standard reduction, in which an expanded instance is generated by copying each element $x_j \in X$ a number of times roughly proportional to its weight $\mu^*(x_j)$ [3,4]. This can generate $\Omega(m)$ copies of each element, which can

---

[3] In addition, it is worth noting that this algorithm can solve the more general *weighted* hitting set problem, in which each element has a given weight, and the goal is to find the minimum weight hitting set.

have notable consequences. First, to achieve a weighted ϵ^*-net in the original instance, one must use a smaller value ϵ' for the expanded instance, on the order of $\mathcal{O}(\epsilon^*/m)$. This results in an approximation ratio of $\mathcal{O}(\log\varphi(\mathcal{O}(m)))$. Secondly, generating copies can increase the number of elements from m to $\Omega(m^2)$. This can increase the runtime considerably. In particular, repeatedly sampling from sets of size $\Theta(m^2)$ can become prohibitive on large instances such as the wireless coverage problem motivating our work.

1.2   Our Contributions

This paper generalizes the elegant net-finder algorithm of Mustafa [15] to the setting of weighted ϵ-nets, in order to produce a fast and simple algorithm for the hitting set problem, which attains asymptotically optimal approximation ratios with respect to the shallow cell complexity. The algorithm enjoys a faster runtime that makes solving larger instances, such as LoRaWAN receiver placement at scale, feasible. This is achieved by combining the weighted ϵ-net finder with the reduction of Even et al. [7]. In doing so, we also improve on the asymptotic approximation ratio from $\max\{1, \log\varphi(m)\}$ to $\max\{1, \mathcal{O}(\log\varphi(\mathcal{O}(z^*)))\}$ where $z*$ is the optimal value to the linear relaxation of the hitting set program (1). While in the worst case $z^* = m$, it is often the case that $z^* \ll m$. However, the multiplicative constants in our analysis are relatively large, matching those of Mustafa [15]. In addition to the algorithm, our analysis generalizes the classic Packing Lemma of Haussler [10], as well as the Shallow Packing Lemma of Mustafa et al.. [17], to the weighted setting, which may be of independent interest.

Key to our approach are adaptations of Mustafa's [17] Shallow Packing Lemma and Haussler's [10] classic Packing Lemma that accommodate non-uniform weights. Our main technical contribution is to allow a notion of *weighted packings*. Consider any non-negative weights $\mu : X \to \mathbb{R}_{\geq 0}$ with $\sum_{x \in X} \mu(X) = 1$, and extend it to element subsets via $\mu(S) = \sum_{x \in S} \mu(S).^4$ A (k,δ)-*packing with respect to weights* μ is a collection of sets $\mathcal{P} \subseteq \mathcal{R}$ in which (i) all sets R in $\mathcal{P}$ are at most k-*heavy*, i.e., have bounded weight $\mu(R) \leq k$; and (ii) all pairs of sets have symmetric differences of weight at least δ. (See Definition 1). Our weighted shallow packing lemma upper bounds the number of sets in $\mathcal{P}$ as a function of the SCC. Our approach accommodates weights μ by sampling elements from a distribution with probability mass proportional to the weights, rather than from a uniform distribution as in the original proofs. Moreover, our proof uses sampling *with replacement* rather than *without replacement* to simplify the analysis. While more generally applicable, our result yields the same bound on the size of $\mathcal{P}$ as in the unweighted setting. An analogous sampling approach is used in proving Theorem 1 [13]. Equipped with our generalized lemma, it is straightforward to adapt Mustafa's [15] analysis to a weighted net-finder. A proof of our Weighted Packing Lemma is included in the extended online version.

---

[4] Any non-negative weights $w : X \to \mathbb{R}_{\geq 0}$ with $w(X) > 0$ can be normalized as $\mu(x) = w(x)/w(X)$.

2 Algorithm and Main Result

Our algorithm combines the LP-relaxation of Even *et al.* [7] with the generalized sampling approach of Mustafa [15]. Our procedure is summarized in Algorithm 1. The algorithm makes use of two global constants, β and γ. These are assumed to be positive, and to satisfy $\gamma \leq 1/4$ and $\beta + \gamma \leq 1$.

Algorithm 1: A simple hitting set algorithm with details

Data: A matrix A with VC-dim$(A) \leq d$ and SCC $\varphi(\cdot, \cdot)$, constants $\gamma, \beta > 0$
$\epsilon^*, (\mu_1^*, \ldots, \mu_n^*) \leftarrow$ solve LP $\{\max \epsilon : A\mu \geq \epsilon, \mu^T \mathbf{1} = 1, \mu \geq \mathbf{0}\}$;
$H \leftarrow \emptyset$;
for $x_j \in X$ **do**
  $\quad H \leftarrow H \cup \{x_j\}$ with probability

$$\min \left\{ 1, \frac{2\mu_j^*}{\left(\frac{3}{4} - \frac{\beta}{2}\right)\epsilon^*} \cdot \max \left\{ \log \left(d^2 \varphi \left(\frac{8d}{\beta\epsilon^*}, \frac{48d}{\beta} \right)^2 \right), d \log \left(\frac{1}{\left(\frac{3}{4} - \frac{\beta}{2}\right)\epsilon^*} \right) \right\} \right\}$$

end
while *there is a set* $R \in \mathcal{R}$ *not hit by* H **do**
  $\quad$ Independently add each $x_j \in R$ to H with probability

$$\min \left\{ 1, \frac{2\mu_j^*}{\gamma \mu^*(R)} \max\{\log 2, d \log \tfrac{1}{\gamma}\} \right\}$$

end
return H

In the while loop, the weights $\mu^*(R) = \sum_{j:x_j \in R} \mu_j^*$ denote the weight of set R under the LP optimal weights $\mu^* = (\mu_1^*, \ldots, \mu_n^*)$.

Conceptually, the algorithm is simple; it randomly selects an initial set of elements H from X, and proceeds to add additional random subsets of elements to H until this is a hitting set. The algorithm relies on an oracle that returns an arbitrary unhit set. This oracle is treated as a black box. Our main result is twofold: we bound the expected size of the solution hitting set H as a function of the cell complexity, and bound the expected number of oracle calls.

Theorem 3. *Let A be a binary matrix encoding a hitting set instance with shallow cell complexity $\varphi(\cdot, \cdot)$ and VC-dim$(\mathcal{R}) \leq d$. Let z^* be LP optimal value. Then the algorithm returns a hitting set of expected size*

$$\mathcal{O}\left(z^* \cdot \max\left\{1, \log \varphi\left(\mathcal{O}(z^*), \mathcal{O}(d)\right)\right\}\right).$$

Furthermore it makes at most $\mathcal{O}(z^*)$ oracle calls in expectation.

Note that the algorithm always returns a hitting set; the randomness is in the size of the solution and the runtime. This is in contrast with the net-finder in Theorem 1. Both algorithms require knowing the VC dimension d; ours must additionally know the shallow cell complexity $\varphi(\cdot, \cdot)$. If unknown, these can be searched for using a standard doubling trick [15].

3   The Weighted Shallow Packing Lemma

The Weighted Shallow Packing Lemma is key to proving Theorem 3. This section formally defines weighted shallow packings, states the lemma, and proves it. To this end, fix non-negative weights μ over X, and define the weight of a subset of elements $S \subseteq X$ as $\mu(S) = \sum_{j \in S} \mu_j$. Assume that $\mu(X) = 1$. To contrast, let card(S) denote the cardinality of any set S. Note that the weights μ induce a probability distribution over the elements X. Throughout, whenever an element u of X is randomly sampled, it is assumed to follow a distribution proportional to $\mu(\cdot)$, in which case we say u is sampled from $\mu(\cdot)$, and denote this by $u \sim \mu(\cdot)$. Note that an element $u \sim \mu(\cdot)$ sampled this way lies in subset $S \subseteq X$ with probability $\mu(S)$.

The main purpose of the weighted shallow packing lemma is to bound the number of sets in a set system in terms of its shallow cell complexity. Clearly, an arbitrary set systems can contain large numbers of sets. Instead, we focus on a particular kind of set system called a *weighted packing*. A set system is a packing if all its sets are "light", and each pair of sets are sufficiently different from each other. Critically, we define "light" and "different" in reference to the weights.

Definition 1. *Let $(X, \mathcal{P})$ be a set system with weights μ, and let $k, \delta \in (0, 1)$ be constants. If all sets S in $\mathcal{P}$ satisfy $\mu(S) \leq k$, and all pairs of distinct sets S, R in $\mathcal{P}$ have symmetric difference of weight at least δ, i.e.*

$$\mu(\Delta(S, R)) = \mu((S \backslash R) \cup (R \backslash S)) \geq \delta, \tag{4}$$

then we say $(X, \mathcal{P})$ is a weighted (k, δ)-packing with respect to μ.

We omit the "with respect to μ"-statement whenever this is clear from context.

The shallow packing lemma bounds the number of sets in a packing as a function of the constants (k, δ), the VC-dimension, and the shallow cell complexity.

Lemma 1 (Weighted shallow packing lemma). *Let $(X, \mathcal{P})$ be a set system on m elements, equipped with weights μ, and let $(X, \mathcal{P})$ be a (k, δ)-packing with respect to μ for constants $k, \delta > 0$. Assume the set system has VC-dim$(\mathcal{P}) \leq d$, and shallow cell complexity $\varphi(\cdot, \cdot)$. Then*

$$card(\mathcal{P}) \leq \frac{24d}{\delta} \cdot \varphi\left(\frac{8d}{\delta}, \frac{48dk}{\delta}\right).$$

The proof to this lemma makes use of our weighted Packing Lemma. The unweighted Packing Lemma is a classic result by Haussler [10] that bounds the number of sets in a packing. We generalize this to nonuniform weights.

Lemma 2 (Weighted packing lemma). *Let $(X, \mathcal{P})$ be a set system with n sets and m elements, equipped with weights μ. Let VC-dim$(\mathcal{P}) \leq d$ for some integer $d \geq 1$ and assume there is a constant $\delta \in (0, 1)$ such that $\mu(\Delta(S_i, S_k)) \geq \delta$ for all $1 \leq i < k \leq n$. Then*

$$card(\mathcal{P}) \leq 2\mathbb{E}\left[card(\mathcal{P}|_Y)\right],$$

where Y is the set of unique elements in a random sample $U = (u_1, u_2, \ldots, u_s)$ of size $s = \lceil \frac{8d}{\delta} \rceil - 1$, in which each element u_k is sampled iid $u_k \sim \mu(\cdot)$ with replacement.

The proof of the latter lemma is in the appendix to the extended online version of the paper; Lemma 1 is proved next.

3.1   Proof of the Weighted Shallow Packing Lemma

Proof. Fix a (k, δ)-packing $\mathcal{P}$ and let $U = (u_1, u_2, \ldots, u_s)$ be a random sample of length s, in which each element is sampled $u_k \sim \mu(\cdot)$, $k = 1, \ldots, s$, independently and with replacement. The number of elements sampled is $s = \lceil \frac{8d}{\delta} \rceil - 1$. Let $Y \subseteq X$ be the set of unique elements in U. For every set $R \in \mathcal{P}$, let $M(R, U) := \sum_{i=k}^{s} \mathbb{1}[u_k \in R]$ denote number of (copies of) elements in U that are in R. Define $\mathcal{P}_L \subseteq \mathcal{P}$ as the sub-collection of "large" sets in packing $\mathcal{P}$ that contain at least $6 \left(\frac{8dk}{\delta} \right)$ (copies of) elements in the random sample U:

$$\mathcal{P}_L = \left\{ R \in \mathcal{P} : M(R, U) \geq 6 \cdot \frac{8dk}{\delta} \right\}.$$

It follows that the probability of a given range R in $\mathcal{P}$ being a member of $\mathcal{P}_L$ is

$$\mathbb{P}[R \in \mathcal{P}_L] = \mathbb{P}\left[M(R, U) \geq 6 \cdot \frac{8dk}{\delta} \right].$$

Our goal is to show that the collection of large sets $\mathcal{P}_L$ has few members in expectation. To do so, it suffices to bound the probability that a fixed set R is a member of $\mathcal{P}_L$. This is achieved using Markov's inequality. Recalling that all sets $R \in \mathcal{P}$ have bounded weight $\mu(R) \leq k$ gives

$$\mathbb{E}[M(R, U)] = \sum_{k=1}^{s} \mathbb{P}[u_k \in R] = \sum_{k=1}^{s} \mu(R) \leq s \cdot k \leq \frac{8dk}{\delta},$$

where we used the fact that we sample from $\mu(\cdot)$, which implies that $\mathbb{P}[u_k \in R] = \mu(R)$. Now, Markov's inequality bounds the probability of R being in $\mathcal{P}_L$:

$$\mathbb{P}[R \in \mathcal{P}_L] = \mathbb{P}\left[M(R, U) \geq 6 \cdot \frac{8dk}{\delta} \right]$$
$$\leq \mathbb{P}\left[M(R, U) \geq 6 \cdot \mathbb{E}[M(R, U)] \right] \leq 1/6.$$

Finally, because $\mathcal{P}_L \subseteq \mathcal{P}$, we conclude that

$$\mathbb{E}[\mathrm{card}(\mathcal{P}|_Y)] \leq \mathbb{E}[\mathrm{card}(\mathcal{P}_L)] + \mathbb{E}[\mathrm{card}((\mathcal{P} \backslash \mathcal{P}_L)|_Y)]$$
$$\leq \sum_{R \in \mathcal{P}} \mathbb{P}[R \in \mathcal{P}_L] + \mathrm{card}(Y) \cdot \varphi \left(\mathrm{card}(Y), 6 \cdot \frac{8dk}{\delta} \right)$$
$$\leq \tfrac{1}{6} \mathrm{card}(\mathcal{P}) + \tfrac{8d}{\delta} \cdot \varphi \left(\tfrac{8d}{\delta}, \tfrac{48dk}{\delta} \right),$$

where the second-to-last inequality uses the shallow cell complexity of $\mathcal{P}$; the system $(Y, (\mathcal{P} \backslash \mathcal{P}_L)|_Y)$ has at most card$(Y) \leq s$ elements, and sets have depth at most $\left(6 \cdot \frac{8dk}{\delta}\right)$, as the system consists only of cells that are not "large". The final inequality holds because $\mathbb{P}[R \in \mathcal{P}_L] \leq 1/6$. Finally, applying Lemma 2 completes the proof.   □

4   Proof of the Main Theorem

Equipped with the Weighted Shallow Packing Lemma, we follow a similar strategy as Mustafa [15]. We state and prove three key lemmas, and finally prove Theorem 3.

4.1   Key Lemmas

The proof of our main theorem relies on all sets having similar weight. Let ϵ and $\mu = (\mu_1, \ldots, \mu_n)$ be a feasible solution to the LP relaxation (3). By the constraints of the LP, each set $R \in \mathcal{R}$ has weight $\mu(R) = \sum_{j:x_j \in R} \mu_j \geq \epsilon$. Partition the collection of sets $\mathcal{R}$ into groups $\ell = 0, 1, \ldots, \lceil \log \epsilon \rceil$ of sets of similar weight; set R belongs to group ℓ if and only if $2^{-\ell-1}\epsilon \leq \mu(R) < 2^{-\ell}\epsilon$. Because the algorithm exclusively takes independent samples, we can view one run of the algorithm as multiple parallel, independent runs on each group of sets. All our bounds scale on the order $\mathcal{O}\left(1/(2^{-\ell}\epsilon)\right)$, so summing over the groups gives a final bound on the order of $\mathcal{O}\left(1/\epsilon\right)$. Hence, we assume henceforth that all sets $R \in \mathcal{R}$ have weight $\epsilon \leq \mu(R) \leq 2\epsilon$.

The key idea of the proof is to amortize the elements added from each processed unhit set throughout the run of the algorithm. We say a set is *processed* each time it is flagged as unhit by the oracle, and a sample is taken from it. We bound the total number of elements sampled using weighted (k, δ)-packings on two levels. The first-level packing is an arbitrary *maximal* packing $\mathcal{P}$ of sets in $\mathcal{R}$. There are a bounded number of sets in $\mathcal{P}$. Next, each processed set R_i is assigned to a set in the first-level packing $\mathcal{P}$. For a fixed set P^j in the first-level packing, *given that it has been assigned processed sets*, we show that the collection of sets R_i assigned to P^j forms a second-level packing. Each second-level packing also has a bounded number of sets. Finally, by bounding the probability that a set in the first-level packing has any sets assigned to it, the total expected number of times the algorithm processes a set is bounded. Note that the assignments are only a tool for analysis; they need not be computed by the algorithm.

We begin by defining the first-level packing. Fix a *maximal* $(2\epsilon, \beta\epsilon)$-packing $\mathcal{P} = \{P^1, \ldots, P^p\}$, where p denotes the number of sets in the packing. The Shallow Packing Lemma 1 upper bounds the number of sets in the packing by

$$p \leq \frac{24d}{\beta\epsilon} \cdot \varphi\left(\frac{8d}{\beta\epsilon}, \frac{96d}{\beta}\right). \tag{5}$$

Now, suppose the algorithm runs for T steps, processing sets $(R_1, \ldots, R_T)$ in sequence. One given set may be processed multiple times. Denote the sets of sampled elements $H_{R_1}, \ldots, H_{R_T}$. The processed sets R_i are assigned to sets P^j in the first-level packing $\mathcal{P}$ as follows. Arbitrarily assign each set R_i to any index $j \in \{1, 2, \ldots, p\}$ satisfying $\mu\left(\Delta(R_i, P^j)\right) < \beta\epsilon$. Such an index j exists because $\mathcal{P}$ is a maximal $(2\epsilon, \beta\epsilon)$-packing. It may be the case that $R_i = P^j$. The next task is to bound the number of sets R_i assigned to any set P^j in the first-level packing.

Let n_j denote the number of processed sets in $(R_1, \ldots, R_T)$ assigned to $P^j \in \mathcal{P}$. For now, condition on first-level packing set P^j having at least one set assigned to it, i.e. $n_j \geq 1$. We study the probability of this event later. Relabel the sets and consider them in the order in which they were processed by the algorithm,

$$\mathcal{S}^j = (R_1^j, \ldots, R_{n_j}^j).$$

Claim. For all $j \in \{1, 2, \ldots, p\}$, $i \in \{1, 2, \ldots, n_j\}$ we have

$$\mu\left(P^j \cap R_i^j\right) > \frac{\mu\left(P^j\right) + \mu\left(R_i^j\right) - \beta\epsilon}{2}. \tag{6}$$

Proof. Fix $j \in \{1, 2, \ldots, p\}$. For all $i \in \{1, 2, \ldots, n_j\}$ we have

$$\mu\left(P^j\right) + \mu\left(R_i^j\right) = \mu\left(P^j \backslash R_i^j\right) + \mu\left(R_i^j \backslash P^j\right) + 2\mu\left(P^j \cap R_i^j\right)$$
$$= \mu\left(\Delta(P^j, R_i^j)\right) + 2\mu\left(P^j \cap R_i^j\right) < \beta\epsilon + 2\mu\left(P^j \cap R_i^j\right).$$

The first equality follows from straightforward accounting, and the second from the definition of symmetric difference. The inequality follows from the manner in which set R_i^j is matched to the packing set P^j. Finally, a simple rearrangement of terms yields the result. □

This proves that the intersection of each set R_i^j with its corresponding first-level packing set P^j is heavy. This lets us define a second-level packing using the intersections $R_i^j \cap P^j$.

Rather than directly bounding the number of processed sets assigned to a first-level packing set, it is easier to first bound the length of a random subsequence of the assigned sets $\mathcal{S}^j$. For any $j \in \{1, \ldots, p\}$, define the subsequence $\mathcal{S}'^j$ as the subsequence of processed sets R in $\mathcal{S}^j$ whose corresponding samples H_R form γ-nets for the system $(R, \mathcal{R}|_R)$:

$$\mathcal{S}'^j = \left(R \in \mathcal{S}^j : H_R \text{ is a } \gamma\text{-net for } (R, \mathcal{R}|_R)\right).$$

We proceed by bounding the length of the subsequence $\mathcal{S}'^j$, and by choosing γ so as to make it likely for a set R in $\mathcal{S}^j$ to be in $\mathcal{S}'^j$, using the ϵ-net Theorem 1. We use this to upper bound the expected number of sets in $\mathcal{S}^j$. Let $\text{len}(\mathcal{S})$ denote the length of a sequence $\mathcal{S}$.

The following claim bounds the length of the subsequence above.

Claim. For any $j \in \{1, 2, \ldots, p\}$:

$$\text{len}(\mathcal{S}'^j) \leq \begin{cases} \frac{24d}{3/2 - \beta - \gamma} \cdot \varphi\left(\frac{8d}{3/2 - \beta - \gamma}, \frac{48d}{3/2 - \beta - \gamma}\right), & \text{if } \beta + \gamma \geq 1/2; \\ \mathcal{O}(1), & \text{otherwise.} \end{cases} \tag{7}$$

Proof. Let $n'_j = \text{len}(\mathcal{S}'^j)$ and relabel the sets so that $\mathcal{S}'^j = \left(R_1^j, \ldots, R_{n'_j}^j \right)$. Now consider an auxiliary sequence of sets based on intersecting the entries R_i^j in $\mathcal{S}'^j$ with P^j:

$$T'^j = \left(S_1^j, \ldots, S_{n'_j}^j \right) \quad \text{with} \quad S_i^j = R_i^j \cap P^j \text{ for each } i \in \{1, \ldots, n'_i\}.$$

This sequence of sets is used to generate a second-level packing. To do this, consider two distinct set-indices $1 \le k < l \le n'_j$. The points $H_{R_k^j}$ are added before set R_l^j is considered, so $H_{R_k^j}$ is a γ-net for $\left(R_k^j, \mathcal{R}|_{R_k^j} \right)$, whereas the set R_l^j – because it is subsequently considered by the algorithm – is not hit by this net. This implies that the intersection of R_k^j and R_l^j is of bounded weight, as it would be hit by the γ-net otherwise:

$$\mu \left(R_k^j \cap R_l^j \right) < \gamma \cdot \mu \left(R_k^j \right).$$

This implies that the weight of the intersection of S_k^j and S_l^j is bounded above:

$$\mu \left(S_k^j \cap S_l^j \right) = \mu \left(R_k^j \cap R_l^j \cap P^j \right) \le \mu \left(R_k^j \cap R_l^j \right) < \gamma \cdot \mu \left(R_k^j \right). \tag{8}$$

The fact that sets in T'^j have pairwise intersections of small weight implies that their symmetric differences are heavy:

$$\begin{aligned}
\mu \left(\Delta(S_k^j, S_l^j) \right) &= \mu \left(S_k^j \right) + \mu \left(S_l^j \right) - 2\mu \left(S_k^j \cap S_l^j \right) \\
&= \mu \left(R_k^j \cap P^j \right) + \mu \left(R_l^j \cap P^j \right) - 2 \cdot \mu \left(S_k^j \cap S_l^j \right) \\
&> \frac{\mu \left(P^j \right) + \mu \left(R_k^j \right) - \beta\epsilon}{2} + \frac{\mu \left(P^j \right) + \mu \left(R_l^j \right) - \beta\epsilon}{2} - 2 \cdot \mu \left(S_k^j \cap S_l^j \right) \\
&> \frac{\mu \left(P^j \right) + \mu \left(R_k^j \right) - \beta\epsilon}{2} + \frac{\mu \left(P^j \right) + \mu \left(R_l^j \right) - \beta\epsilon}{2} - 2\gamma \cdot \mu \left(R_k^j \right) \\
&= \mu \left(P^j \right) - \beta\epsilon + \frac{1}{2}\mu \left(R_l^j \right) + (1/2 - 2\gamma) \mu \left(R_k^j \right) \\
&\ge (3/2 - \beta - \gamma) \cdot \mu \left(P^j \right),
\end{aligned}$$

where the first inequality uses Eq. (6), the second Eq. (8), and the last exploits the fact that sets R_k^j, R_l^j, and P^j are each of measure at least ϵ and at most 2ϵ, and that $\gamma \le 1/4$. Thus, depending on the constants, the sequence T'^j may form a weighted packing.

Finally, reviewing two cases for the constants β and γ makes the above more precise. First, if $\beta + \gamma < 1/2$, the inequality above implies that the symmetric difference of S_k^j and S_l^j is strictly larger than $\mu \left(P^j \right)$. This cannot be the case as both sets are subsets of P^j. Thus, the only sequence $\mathcal{S}'^j$ for which $\beta + \gamma$ can be less than a half is if there are no two unique indices, implying that $\text{len}(\mathcal{S}'^j) \le 1$. Secondly, if $\beta + \gamma \ge 1/2$, the sets in T'^j form a $\left(\mu \left(P^j \right), (3/2 - \beta - \gamma)\mu \left(P^j \right) \right)$-packing over P^j; all sets have measure at most $\mu \left(P^j \right)$, and every symmetric difference is at least $(3/2 - \beta - \gamma)\mu\left(P^j\right)$. This is our second-level packing. Now, the Shallow Packing Lemma 1 implies:

$$\text{len}\left(\mathcal{S}'^j\right) = \text{len}\left(\mathcal{T}'^j\right) \leq \frac{24d}{3/2 - \beta - \gamma} \cdot \varphi\left(\frac{8d}{3/2 - \beta - \gamma}, \frac{48d}{3/2 - \beta - \gamma}\right),$$

where we have used the fact that $\varphi\left(\cdot, \cdot\right)$ is non-decreasing and that $\mu(P^j) \leq 1$.
□

We can now bound the length of the full sequence of sets assigned to the packing set P^j. Taking expectations sidesteps any dependencies in the sequences. For instance, a set R can only be in $\mathcal{S}^j$ if previous samples failed to hit it. However, for each fixed set $R \in \mathcal{R}$, the probability of the sampled points H_S forming a γ-net for $(R, R|_{\mathcal{R}})$ is independent of previous sampling. Indeed, by Theorem 1, the probability that H_R is a γ-net is at least $1 - \gamma \geq 1/2$.

Lemma 3 (Mustafa, Lemma 5 [15])

$$\mathbb{E}\left[\text{len}(\mathcal{S}^j)\,\big|\,n_j \geq 1\right] \leq \frac{48}{3/2 - \beta - \gamma} \cdot \varphi\left(\frac{8d}{3/2 - \beta - \gamma}, \frac{48d}{3/2 - \beta - \gamma}\right)$$

Proof. We use a simple application of linearity of expectation, and Theorem 1:

$$\mathbb{E}[\text{len}\left(\mathcal{S}'^j\right)\,\big|\,n_j \geq 1] = \sum_{R \in \mathcal{S}^j} \mathbb{P}[H_R \text{ is a } \gamma\text{-net for } (R, \mathcal{R}|_R)] \geq \tfrac{1}{2} \cdot \text{len}(\mathcal{S}^j),$$

where we drop the conditioning on $n_j \geq 1$ because the event that a particular sample H_R is a γ-net is independent of the number of previous samples. On the other hand, Eq. (7) upper bounds the size of $\text{len}(\mathcal{S}'^j)$. Piecing these together yields the inequality:

$$\tfrac{1}{2}\text{len}(\mathcal{S}^j) \leq \mathbb{E}\left[\text{len}(\mathcal{S}'^j)\,\big|\,n_j \geq 1\right] \leq \text{len}(\mathcal{S}'^j) \leq \frac{24d}{3/2 - \beta - \gamma}\varphi\left(\frac{8d}{3/2 - \beta - \gamma}, \frac{48d}{3/2 - \beta - \gamma}\right).$$

□

Thus far we have conditioned on a set in the first-level packing being assigned at least one processed set. We now bound the probability of this being the case. Later, this probability is used to compute the expected number of processed sets assigned to a first-level packing set.

Lemma 4. *Let H_0 be the initial sample taken by the algorithm. Then for any $j \in \{1, \ldots, p\}$:*

$$\mathbb{P}[n_j \geq 1] = \mathcal{O}\left(\frac{1}{d^2\varphi\left(\frac{8d}{\beta\epsilon}, \frac{48d}{\beta}\right)^2}\right).$$

Proof. Fix an index $j \in \{1, \ldots, p\}$. Suppose that $n_j \geq 1$. By Eq. (6), for any $i \in \{1, \ldots, n_j\}$:

$$\mu\left(P^j \cap R_i^j\right) > \frac{\mu\left(P^j\right) + \mu\left(R_i^j\right) - \beta\epsilon}{2}$$

$$\geq \frac{\mu\left(P^j\right) + \mu\left(P^j\right)/2 - \beta\mu\left(P^j\right)}{2} = \left(\frac{3}{4} - \frac{\beta}{2}\right) \cdot \mu\left(P^j\right)$$

The second inequality above follows from the assumption that all sets have weights within a factor 2 of each other. The above implies that, if H_0 is a $\left(\frac{3}{4} - \frac{\beta}{2}\right)$-net for $\left(P^j, \mathcal{R}|_{P^j}\right)$, then any $R \in \mathcal{S}^j$ would be hit by H_0. In other words, $n_j \geq 1$ only if H_0 is not a $\left(\frac{3}{4} - \frac{\beta}{2}\right)$-net for $\left(P^j, \mathcal{R}|_{P^j}\right)$:

$$\mathbb{P}[n_j \geq 1] \leq \mathbb{P}\left[H_0 \text{ is not a } \left(\frac{3}{4} - \frac{\beta}{2}\right)\text{-net for } \left(P^j, \mathcal{R}|_{P^j}\right)\right].$$

Because $\frac{\mu_j}{\epsilon} \geq \frac{\mu_j}{\mu(R)}$, the initial sample includes each element with sufficient probability to apply Theorem 1 to the RHS above, completing the proof.   □

4.2   Proof of Theorem 3

Proof of Theorem 3. At this stage, the analysis closely follows Mustafa's [15]. Clearly, the algorithm proceeds until H is an ϵ^*-net with respect to measure μ^*, i.e., a hitting set. It suffices to bound the expected size of the hitting set H, as well as the expected number of oracle calls. These quantities are related, since the number of points added depends on the number of times a set is processed.

First, consider the expected size of the hitting set. There are two contributions to the set: the initial sample H_0, and the samples from the processed sets $H_{R_1}, \ldots, H_{R_T}$. We bound the expected size of the initial sample first.

Claim. The expected size of the initial sample, $\mathbb{E}[\text{card}(H_0)]$ is bounded by

$$\mathcal{O}\left(\frac{1}{\left(\frac{3}{2} - \frac{\beta}{2}\right)\epsilon} \max\left\{\log\left(d\varphi\left(\frac{8d}{\beta\epsilon}, \frac{48d}{\beta}\right)\right), \frac{d}{\left(\frac{3}{2} - \frac{\beta}{2}\right)} \log \frac{1}{\left(\frac{3}{2} - \frac{\beta}{2}\right)\epsilon}\right\}\right). \quad (9)$$

This follows by summing the probability of sampling x for each $x \in X$. An analogous result is used for the number of points added during the processing of a set $R \in \mathcal{R}$, provided it is processed:

Claim. For any fixed set $R \in \mathcal{R}$, conditional on being processed, the expected number of points added each time it is processed is

$$\mathbb{E}\left[\text{card}(H_R)\right] \leq 2\left(\frac{\log 2}{\gamma} + \frac{d}{\gamma} \log \frac{1}{\gamma}\right) = \mathcal{O}(1). \quad (10)$$

This bound applies irrespective of whether or not a set was processed previously.

The number of points added during processing, and the number of oracle calls, can be bounded together. Recalling that $R_1, \ldots, R_T$ are the processed sets, and using the claim above, the number of added elements is at most

$$\mathbb{E}\left[\sum_{i=1}^{T} \mathrm{card}(H_{R_i})\right] \leq \mathbb{E}\left[\sum_{i=1}^{T} 2\left(\frac{\log 2}{\gamma} + \frac{d}{\gamma}\log\frac{1}{\gamma}\right)\right] = \mathbb{E}[T]\cdot 2\left(\frac{\log 2}{\gamma} + \frac{d}{\gamma}\log\frac{1}{\gamma}\right). \tag{11}$$

Thus, it suffices to bound the expected number of oracle calls $\mathbb{E}[T]$. This is where we employ both the first-, and second-level packings. In particular

$$\mathbb{E}[T] = \mathbb{E}\left[\underbrace{\sum_{j=1}^{p} \mathrm{len}(\mathcal{S}^j)}_{(i)}\right] = \sum_{j=1}^{p} \cdot \underbrace{\mathbb{E}[\mathrm{len}(\mathcal{S}^j)\,|\,n_j \geq 1]}_{(ii)} \cdot \underbrace{\mathbb{P}[n_j \geq 1]}_{(iii)}. \tag{12}$$

The terms (i), (ii) and (iii) are bounded using Eq. (5), Lemma 3, and Lemma 4, respectively. In addition, using $\frac{3}{2} - \beta - \gamma \geq \frac{1}{2} \geq \max\{\beta\epsilon, \beta/2\}$:

(i) $p \leq \frac{24d}{\beta\epsilon}\varphi\left(\frac{8d}{\beta\epsilon}, \frac{24d}{\beta}\right)$

(ii) $\mathbb{E}\left[\mathrm{len}(\mathcal{S}^j)\,|\,n_j \geq 1\right] \leq \frac{48d}{3/2-\beta-\gamma}\varphi\left(\frac{8d}{3/2-\beta-\gamma}, \frac{24d}{3/2-\beta-\gamma}\right) \leq \frac{48d}{3/2-\beta-\gamma}\varphi\left(\frac{8d}{\beta\epsilon}, \frac{24d}{\beta}\right)$;

(iii) $\mathbb{P}[n_j \geq 1] \leq \left(d^2\varphi\left(\frac{8d}{\beta\epsilon}, \frac{24d}{\beta}\right)^2\right)^{-1}$.

Combining the right-hand-side terms, we obtain the bound

$$\mathbb{E}[T] \leq \frac{24 \cdot 48}{\beta \cdot (3/2 - \beta - \gamma)}\frac{1}{\epsilon} = \mathcal{O}\left(\frac{1}{\epsilon}\right). \tag{13}$$

This is minimized by choosing a small γ, e.g. $\gamma = 1/100$, and setting $\beta = 3/4$.

Finally, summing over the ℓ groups of sets, and adding the expected number of initial samples to the expected number added points completes the proof. Note that Eq. (13) also bounds the expected number of oracle calls made during the run of the algorithm.   □

References

1. Aronov, B., Ezra, E., Sharir, M.: Small-size ε-nets for axis-parallel rectangles and boxes. In: Proceedings of the Forty-First Annual ACM Symposium on Theory of Computing, pp. 639–648 (2009)
2. Bar-Yehuda, R., Even, S.: A linear-time approximation algorithm for the weighted vertex cover problem. J. Algorithms **2**(2), 198–203 (1981)
3. Brönnimann, H., Goodrich, M.T.: Almost optimal set covers in finite VC-dimension. Discret. Comput. Geom. **14**, 263–279 (1995)

4. Chan, T.M., Grant, E., Könemann, J., Sharpe, M.: Weighted capacitated, priority, and geometric set cover via improved quasi-uniform sampling. In: Proceedings of the Twenty-Third Annual ACM-SIAM Symposium on Discrete Algorithms, SODA 2012, pp. 1576–1585. Society for Industrial and Applied Mathematics (2012)

5. Chvatal, V.: A greedy heuristic for the set-covering problem. Math. Oper. Res. **4**(3), 233–235 (1979)

6. Clarkson, K.L.: A randomized algorithm for closest-point queries. SIAM J. Comput. **17**(4), 830–847 (1988)

7. Even, G., Rawitz, D., Shahar, S.M.: Hitting sets when the VC-dimension is small. Inf. Process. Lett. **95**(2), 358–362 (2005)

8. Feige, U.: A threshold of ln n for approximating set cover. J. ACM (JACM) **45**(4), 634–652 (1998)

9. Garey, M.R., Johnson, D.S.: Computers and Intractability, vol. 174. Freeman San Francisco (1979)

10. Haussler, D.: Sphere packing numbers for subsets of the Boolean n-cube with bounded Vapnik-Chervonenkis dimension. J. Comb. Theory Ser. A **69**(2), 217–232 (1995)

11. Haussler, D., Welzl, E.: Epsilon-nets and simplex range queries. In: Proceedings of the Second Annual Symposium on Computational Geometry, pp. 61–71 (1986)

12. Johnson, D.S.: Approximation algorithms for combinatorial problems. In: Proceedings of the Fifth Annual ACM Symposium on Theory of Computing, pp. 38–49 (1973)

13. Komlós, J., Pach, J., Woeginger, G.J.: Almost tight bounds for epsilon-nets. Discret. Comput. Geom. **7**, 163–173 (1992)

14. Mustafa, N.H.: A simple proof of the shallow packing lemma. Discret. Comput. Geom. **55**(3), 739–743 (2016)

15. Mustafa, N.H.: Computing optimal epsilon-nets is as easy as finding an unhit set. In: Baier, C., Chatzigiannakis, I., Flocchini, P., Leonardi, S. (eds.) 46th International Colloquium on Automata, Languages, and Programming, ICALP 2019, Patras, Greece, 9–12 July 2019. LIPIcs, vol. 132, pp. 87:1–87:12. Schloss Dagstuhl - Leibniz-Zentrum für Informatik (2019)

16. Mustafa, N.H.: Sampling in Combinatorial and Geometric Set Systems, vol. 265. American Mathematical Society (2022)

17. Mustafa, N.H., Dutta, K., Ghosh, A.: A simple proof of optimal epsilon nets. Combinatorica **38**(5), 1269–1277 (2018)

18. Mustafa, N.H., Varadarajan, K.: Epsilon-approximations & epsilon-nets. In: Handbook of Discrete and Computational Geometry, pp. 1241–1267. Chapman and Hall/CRC (2017)

19. Varadarajan, K.: Epsilon nets and union complexity. In: Proceedings of the Twenty-Fifth Annual Symposium on Computational Geometry, pp. 11–16 (2009)

20. Yousuf, A.M., Rochester, E.M., Ghaderi, M.: A low-cost LoRaWAN testbed for IoT: implementation and measurements. In: 2018 IEEE 4th World Forum on Internet of Things (WF-IoT), pp. 361–366 (2018)

Any-Order Online Interval Selection

Allan Borodin and Christodoulos Karavasilis$^{(\boxtimes)}$

University of Toronto, Toronto, Canada
{bor,ckar}@cs.toronto.edu

Abstract. We consider the problem of online interval scheduling on a single machine, where intervals arrive online in an order chosen by an adversary, and the algorithm must output a set of non-conflicting intervals. Traditionally in scheduling theory, it is assumed that intervals arrive in order of increasing start times. We drop that assumption and allow for intervals to arrive in any possible order. We call this variant *any-order interval selection* (AOIS). We assume that some online acceptances can be revoked, but a feasible solution must always be maintained. For unweighted intervals and deterministic algorithms, this problem is unbounded. Under the assumption that there are at most k different interval lengths, we give a simple algorithm that achieves a competitive ratio of $2k$ and show that it is optimal amongst deterministic algorithms, and a restricted class of randomized algorithms we call *memoryless*, contributing to an open question by Adler and Azar [1]; namely whether a randomized algorithm without memory or with only "bounded" access to history can achieve a constant competitive ratio. We connect our model to the problem of *call control* on the line, and show how the algorithms of Garay et al. [22] can be applied to our setting, resulting in an optimal algorithm for the case of proportional weights. We also discuss the case of intervals with arbitrary weights, and show how to convert the single-length algorithm of Fung et al. [20] into a *classify and randomly select* algorithm that achieves a competitive ratio of $2k$. Finally, we consider the case of intervals arriving in a *random order*, and show that for single-lengthed instances, a *one-directional* algorithm (i.e. replacing intervals in one direction), is the only deterministic memoryless algorithm that can possibly achieve a strict competitive ratio less than 2.

Keywords: interval selection · scheduling · online algorithms · call control

1 Introduction

We consider the problem of scheduling intervals online with revoking[1]. Intervals arrive with a fixed start time and fixed end time, and have to be taken right away, or be discarded upon arrival, while no intervals in the solution conflict.

---

[1] Displacing one or more previously scheduled intervals with a conflicting new interval.

© The Author(s), under exclusive license to Springer Nature Switzerland AG 2023
J. Byrka and A. Wiese (Eds.): WAOA 2023, LNCS 14297, pp. 175–189, 2023.
https://doi.org/10.1007/978-3-031-49815-2_13

The algorithm has to decide which intervals to include in the final schedule, so as to optimize some objective.

In the unweighted case, the goal is to maximize the number of intervals in the final solution. In the weighted case, we want an interval-set of maximum weight. Following previous work, we allow some revoking of online decisions, which is often considered even in the conventional start-time-ordered scheduling model. More precisely, if a newly arrived interval conflicts with other intervals already taken by the algorithm, we are able to take the new interval and discard the conflicting intervals. We are able to displace multiple existing intervals at once, although this won't be needed in the unweighted case. To avoid confusion, we should note that *preemption*[2] is often used in the interval selection literature to mean precisely this revoking of previous decisions we just described. Under this definition, preemption is allowed in our model. When we discard an interval it is final and it cannot be taken again.

We focus mainly on the unweighted case, where all intervals have the same weight. We discuss the competitive ratio of the problem in terms of k, the number of distinct interval lengths. However our algorithm does not need a priori knowledge of k. We show that a simple, deterministic, "memoryless" algorithm that only replaces when the new interval is entirely subsumed by an existing one, achieves the optimal competitive ratio in terms of the parameter k. We also show that "memoryless" randomized algorithms can not do any better. The main difference between our model and most of the interval selection literature, is allowing intervals to arrive in any order, a strict generalization of the ordered case. Bachmann et al. [5] have studied the any-order input model in the context of "t-intervals" (we are concerned with $t = 1$). They consider randomized algorithms, and don't allow revoking. In that model, they get a lower bound of $\Omega(N)$, with N being the number of intervals in a given input instance. The next most closely related problem is that of call admission [21] on the line graph, with online intervals corresponding to paths of a given line graph. The connection between call control on the line graph and interval selection has been noted before, but has not been carefully defined. We wish to clarify this connection by explaining the similarities as well as the differences, and how results correspond. We note that the parameter $k \leq N$ (respectively, $k \leq n - 1$) is an obvious refinement of the number of intervals (respectively, the number of vertices for call admission on a line graph with n vertices).

The applications of interval selection problems are plentiful. Some examples are resource allocation, network routing, transportation, and computer wiring. We refer the reader to the surveys by Kolen et al. [27], and Kovalyov et al. [30] for an overview of results and applications in the area of interval scheduling.

Related Work. Lipton and Tomkins [31] introduced the online interval scheduling problem. In our terminology, they consider the arrival of intervals with increasing start times (ordered), and interval weights that are proportional to

---

[2] In contrast to revoking, preemption in much of the scheduling literature means the pausing of a scheduled job, and resuming it later.

the lengths. They don't allow displacement of existing intervals, and give a randomized algorithm with competitive ratio $O((log\Delta)^{1+\epsilon})$, where Δ is the ratio of the longest to shortest interval.

In the unweighted case with increasing starting times, Faigle and Nawijn [17] give an optimal 1-competitive algorithm that is allowed to revoke previous decisions (replace intervals). In the weighted case with increasing starting times, Woeginger [37] shows that for general weights, no deterministic algorithm can achieve a constant competitive ratio. Canetti and Irani [11] extend this and show that even randomized algorithms with revocable decisions cannot achieve a constant ratio for the general weighted case. For special classes of weight functions based on the length (including proportional weights), Woeginger [37] gives an optimal deterministic algorithm with competitive ratio 4. Seiden [34] gives a randomized $(2+\sqrt{3})$-competitive algorithm when the weight of an interval is given by a continuous convex function of the length. Epstein and Levin [16] give a 2.45-competitive randomized algorithm for weights given by functions of the length that are monotonically decreasing, and they also give an improved $1 + \ln(2) \approx 1.693$ upper bound for the weight functions studied by Woeginger [37]. Fung et al. [20] provide the best known upper bounds, giving *barely random* algorithms that achieve a competitive ratio of 2 for all the Woeginger weight functions. These algorithms randomly choose one of two deterministic algorithms at the beginning. More generally, barely random algorithms have access to a small number of deterministic algorithms, and randomly choose one.

Restricting interval lengths has previously been considered in the literature, e.g. Lipton and Tomkins [31] study the case of two possible lengths, and Bachmann et al. [5] consider single and two-length instances. For the related offline problem of throughput maximization, Hyatt-Denesik et al. [25] consider c distinct processing times. The special case of single-length jobs has been studied in the job scheduling [6,13,35], sum coloring [9], and the interval selection literature [19,32]. Woeginger [37] also points out how his results can be extended to the case of equal lengths and arbitrary weights. Miyazawa and Erlebach [32] point out the equivalence between fixed length (w.l.o.g. unit) instances, and proper interval instances, i.e. instances where no interval is contained within another. This is because of a result by Bogart and West [8], showing the equivalence of the corresponding interval graphs in the offline setting.

There has also been some work on multiple identical machines. For the case of equal-length, arbitrary-weight intervals, Fung et al. [19] give an algorithm that is 2-competitive when m, the number of machines, is even, and $(2 + \frac{2}{2m-1})$ when m is odd. Yu & Jacobson [38] consider C-benevolent (weight function is convex increasing) jobs and get an algorithm that is 2-competitive when m is even, and $(2 + \frac{2}{m})$-competitive when m is odd.

In the problem of call control, a graph is given, and requests that correspond to pairs of nodes of the graph arrive online. The goal is to accept as many requests as possible, with the final set consisting of disjoint paths. When the underlying graph is a line, this problem is closely related to ours. For call control on the line, Garay et al. [22] give optimal deterministic algorithms. In the unweighted case, they achieve a $O(\log(n))$ competitive ratio, where n is the number of the vertices of the graph. In the case of proportional weights (weight is equal to the length of the path), they give an optimal algorithm that is $(\sqrt{5}+2) \approx 4.23$-competitive (its optimality was shown by Furst and Tomkins [36]). Adler and Azar [1] use randomization to overcome the $\log(n)$ lower bound, and give a 16-competitive algorithm. Emek et al. [15] study interval selection in the streaming model, and show how to modify their streaming algorithm to work online, achieving a competitive ratio of 6, improving upon the 16-competitive algorithm of Adler and Azar. It is noteworthy that the Adler and Azar algorithm uses memory proportional to the entire input sequence. In contrast, the Emek et al. algorithm only uses memory that is within a constant factor of a current OPT solution. It is still an open question if a randomized algorithm using only constant bounded memory can get a constant ratio in the unweighted case. We show that for a strict, but natural definition of memoryless randomized algorithms, a constant ratio cannot be obtained. The algorithms presented in this paper, along with the optimal algorithms by Garay et al. [22] and Woeginger [37], fall under our definition of memoryless. It is worth noting that similar notions of memoryless algorithms, and comparison between randomized memoryless and deterministic, have appeared in the k-server, caching, and facility location literature [12,14, 18,26,28,29,33]. We would note that barely random algorithms as described earlier (i.e. algorithms that initially generate some random bits, which are used in every online step), are not memoryless but usually satisfy bounded memory. The algorithms by Fung et al. [20] are an example of this. More generally, this use of initial random bits are the *classify and randomly select* algorithms (e.g. Lipton and Tomkins [31] and Awerbuch et al. [3]). It's important to note that such algorithms may require prior knowledge of bounds on lengths of intervals. In the full version of the paper [10] we discuss our meaning of memoryless and bounded memory online algorithms, and the relation to randomness, advice, and the Adler and Azar question.

The problem of admission control has also been studied under the model of minimizing rejections [2,7] instead of maximizing acceptances. An alternative input model for interval selection is that of arriving conflicts [23] instead of single intervals, with the algorithm being able to choose at most one item from each conflict. We also note that an instance of interval selection can be represented as an interval graph, with intervals corresponding to vertices, and edges denoting a conflict between two intervals. Generally, interval graphs reveal much less about the instance compared to receiving the actual intervals. In the interval graph representation, arriving vertices may have an adjacency list only in relation to already arrived vertices, or they may show adjacency to future vertices as well.

Our Results. For the unweighted adversarial case, we know that no deterministic algorithm is bounded (follows from [22]). Assuming there are at most k different lengths, we show how a simple greedy algorithm achieves a competitive ratio of $2k$. We also give a matching lower bound that holds for all deterministic algorithms, as well as "memoryless" randomized algorithms. We note that an instance with k different lengths can have a nesting depth of at most $k-1$. Alternatively, we can state our results in terms of d, the nesting depth (see Fig. 1), noting that $d \leq (k-1)$. This implies that our $2k$ bounds can be restated as $2(d+1)$. We also show how to extend the classify and randomly select paradigm used by Fung et al. [20] to obtain a randomized algorithm that is $2k$-competitive for the case of arbitrary weights and k different interval lengths. It's worth noting that the analysis by Canetti and Irani [11] implies an $\Omega(\sqrt{k})$ lower bound for randomized algorithms with arbitrary weights.

We show how the problem of call control on the line [22] relates to interval selection, and in particular how their $\log n$-competitive algorithm for the unweighted case and their $(2 + \sqrt{5})$-competitive algorithm for proportional weights carries over to interval selection. In doing so, we explain why there is no contradiction between our optimal $2k$ bound, and their optimal $\log n$ bound. Lastly, we consider deterministic memoryless algorithms for the problem of any-order, unweighted, single-lengthed (i.e. unit) intervals with random order arrivals. We show that the only deterministic memoryless algorithm that can possibly perform better than the adversarial bound is one-directional, only replacing intervals if they overlap in that particular direction.

Organization of the Paper. Section 2 has some definitions to clarify the model. Section 3 has our upper and lower bounds in the adversarial case. Section 4 discusses arbitrary weights. Section 5 is about interval selection in the random order model. We end with some conclusions and open problems. The connection to call control, and the application of the proportional weights algorithm to our model can be found in the full version of the paper.

2   Preliminaries

Our model consists of intervals arriving on the real line. An interval I_i is specified by a starting point s_i, and an end point f_i, with $s_i < f_i$. It occupies space $[s_i, f_i)$ on the line, and the conventional notions of intersection, disjointness, and containment apply. This allows two adjacent intervals $[s_1, f)$ and $[f, f_2)$ to not conflict, although our results would apply even if we considered closed intervals $[s_i, f_i]$ with $[s_1, f]$ and $[f, f_2]$ conflicting. There are two main ways two intervals can conflict, and they are shown in Fig. 1. In the case of containment, we say that the smaller intervals are *subsumed* by the larger one.

We use the notion of competitive ratio to measure the performance of our online algorithms. Given an algorithm A, let ALG denote the objective value of the solution achieved by the algorithm, and let OPT denote the optimal value achieved by an offline algorithm. The competitive ratio of A is defined as follows: $CR(A) = \frac{OPT}{ALG} \geq 1$. We should note that we can repeat disjoint copies of our nemesis sequences, and get the corresponding tight lower bounds. As a result, we can omit the standard additive term in our definition of competitive ratio. We will sometimes abuse notation and use ALG and OPT to denote the sets of intervals maintained by the algorithm at some given point, and the set of intervals of an optimal solution respectively. In the case of deterministic algorithms and random arrival of intervals, the performance of an algorithm is a random variable, and the competitive ratios hold w.h.p. (definition of competitive ratio remains unchanged). The algorithm we present in the case of arbitrary weights is randomized, and its expected competitive ratio is defined as $CR(A) = \frac{OPT}{\mathbb{E}[ALG]}$.

(a)Partial Conflict.

(b)Containment with nesting depth 1.

Fig. 1. Types of conflicts.

We sometimes refer to a *chain* of intervals (Fig. 2). This is a set of intervals where each interval partially conflicts with exactly two other intervals, except for the two end intervals that partially conflict with only one.

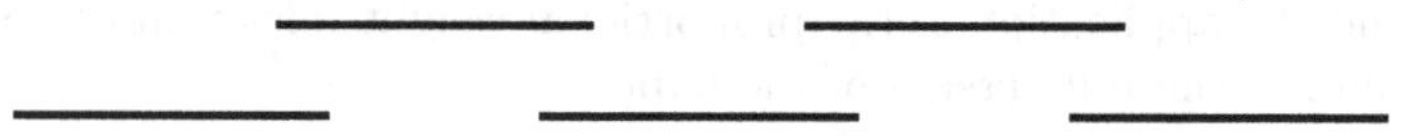

Fig. 2. Interval chain.

3   Adversarial Order

3.1   Unweighted

In this section, we assume an adversary chooses the instance configuration, along with the arrival order of all intervals. Lemma 1 shows that revocable decisions are necessary even in the case of two different lengths. Algorithm 1 is the greedy algorithm that achieves the optimal competitive ratio of $2k$ in the unweighted case, and it works as follows: On the arrival of a new interval, take it if there's no conflict. If there's a conflict, take the new interval only if it is properly contained inside an existing interval.

Lemma 1. *The problem of any-order unweighted interval scheduling with two different lengths and irrevocable decisions is unbounded.*

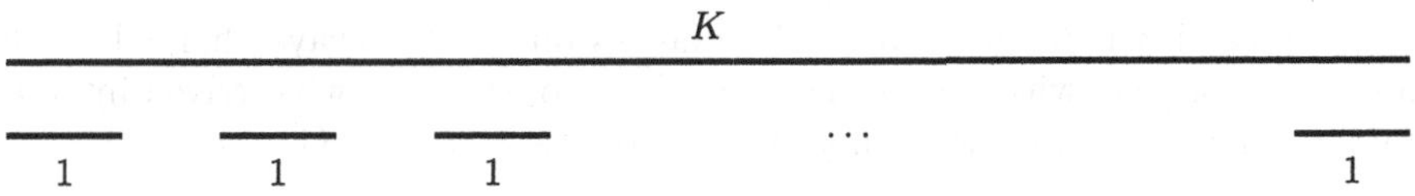

$$K$$

1   1   1   . . .   1

Fig. 3. Unweighted instance with two different lengths.

Algorithm 1.

On the arrival of I:
$I_s \leftarrow$ Set of intervals currently in the solution conflicting with I
for $I' \in I_s$ **do**
    if $I \subset I'$ **then**
        Take I and discard I'
        return
    end if
end for
Discard I

Proof. Consider two possible interval lengths of 1 and K (Fig. 3). Let an interval of length K arrive first. W.l.o.g. the algorithm takes it (otherwise no smaller intervals arrive). Then K 1-length non-overlapping intervals arrive next, all of them overlapping with the first K-length interval. The algorithm cannot take any of the 1-length intervals, achieving a competitive ratio of $\frac{1}{K}$. This construction can be repeated multiple times.

Theorem 1. *Algorithm 1 achieves a competitive ratio of 2k for the problem of any-order unweighted interval scheduling with k different lengths.*

Proof. We define a mapping of intervals $f : OPT \longrightarrow ALG$, where every interval in ALG has at most $2k$ intervals in OPT mapped to it. Because intervals taken by the algorithm might be replaced during the execution, the mapping f might be redefined multiple times. What follows is the way optimal intervals $I \in OPT$ are charged, as soon as they arrive, to intervals $I' \in ALG$. There are four cases of interest:

Case 1: The newly arrived optimal interval is taken by the algorithm.
This can happen either because this interval did not conflict with any other intervals taken by the algorithm, or because it was entirely subsumed by a larger interval in ALG, in which case the algorithm would have replaced the large interval with the new small one. In this case, this optimal interval is mapped onto itself.

Case 2: The newly arrived optimal interval partially conflicts with one interval currently in ALG. In this case, this optimal interval is charged to the interval it conflicts with.

Case 3: The newly arrived optimal interval partially conflicts with two intervals currently in ALG. In this case, this optimal interval can be charged to either of these two intervals arbitrarily. We may assume it is always charged to the interval it conflicts with on the right. Notice also, that a newly arrived interval, cannot partially conflict with more than two intervals in ALG.

Case 4: The newly arrived optimal interval subsumes an interval currently in ALG. W.l.o.g. we can assume this never happens. Any such optimal solution OPT can be turned into an optimal solution OPT', with the smaller interval in place of the larger one. We can restrict ourselves to only look at optimal solutions where no such transformation can take place. This case also encapsulates the case of an optimal interval perfectly coinciding with an interval taken by the algorithm.

An interval (I_l) taken by the algorithm can later be replaced, if a smaller one (I_s) comes along and is subsumed by it. When this happens, all intervals in OPT charged to I_l up to that point, will be transferred and charged to I_s. As a result, there are two ways an interval taken by the algorithm can be charged by intervals in OPT. The first way is when an interval $I \in OPT$ is directly charged to an interval $I' \in ALG$ when I arrives (Cases 1–4). This will be referred to as *direct charging*. The second way is when a new interval, I_n, arrives, and replaces an existing interval I_e, in which case all optimal intervals previously charged to I_e, will now be charged to I_n. This will be referred to as *transfer charging*.

Proposition 1. *An interval taken by the algorithm (even temporarily), can be charged by at most two optimal intervals through direct charging.*

To see why this proposition is true, we consider the three main cases of direct charging explained earlier. In *Case* 1, the optimal interval is taken by the algorithm and is charged to itself. Because no other optimal interval can conflict with it, we know this interval will never be directly charged again.

In *Cases* 2 and 3, direct charging happens because of the optimal interval partially conflicting with one or two intervals currently taken by the algorithm. Because an interval taken by the algorithm can partially conflict with at most two optimal intervals (one on each side), it can be charged twice at most.

Proposition 2. *An interval taken by the algorithm can be charged by at most $2k - 2$ optimal intervals through transfer charging.*

Consider a sequence of interval replacements by the algorithm, where all optimal intervals charged to an interval in the sequence are passed down to the next interval in the sequence. The last interval in that sequence will have accumulated all the optimal intervals charged to the previous intervals in that sequence. Because we consider k different lengths, such a sequence can have up to k intervals, participating in $k - 1$ transfer charging events. We also know that every interval in that sequence can be charged at most two optimal intervals through direct charging (Proposition 1) before being replaced. Consequently, assuming two additional charges are added to each interval in that sequence, the last (smallest) interval will be charged $2(k - 1)$ optimal intervals through transfer charging.

We have described a process during which every optimal interval is charged to an interval in ALG. By Propositions 1 & 2, we know that an interval in ALG, can be charged by $2k$ intervals in OPT at most. Therefore, our algorithm has a competitive ratio of $2k$ for the problem of unweighted interval selection with revocable decisions and k different possible interval lengths.

We now provide a matching lower bound, showing that no deterministic algorithm can do better.

Theorem 2. *No deterministic algorithm can achieve a competitive ratio better than $2k$ for the problem of unweighted interval selection with revocable decisions and k different lengths.*

Proof. At any point during the execution, the algorithm will have exactly one interval in its solution, while the size of the optimal solution will keep growing. We begin by describing how the main component of the instance is constructed, using intervals of the same length. First, the adversary must decide on an overlap amount v, which can be arbitrarily small. All partially conflicting intervals will overlap by exactly this amount. Consider now the instance of Fig. 4. Intervals I_1 and I_2 arrive first in that order. If I_1 is taken by the algorithm and is then replaced by I_2, then I_4 arrives. If I_1 was taken by the algorithm but was not replaced by I_2, then I_3 would arrive. Because this case is symmetrical, we only consider the former case of I_2 replacing I_1. What happens is that this chain keeps growing in the same direction, until the algorithm decides to stop replacing. When that happens, we look at the last three intervals of the chain. For example, when I_4 arrived, if the algorithm chose to not select I_4 and instead maintain I_2, we stop growing the chain and consider the intervals (I_1, I_2, I_4). If the algorithm never stops replacing, it will end up with I_5 in its solution. Notice that if the algorithms keeps replacing a growing chain, this will hurt the competitive ratio. In all cases, there exists an optimal solution of at least two intervals, with neither of them being the one taken by the algorithm. Note also that this construction requires at most four intervals of length L, occupying space at most $(4L - 3v)$ in total.

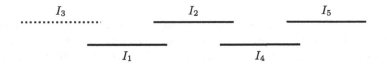

Fig. 4. Base adversarial construction

A small detail is that w.l.o.g. we can assume I_1 is always taken by the algorithm when it first arrives. Because this construction will take place a number of times during the execution, when the algorithm will already have an interval in its solution, it is useful to consider the case when I_1 is not taken by the algorithm. In this case, we start growing the chain regardless. If I_2 or I_4 are taken by the algorithm, we treat it similarly to when I_1 was taken and the algorithm kept replacing. If the algorithm hasn't taken any interval even after I_4 has arrived, the chain stops growing and we consider the intervals (I_1, I_2, I_4).

Let I_{alg} be the interval taken by the algorithm (or I_2 if no intervals were taken). All remaining intervals to arrive will be subsumed by I_{alg}, and thus will not conflict with the two neighboring intervals taken by OPT. Assuming I_{alg} conflicts with one interval on the left and one on the right, that leaves space of length $(L - 2v)$ for all remaining intervals. Inside that space, the exact same construction described will take place, only when the algorithm takes a new interval, it implies I_{alg} is replaced. This can be thought of as going a level deeper, and using a sufficiently smaller interval length. More precisely, if L' is the new (smaller) length that will be used, it must hold that $L' \leq \frac{L+v}{4}$.

After each such construction is completed, the size of the optimal solution grows by at least 2. Because there are at most k different lengths, this can be repeated at most k times. Finally, because the algorithm only ever keeps a single interval in its solution, it will achieve a competitive ratio of $2k$.

We now extend Theorem 2 and show that the $2k$ lower bound also holds for a class of randomized algorithms we call *memoryless*. Intuitively, memoryless algorithms decide on taking or discarding the newly arrived interval, only by looking at the new interval, and all the intervals currently in the solution, using no information from previous online rounds. Although not randomized, it is worth noting that Algorithm 1, along with the optimal deterministic algorithms for call control [22], are memoryless.

Definition 1 (*Memoryless randomized algorithm*). *We call a randomized algorithm memoryless, if a newly arrived interval I_{new} is taken with probability $F(I_{new}, S)$, where $S = \{I_1, I_2, ...\}$ is the set of intervals currently in the solution, and each interval is a tuple of the form (s_i, f_i).*

Notice that Definition 1 only allows us to make use of random bits of this current step, and it does not allow access to random bits from previous rounds. In particular, this definition does not capture barely random algorithms (as mentioned in the introduction), or algorithms that fall under the *classify and randomly select* paradigm.

Theorem 3. *No memoryless randomized algorithm can achieve a competitive ratio better than $2k$ for the problem of unweighted interval selection with revocable decisions and k different lengths. More specifically, for all $p \in (0, 1]$, there exists an $\epsilon_p > 0$, such that the competitive ratio is greater than $2k - \epsilon_p$ with probability p.*

Proof. The proof is very similar to the proof of Theorem 2. The instance has the same structure as the one described in the proof of Theorem 2, with the difference that whenever a new interval is taken with probability $p > 0$, the adversary will have to add as many copies of that interval as necessary, so that it is taken w.h.p. Fig. 5 shows an example of multiple copies of a new interval, ensuring that a replacement happens w.h.p.

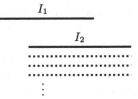

Fig. 5. Replacing I_1 w.h.p when $F(I_2, \{I_1\}) > 0$.

It is worth mentioning that similar to how we extend our lower bound to hold for memoryless randomized algorithms, one can extend the $\log(n)$ lower bound for call control [22] to also hold for memoryless randomized algorithms.

4 Arbitrary Weights

The case of intervals having an arbitrary weights has previously been considered for the case of single-length instances and ordered arrivals. Woeginger [37] gives an optimal deterministic algorithm that is 4-competitive. Fung et al. [20] give a barely random algorithm that is 2-competitive, and show that it is optimal amongst barely random algorithms that choose between two deterministic algorithms. Woeginger [37] shows that in the case of two different lengths, there does not exist a deterministic algorithm with finite competitive ratio. We show how to combine the barely random algorithm of Fung et al., with a classify and randomly select algorithm, to obtain a randomized algorithm for the any-order case, that achieves a competitive ratio of $2k$, when there are k different lengths.

First, one can observe that the 2-competitive single-length algorithm by Fung et al. [20] (Theorem 3.1), works even in the case of any-order arrivals. Our algorithm (denoted as ARB) works as follows: Choose one of k lengths, uniformly at random. Then execute the algorithm of Fung et al., looking only at intervals of the chosen length.

Theorem 4. *Algorithm ARB achieves a competitive ratio of $2k$ for the problem of any-order interval selection with k different lengths and arbitrary weights.*

Proof. Let $L_1, L_2, ..., L_k$ be all the different lengths of an instance. Associated with length L_i, is a sub-instance C_i, comprised only of the intervals of length L_i. Let OPT_i denote the weight of an optimal solution on sub-instance C_i. The expected performance of the algorithm can be bounded as follows:

$$\mathbb{E}[ALG] \geq \frac{1}{k}\frac{OPT_1}{2} + \frac{1}{k}\frac{OPT_2}{2} + ... + \frac{1}{k}\frac{OPT_k}{2} \geq \frac{OPT}{2k}$$

The first inequality holds because applying Fung et al. [20] on C_i gives a solution of weight at least $\frac{OPT_i}{2}$. The second inequality holds because for every length L_j, the total weight of the intervals of length L_j in the final solution, is at most OPT_j.

We note that the algorithm does not need to know the actual lengths beforehand, or even k. The algorithm can start working with the first length that appears. When a second length arrives, the algorithm discards its current solution and chooses the new length with probability $\frac{1}{2}$. More generally, when the ith length arrives, the algorithm starts over using the new length with probability $\frac{1}{i}$. One can see that the probability that any length is chosen is $\frac{1}{k}$. This procedure can be viewed as a form of reservoir sampling. Moreover, by replacing the 2-competitive arbitrary weights algorithm with a simple greedy algorithm, we get a randomized algorithm for the unweighted case that is $2k$-competitive and does not use revoking.

5   Random Order

In this section, we assume the adversary chooses the instance configuration, but the intervals arrive in a random order. We consider unweighted, single-lengthed instances, and deterministic memoryless algorithms with revocable acceptances. We consider various cases and show that the only type of algorithm that can beat the adversarial bound is a *one-directional* algorithm, namely an algorithm that only replaces intervals on the left side, or only on the right side, regardless of the amount of overlap. For any other algorithm, we show how the adversary can enforce a competitive ratio of 2, resulting in no benefit over adversarial arrivals for single-lengthed instances.

The argument works as follows. We first consider two simple algorithms, an algorithm that always replaces, and an algorithm that never replaces, and give a class of instances where these algorithms are no better than 2-competitive. We then show that for any algorithm that isn't one-directional, we can construct an instance on which the algorithm's behavior is the same as that of an always-replace, or never-replace algorithm on that class of instances. As a result, we get the following, arguably surprising, theorem. The proof of Theorem 5 is presented in the full version of the paper.

Theorem 5. *Every deterministic memoryless algorithm that isn't one-directional, can be forced to a strict competitive ratio of at least 2 for the problem of online unweighted single-lengthed interval selection under random order arrivals.*

We note that in ongoing work with additional authors, we have shown that for chain instances, the one-directional algorithm is substantially better than 2-competitive.

6   Conclusions and Open Problems

There are a number of possible directions for future work. A very natural direction is looking at specific weighted cases. Deterministically, Garay et al. [22] have settled the case of proportional weights with an optimal, constant-competitive algorithm. It's interesting to see if a similar constant can be achieved for the more general weight functions studied by Woeginger [37], with or without randomness. We considered the case of arbitrary weights in Sect. 4.

It is fair to say that we have a limited understanding of randomized algorithms for interval selection. In the unweighted adversarial setting, we have shown that no memoryless randomized algorithm can be constant-competitive. With memory, the best known algorithm is 6-competitive, and we know of no better lower bound than $\frac{4}{3}$ (using a chain of 3 intervals). For some weighted cases (including proportional weights), Fung et al. show that with one random bit, their 2-competitive algorithm is optimal. However, these upper bounds don't necessarily hold in the any-order model. We also note that for arbitrary weights, there is a gap between our $2k$ upper bound, and the $\Omega(\sqrt{k})$ lower bound by Canetti and Irani [11]. We would like to extend the memoryless model to algorithms with constant memory beyond the current solution. In particular, we would want to allow access to a few initial random bits which would also capture algorithms that fall under the *classify and randomly select* paradigm.

A natural extension of revocable decisions is assigning a cost to the removal and replacement of previously accepted items, a model arguably more relevant in practice. This has been applied to problems such as online knapsack [24], and online advertising [4]. We find it interesting to consider interval selection with costs, and devise algorithms that aim to optimize the solution while limiting the total cost of revoking.

Finally, to the best of our knowledge, we have initiated the study of this model under random order arrivals, where there are many open questions for future work. We have only looked at single-lengthed instances, a special case that, in the adversarial setting, doesn't even require revoking. Looking at multiple lengths under random arrivals is a natural next step. Lastly, we have shown that one-directional algorithms for single-lengthed instances, are the only type of deterministic memoryless algorithms that can achieve better than $2-$competitiveness. We don't have any provable upper bounds on the performance of a one-directional algorithm, but we have conducted experiments that suggest it may achieve much better than 2-competitiveness. This is an interesting contrast with the adversarial model, where a one-directional algorithm would perform arbitrarily bad.

Acknowledgements. We would like to thank Denis Pankratov, Adi Rosén and Omer Lev for many helpful comments.

References

1. Adler, R., Azar, Y.: Beating the logarithmic lower bound: randomized preemptive disjoint paths and call control algorithms. J. Scheduling **6**(2), 113–129 (2003)
2. Alon, N., Azar, Y., Gutner, S.: Admission control to minimize rejections and online set cover with repetitions. In: Proceedings of the Seventeenth Annual ACM Symposium on Parallelism in Algorithms and Architectures, pp. 238–244 (2005)

3. Awerbuch, B., Bartal, Y., Fiat, A., Rosén, A.: Competitive non-preemptive call control. In: SODA, vol. 94, pp. 312–320. Citeseer (1994)
4. Babaioff, M., Hartline, J.D., Kleinberg, R.D.: Selling ad campaigns: online algorithms with cancellations. In: Proceedings of the 10th ACM Conference on Electronic Commerce, pp. 61–70 (2009)
5. Bachmann, U.T., Halldórsson, M.M., Shachnai, H.: Online selection of intervals and t-intervals. Inf. Comput. **233**, 1–11 (2013)
6. Baptiste, P.: Scheduling equal-length jobs on identical parallel machines. Discret. Appl. Math. **103**(1–3), 21–32 (2000)
7. Blum, A., Kalai, A., Kleinberg, J.: Admission control to minimize rejections. In: Dehne, F., Sack, J.-R., Tamassia, R. (eds.) WADS 2001. LNCS, vol. 2125, pp. 155–164. Springer, Heidelberg (2001). https://doi.org/10.1007/3-540-44634-6_15
8. Bogart, K.P., West, D.B.: A short proof that "proper= unit." Discret. Math. **201**(1–3), 21–23 (1999)
9. Borodin, A., Ivan, I., Ye, Y., Zimny, B.: On sum coloring and sum multi-coloring for restricted families of graphs. Theor. Comput. Sci. **418**, 1–13 (2012)
10. Borodin, A., Karavasilis, C.: Any-order online interval selection. arXiv preprint arXiv:2303.06127 (2023)
11. Canetti, R., Irani, S.: Bounding the power of preemption in randomized scheduling. In: Proceedings of the Twenty-Seventh Annual ACM Symposium on Theory of Computing, pp. 606–615 (1995)
12. Christou, D., Fotakis, D., Koumoutsos, G.: Memoryless algorithms for the generalized k-server problem on uniform metrics. In: Kaklamanis, C., Levin, A. (eds.) WAOA 2020. LNCS, vol. 12806, pp. 143–158. Springer, Cham (2021). https://doi.org/10.1007/978-3-030-80879-2_10
13. Chrobak, M., Dürr, C., Jawor, W., Kowalik, Ł, Kurowski, M.: A note on scheduling equal-length jobs to maximize throughput. J. Scheduling **9**(1), 71–73 (2006)
14. Coester, C., Koutsoupias, E.: The online k-taxi problem. In: Proceedings of the 51st Annual ACM SIGACT Symposium on Theory of Computing (2019)
15. Emek, Y., Halldórsson, M.M., Rosén, A.: Space-constrained interval selection. ACM Trans. Algorithms (TALG) **12**(4), 1–32 (2016)
16. Epstein, L., Levin, A.: Improved randomized results for that interval selection problem. In: Halperin, D., Mehlhorn, K. (eds.) ESA 2008. LNCS, vol. 5193, pp. 381–392. Springer, Heidelberg (2008). https://doi.org/10.1007/978-3-540-87744-8_32
17. Faigle, U., Nawijn, W.M.: Note on scheduling intervals on-line. Discret. Appl. Math. **58**(1), 13–17 (1995)
18. Fotakis, D.: On the competitive ratio for online facility location. Algorithmica **50**(1), 1–57 (2008)
19. Fung, S.P., Poon, C.K., Yung, D.K.: On-line scheduling of equal-length intervals on parallel machines. Inf. Process. Lett. **112**(10), 376–379 (2012)
20. Fung, S.P., Poon, C.K., Zheng, F.: Improved randomized online scheduling of intervals and jobs. Theory Comput. Syst. **55**(1), 202–228 (2014)
21. Garay, J.A., Gopal, I.S.: Call preemption in communication networks. In: Proceedings IEEE INFOCOM 1992: Conference on Computer Communications (1992)
22. Garay, J.A., Gopal, I.S., Kutten, S., Mansour, Y., Yung, M.: Efficient on-line call control algorithms. J. Algorithms **23**(1), 180–194 (1997)
23. Halldórsson, M.M., Patt-Shamir, B., Rawitz, D.: Online scheduling with interval conflicts. Theory Comput. Syst. **53**(2), 300–317 (2013)

24. Han, X., Kawase, Y., Makino, K.: Online Knapsack problem with removal cost. In: Gudmundsson, J., Mestre, J., Viglas, T. (eds.) COCOON 2012. LNCS, vol. 7434, pp. 61–73. Springer, Heidelberg (2012). https://doi.org/10.1007/978-3-642-32241-9_6

25. Hyatt-Denesik, D., Rahgoshay, M., Salavatipour, M.R.: Approximations for throughput maximization. arXiv preprint arXiv:2001.10037 (2020)

26. Kleinberg, J.M.: A lower bound for two-server balancing algorithms. Inf. Process. Lett. **52**(1), 39–43 (1994)

27. Kolen, A.W., Lenstra, J.K., Papadimitriou, C.H., Spieksma, F.C.: Interval scheduling: a survey. Nav. Res. Logist. (NRL) **54**(5), 530–543 (2007)

28. Koutsoupias, E.: The k-server problem. CS Rev. **3**(2), 105–118 (2009)

29. Koutsoupias, E., Taylor, D.S.: The CNN problem and other k-server variants. Theor. Comput. Sci. **324**(2–3), 347–359 (2004)

30. Kovalyov, M.Y., Ng, C.T., Cheng, T.E.: Fixed interval scheduling: models, applications, computational complexity and algorithms. Eur. J. OR (2007)

31. Lipton, R.J., Tomkins, A.: Online interval scheduling. In: SODA, vol. 94 (1994)

32. Miyazawa, H., Erlebach, T.: An improved randomized on-line algorithm for a weighted interval selection problem. J. Sched. **7**(4), 293–311 (2004)

33. Raghavan, P., Snir, M.: Memory versus randomization in on-line algorithms. In: Ausiello, G., Dezani-Ciancaglini, M., Della Rocca, S.R. (eds.) ICALP 1989. LNCS, vol. 372, pp. 687–703. Springer, Heidelberg (1989). https://doi.org/10.1007/BFb0035792

34. Seiden, S.S.: Randomized online interval scheduling. Oper. Res. Lett. **22**(4–5), 171–177 (1998)

35. Sgall, J.: On-line scheduling. Online Algorithms, pp. 196–231 (1998)

36. Tomkins, A.: Lower bounds for two call control problems. Inf. Process. Lett. **56**(3), 173–178 (1995)

37. Woeginger, G.J.: On-line scheduling of jobs with fixed start and end times. Theor. Comput. Sci. **130**(1), 5–16 (1994)

38. Yu, G., Jacobson, S.H.: Online c-benevolent job scheduling on multiple machines. Optim. Lett. **12**(2), 251–263 (2018)

Improved Approximations for Relative Survivable Network Design

Michael Dinitz[1], Ama Koranteng[1(✉)], Guy Kortsarz[2], and Zeev Nutov[3]

[1] Johns Hopkins University, Baltimore, USA
mdinitz@cs.jhu.edu, akorant1@jhu.edu
[2] Rutgers University, Camden, USA
guyk@camden.rutgers.edu
[3] The Open University, Ra'anana, Israel
nutov@openu.ac.il

Abstract. One of the most important and well-studied settings for network design is edge-connectivity requirements. This encompasses uniform demands (e.g. the Minimum k-Edge-Connected Spanning Subgraph problem), as well as nonuniform demands (e.g. the Survivable Network Design problem (SND)). In a recent paper [Dinitz, Koranteng, Kortsarz APPROX '22], the authors observed that a weakness of these formulations is that we cannot consider fault-tolerance in graphs that have small cuts but where *some* large fault sets can still be accommodated. To remedy this, they introduced new variants of these problems under the notion *relative* fault-tolerance. Informally, this requires not that two nodes are connected if there are a bounded number of faults (as in the classical setting), but that they are connected if there are a bounded number of faults *and the nodes are connected in the underlying graph post-faults*.

Due to difficulties introduced by this new notion of fault-tolerance, the results in [Dinitz, Koranteng, Kortsarz APPROX '22] are quite limited. For the Relative Survivable Network Design problem (RSND) with non-uniform demands, they are only able to give a nontrivial result when there is a single demand with connectivity requirement 3—a non-optimal 27/4-approximation. We strengthen this result in two significant ways: We give a 2-approximation for RSND when *all requirements are at most* 3, and a $2^{O(k^2)}$-approximation for RSND with a single demand of *arbitrary value* k. To achieve these results, we first use the "cactus representation" of minimum cuts to give a lossless reduction to normal SND. Second, we extend the techniques of [Dinitz, Koranteng, Kortsarz APPROX'22] to prove a generalized and more complex version of their structure theorem, which we then use to design a recursive approximation algorithm.

Keywords: Fault tolerance · Network design

See [17] for the full version.
M. Dinitz—Supported in part by NSF awards CCF-1909111 and CCF-2228995.
A. Koranteng—Supported in part by an NSF Graduate Research Fellowship and NSF award CCF-1909111.

© The Author(s), under exclusive license to Springer Nature Switzerland AG 2023
J. Byrka and A. Wiese (Eds.): WAOA 2023, LNCS 14297, pp. 190–204, 2023.
https://doi.org/10.1007/978-3-031-49815-2_14

1   Introduction

Fault-tolerance has been a central object of study in approximation algorithms, particularly for network design problems where the graphs we study represent physical objects which might fail (communication links, transportation links, etc.). In these settings it is natural to ask for whatever object we build to be fault-tolerant. The precise definition of "fault-tolerance" varies in different settings, but a common formulation is edge fault-tolerance, which typically takes the form of edge connectivity. Informally, these look like guarantees of the form "if up to k edges fail, then the nodes I want to be connected are still connected." For example, consider the following classical fault-tolerance problem.

Definition 1. *In the Survivable Network Design problem (SND, sometimes referred to as Generalized Steiner Network) we are given an edge-weighted graph G and demands $\{(s_i, t_i, k_i)\}_{i \in [\ell]}$, and we are supposed to find the minimum-weight subgraph H of G so that there are at least k_i edge-disjoint paths between s_i and t_i for every $i \in [\ell]$. In other words, for every $i \in [\ell]$, if fewer than k_i edges fail then s_i and t_i will still be connected in H even after failures.*

The Survivable Network Design problem is well-studied (see [15,23,25,30] for a sample); notably, Jain [25] gives a 2-approximation algorithm for the problem in a seminal paper. Beyond SND, edge fault-tolerance has been studied in many related network design contexts, with the k-Edge Connected Spanning Subgraph, Fault-Tolerant Group Steiner Tree, Fault-Tolerant Spanner, and Fault-Tolerant Shortest Paths problems being just a few examples (see [4,15,19,26]). These and other classical fault-tolerance problems, including the Survivable Network Design problem, are *absolute* fault-tolerance problems—if up to k objects fail, the remaining graph should function as desired. This differs from the stronger notion of fault-tolerance introduced in [16], called *relative* fault-tolerance. Relative fault-tolerance makes guarantees that rather than being absolute ("if at most k edges fail the network still functions") are relative to an underlying graph or system ("if at most k edges fail, the subgraph functions just as well as the original graph post-failures").

Relative fault-tolerance is therefore a natural generalization of absolute fault-tolerance: If the input graph has the desired connectivity, then the relative fault-tolerance and absolute fault-tolerance definitions are equivalent. However, if the input graph does not have the requested connectivity, then relative fault-tolerance allows us to return a solution with interesting and nontrivial guarantees while absolute fault-tolerance forces us to return nothing. In this way, relative fault-tolerance overcomes a significant weakness of absolute fault-tolerance.

This relative fault-tolerance definition was inspired by a recent line of work on relative notions of fault-tolerance for graph spanners and emulators [5–9,11,18,19]. In these settings, the goal is generally to find existential bounds and algorithms to achieve them, rather than to do optimization. In [16], by contrast, their approach takes the point of view of optimization and approximation algorithms. With this notion of fault-tolerance in network design, the authors of [16] define the relative version of the Survivable Network Design problem.

Definition 2. *In the Relative Survivable Network Design problem (RSND), we are given a graph $G = (V, E)$ with edge weights $w : E \to \mathbb{R}_{\geq 0}$ and demands $\{(s_i, t_i, k_i)\}_{i \in [\ell]}$. A feasible solution is a subgraph H of G where for all $i \in [\ell]$ and $F \subseteq E$ with $|F| < k_i$, if there is a path in $G \setminus F$ from s_i to t_i then there is also a path in $H \setminus F$ from s_i to t_i. Our goal is to find the minimum weight feasible solution.*

Note that if s_i and t_i are k_i-connected in G for every $i \in [\ell]$, then RSND is exactly the same as SND. If in G there exists some $i \in [\ell]$ such that s_i and t_i are *not* k_i-connected, then although there is no solution for SND, there is a meaningful RSND solution.

There has been recent work on a related network design model introduced by Adjiashvili [1–3,10,13]. In this model, E is partitioned into "safe" and "unsafe" edges. Informally, in the Flex-SNDP problem we are given a graph $G = (V, E)$ with edge costs and with functions $p, q : V \times V \to \mathbb{Z}^+$. We must return a min cost subgraph such that for each vertex pair u, v, they are $p(u, v)$-connected after deleting any subset of up to $q(u, v)$ unsafe edges. Like RSND, Flex-SNDP is a natural generalization of SND. However, it is an absolute fault-tolerance problem since it does not consider the underlying connectivity of the input. No polynomial-time approximation algorithms are known for general Flex-SNDP, though there has been recent work on several special cases [3,10,12,13].

The Results of [16]. Although relative fault-tolerance is a natural and promising generalization of fault-tolerance, the results given in [16] for the RSND problem are quite limited. Outside of a 2-approximation algorithm for the special case in which all demands are identical, [16] is only able to give algorithms for some of the simplest RSND special cases. First, they give an extremely simple 2-approximation for the RSND special case where all demands are in $\{0, 1, 2\}$ (also known as the 2-RSND problem). The algorithm falls out of the observation that there is only a difference between a relative demand of 2 and an absolute demand of 2 when there is a cut of size one separating the vertex demand pair. Cuts of size one are very easy to handle, allowing for a simple and straightforward reduction to SND.

Cuts of size two or larger are significantly more difficult to reason about, and so the 2-RSND algorithm does not extend to larger demands. As a result of this more complex cut structure, [16] is only able to handle demands of value 3 (and reason about the size two cuts between them) when there is only *a single demand*, with value 3 (also known as the SD-3-RSND problem). Despite this being an extremely restricted special case of RSND, the algorithm and analysis given by [16] are quite complex, depending on a careful graph decomposition involving "important separators" (a concept from fixed-parameter tractability [28]). Moreover, this algorithm only achieved a 27/4-approximation for the problem, far from the 2-approximation (or even exact algorithm) that one might hope for.

The limited results of [16] show that while relative fault-tolerance is an attractive notion, applying it to the Survivable Network Design problem significantly changes the structure of the problem and makes it difficult to reason about and develop algorithms for. For example, while [16] only gives a 27/4-approximation for SD-3-RSND, there is an exact polynomial-time algorithm for the SND equivalent (by a simple reduction to the Min-Cost Flow problem). So one might worry that relative fault-tolerance is simply too difficult of a definition, and the results of [16] are limited precisely because nothing is possible for even slightly more general settings.

1.1 Our Results and Techniques

In this paper, we seek to alleviate this worry by providing improved bounds for generalizations of the settings considered in [16]. In particular, we study two natural generalizations of the SD-3-RSND problem (which [16] provided a 27/4-approximation for). First, rather than only a single demand with value at most 3, can we handle an *arbitrary* number of demands that are at most 3? Secondly, in the single demand setting, instead of only handling a demand of at most 3, can we generalize to *arbitrary* values?

3-RSND. We begin with the setting where all demands are at most 3, but there can be an arbitrary number of such demands. We call this the 3-RSND problem. Note that, as discussed, there are no previous results for this setting, and the most related result is a 27/4-approximation if there is only a single such demand [16]. We prove the following theorem.

Theorem 1. *There is a polynomial-time 2-approximation for the 3-RSND problem.*

To obtain this theorem, we use entirely different techniques from those used in [16]. Most notably, we use the *cactus representation* of the global minimum cuts (which in this case are 2-cuts) of the input graph. The cactus representation of global minimum cuts is well studied and has been leveraged in a number of settings (see [20, 21, 24, 27, 29] for a sample). While it can be defined and constructed for more general connectivity values, for our setting we can construct the cactus representation by contracting components with certain connectivity properties. This results in a so-called *cactus graph*, which at a high level is a "tree of cycles": every pair of cycles intersects on at most one component in the construction. This cactus graph now has a simple enough structure that it allows us to reduce the original problem to a simpler problem in each of the contracted components. That is, we are able to show that certain parts of the cactus are essentially "forced", while other parts are not necessary, so the only question that remains is what to do "inside" of each cactus vertex, i.e., each component. This reduction makes the connectivity demands inside each component more complicated, but fortunately we are guaranteed 3-connectivity between special vertices inside the component. Hence we can use Jain's 2-approximation for SND [25] without worrying about the relative nature of the demands.

SD-*k*-RSND. Our second improvement is orthogonal: rather than allowing for more demands of at most 3, we still restrict ourselves to a single demand but allow it to be a general constant k rather than 3. We call this the SD-k-RSND problem. As with the 3-RSND problem discussed earlier, there are no known results for this problem. We prove the following theorem:

Theorem 2. *There is a polynomial-time $2^{O(k^2)}$-approximation for the SD-k-RSND problem.*

To prove this, we extend the technique used by [16] for the $k = 3$ case. They construct a "chain" of 2-separators (cuts of size 2 that are also important separators) so that in each component in the chain, there are no 2-cuts between the incoming separator and the outgoing separator. They are then able to use this structure to characterize the connectivity requirement of any feasible solution restricted to that component. To extend this technique, we use important separators of size up to $k - 1$ to carefully construct a *hierarchy* of chains. The hierarchy has $k - 1$ levels of nested components, so that for each component in the ith level of the hierarchy, there are no cuts of size at most i between the incoming and outgoing separators. There are multiple ways of constructing such a hierarchy, but we prove that a particular construction yields a hierarchy with a number of useful but delicate properties within a single level and between different levels of the hierarchy. With these properties, we can characterize the complex connectivity requirement of any feasible solution when restricted to a component in the hierarchy. Once we have this structure theorem, we approximate the optimal solution in each component of the hierarchy via a recursive algorithm.

Simplification of k-EFTS. The k-Edge Fault Tolerant Subgraph problem (k-EFTS) is the special case of RSND where all demands are identical: every two nodes have a demand of exactly k. A 2-approximation for k-EFTS was recently given in [16] via a somewhat complex proof; in particular, they defined a new property called *local weak supermodularity* and used it to show that Jain's iterative rounding still gave the same bounds in the relative setting. In the full version [17], we give a simplification of this proof. It turns out that local weak supermodularity is not actually needed, and a more classical notion of $\mathcal{F}$-supermodularity suffices. This allows us to reduce to previous work in a more black-box manner.

2   Preliminaries

We will consider the following special cases of RSND (Definition 2):

- The k-RELATIVE SURVIVABLE NETWORK DESIGN problem (k-RSND) is the special case of RSND where $r(s, t) \leq k$ for all $s, t \in V$. In this paper we consider the case $k = 3$, namely, the 3-RSND problem.
- The SINGLE DEMAND k-RELATIVE SURVIVABLE NETWORK DESIGN problem (SD-k-RSND) is the special case of RSND where $r(s, t) = k$ for exactly one vertex pair $s, t \in V$ and there is no demand for any other vertex pairs (equivalently, all other demands are 0). We consider the full SD-k-RSND problem for arbitrary k.

– The k-EDGE-FAULT-TOLERANT-SUBGRAPH problem (k-EFTS) is the special case of RSND where $r(s,t) = k$ for all $s, t \in V$.

For each of the listed RSND problem variants, we will use the following notation and definitions throughout. Let $G = (V, E)$ be a (multi-)graph and H a spanning subgraph (or an edge subset) of G. For $A \subseteq V$, let $\delta_H(A)$ denote the set of edges in H with exactly one endpoint in A, and let $d_H(A) = |\delta_H(A)|$ be their number. Additionally, let $G[A]$ denote the subgraph of G induced by the vertex set A. Let $s, t \in V$. We say that A is an st-set if $s \in A$ and $t \notin A$, and that $\delta_G(A)$ (or $\delta_E(A)$) is an st-cut of G (induced by A). An st-cut $\delta_G(A)$ (or an st-set A) is G-minimal if $\delta_G(A)$ contains no other st-cut of G. Assuming G is connected, it is easy to see that $\delta_G(A)$ is G-minimal if and only if both $G[A]$ and $G[V \setminus A]$ are connected. One can also see that if an st-cut $X \subseteq E$ is G-minimal, then $X = \delta_G(A)$ for some $A \subseteq V$. Finally, let $\lambda_G(s,t)$ denote the size of a min st-cut in G.

By Theorem 17 of [16], we may assume without loss of generality that the input graph G of any RSND instance is 2-edge-connected (or "2-connected").

3   2-Approximation for 3-RSND (and SD-3-RSND)

Given an RSND instance, we say that a vertex pair s, t is a k-**demand** if $r(s,t) = k$. We call a k-demand **relative** if the minimum st-cut has size less than k; that is, if $\lambda_G(s,t) < k$. A k-demand is then **ordinary** if $\lambda_G(s,t) \geq k$. Recall that SD-3-RSND has only one demand st, and that it is a 3-demand. The edges of any size 2 st-cut, or 2-st-cut, belong to any feasible solution so we call them **forced edges**. As a result, we can assume without loss of generality that they have cost 0.

3.1   Overview

We first give an overview of the theorems and proofs in this section. In order to prove Theorem 1, we will show that we can replace a single *relative* 3-demand by an equivalent set of *ordinary* 3-demands. More formally, we will prove the following.

Theorem 3. *Given an* SD-3-RSND *instance, there exists a polynomially computable set of ordinary 3-demands, D, such that for any $H \subseteq E$ that contains all forced edges, H is a feasible* SD-3-RSND *solution if and only if H satisfies all demands in D.*

This theorem reduces SD-3-RSND to the ordinary 3-SND problem (that is, the special case of SND where all demands are at most 3). In fact, this also gives us a lossless reduction from 3-RSND to 3-SND: Given a 3-RSND instance, we include the forced edges of all 3-demands into our solution, replace each relative 3-demand by an equivalent set of ordinary demands, and obtain an equivalent ordinary 3-SND instance. Since SND admits approximation ratio 2, this reduction from 3-RSND to 3-SND implies Theorem 1.

We will also show that SD-3-RSND is approximation equivalent to certain instances of a special case of 3-SND. Before we define this special case, we must give a definition. A vertex subset R is a k-**edge-connected subset** in a graph H if $\lambda_H(u,v) \geq k$ for all vertex pairs $u,v \in R$. Since the relation $\{(u,v) \in V \times V : \text{no } (k-1)\text{-cut separates } u,v\}$ is transitive, this is equivalent to requiring that $\lambda_H(u,v) \geq k$ for pairs u,v that form a tree on R. We will prove that SD-3-RSND is approximation equivalent to special instances of the following problem:

4-SUBSET 3-EC

Input: A graph $J = (V', E')$ with edge costs, and a set $R \subseteq V'$ of at most 4 terminals.

Output: A min-cost subgraph H of J, such that R is 3-edge-connected in H.

More specifically, we will prove the following.

Theorem 4. *Let s and t be vertices in $J = (V', E')$, where J is the input graph to an instance of 4-SUBSET 3-EC. SD-3-RSND admits approximation ratio ρ if and only if 4-SUBSET 3-EC with the following properties (A,B) admits approximation ratio ρ:*

(A) $d_J(s) = d_J(t) = 2$ and R is the set of neighbors of s, t.
(B) If $d_J(A) = 2$ for some st-set A, then $A = \{s\}$ or $A = V' \setminus \{t\}$. Namely, if F is a set of 2 edges of J such that $J \setminus F$ has no st-path, then $F = \delta_J(s)$ or $F = \delta_J(t)$.

The general 4-SUBSET 3-EC problem admits approximation ratio 2, since it is a special case of SND. However, it is not actually known whether 4-SUBSET 3-EC is in P or is NP-hard (see [14,22] for results on a closely related problem). This 2-approximation is the best known for the 4-SUBSET 3-EC problem, so our 2-approximation for 3-RSND is the best we can hope for. In the rest of this section, we prove Theorems 4, 3, and 1. All missing proofs can be found in the full version [17].

3.2   Cactus Representation and Definitions

We first give some definitions and describe the cactus representation. The relation $\{(u,v) \in V \times V : \text{no } (k-1)\text{-cut separates } u,v\}$ is an equivalence, and we will call its equivalence classes k-**classes**. We construct a **cactus** $\mathcal{G}$ by shrinking every nontrivial 3-class (that is, every 3-class with at least 2 nodes) of the input graph G. Note that since G is 2-connected, $\mathcal{G}$ is a connected graph in which every two cycles have at most one node in common. Going forward, we will identify every 3-class with the corresponding node of $\mathcal{G}$. The edge pairs that belong to the same cycle of $\mathcal{G}$ are the 2-cuts of G.

We can assume that vertex pair st is a relative 3-demand. We say that the st-**chain of cycles** of $\mathcal{G}$ consists of all the cycles (and their nodes) in $\mathcal{G}$ that contain a 2-st-cut. We refer to the nodes, 3-classes, on these cycles as st-**relevant** nodes. Note that the set of edges in $\mathcal{G}$ that are in the st-chain of cycles are the forced edges. We also say that an st-relevant 3-class is **central** if it contains s or t, or belongs to two cycles of the st-chain. Additionally, the **attachment nodes** of an st-relevant 3-class are nodes in the 3-class that are either s or t, or are the ends of the edges (the **attachment edges**) that belong to some cycle in the st-chain of cycles. Since G is 2-connected, the number of attachment nodes in a non-central 3-class is exactly 2, while the number of attachment nodes in a central 3-class is between 2 and 4.

3.3   Proof of Theorems 4, 3, and 1

For the proof of Theorems 4 and 3, we associate with each st-relevant 3-class, C, a certain graph G_C which we call the **component of** C, obtained as follows:

– If C is a non-central 3-class then, in the graph obtained from G by removing the two attachment edges of C, G_C is the connected component that contains C.
– If C is a central 3-class, then removing the attachment edges of C results in at least one and at most two connected components that do not contain C – one contains s and the other contains t, if any. We obtain G_C from G by contracting the connected component that contains s into node s, and contracting the connected component that contains t into node t.

We now modify the central components G_C to satisfy properties (A,B) from Theorem 4. Consider some central 3-class C, and consider its component $J = G_C$. If J does not contain one of the original nodes s or t, then it has properties (A,B) and no modification is needed. If J contains the original node s, then we rename s to s', add a new node s, and connect new s by two zero cost edges to s'. The obtained J now has properties (A,B). A similar transformation applies if J contains the original node t.

The following lemma is about both the non-central components and these *modified* central components; in the lemma, we show that for H to be a feasible SD-3-RSND solution, it is necessary and sufficient to satisfy certain connectivity properties within each component.

Lemma 1. *Let H be a subgraph of G, and suppose that H contains all forced edges. Subgraph H is a feasible SD-3-RSND solution if and only if for every component J, the following holds.*

(i) *If J is a non-central component, then $H[J]$ contains two edge-disjoint uv-paths, where u and v are the two attachment nodes of J.*
(ii) *If J is a central component, then $H[J]$ is a feasible solution to the SD-3-RSND instance in J (with demand $r(s,t) = 3$).*

Suppose that for the special SD-3-RSND instances that arise in the central components we can achieve approximation ratio α. Then, we can achieve ratio α for general SD-3-RSND by picking into our solution H three types of edge sets.

1. The forced edges.
2. A min-cost set of 2 edge-disjoint paths between the attachment nodes of each st-relevant non-central component.
3. An α-approximate solution in each st-relevant central component.

Note that edges picked in steps 1,2 do not invoke any cost in the approximation ratio, since by Lemma 1 we actually pick parts of an optimal solution. Thus we get that the approximability of SD-3-RSND is equivalent to the approximability of the very special instances that arise in the central components. We will now show that these special instances from the central components are in fact instances of 4-Subset 3-EC with properties (A,B) from Theorem 4, thus proving Theorem 4. We will consider only central components with 4 attachment nodes; other cases with 3 or 2 attachment nodes are similar.

In what follows, let $\mathcal{I}$ be an SD-3-RSND instance on input graph J with properties (A,B) (just as in our central components). Let $R = \{x, y, z, w\}$ where x, y are the neighbors of s and z, w are the neighbors of t and let H be a subgraph of J that includes the four forced edges sx, sy, zt, and wt. We have the following.

Lemma 2. *Subgraph H is a feasible solution for instance $\mathcal{I}$ if and only if $R = \{x, y, z, w\}$ is a 3-edge-connected subset in H.*

By Lemma 2, H is a feasible solution for $\mathcal{I}$ if and only if H includes all forced edges and R is a 3-edge-connected subset—that is, R forms a feasible solution to 4-Subset 3-EC—in H. This, along with Lemma 1, implies that the approximability of SD-3-RSND is equivalent to that of 4-Subset 3-EC with properties (A,B), concluding the proof of Theorem 4.

Proof of Theorem 3. We will prove that a single relative 3-demand st can be replaced by an equivalent forest of ordinary 3-demands in polynomial time, where the trees in this forest span the sets of attachment nodes of the st-relevant 3-classes.

Recall that by Lemmas 1 and 2, subgraph H is a feasible SD-3-RSND solution for 3-demand st if and only if the following holds for every st-relevant 3-class C:

(i) If C is central, then the set R_C of attachment nodes of C is a 3-connected subset in H.
(ii) If C is non-central, then $H[C]$ contains 2 edge-disjoint uv-paths, where u and v are the attachment nodes of C.

The first condition is equivalent to satisfying a clique of 3-demands on R_C.[1] For the second condition, consider a non-central st-relevant 3-class C with attachment nodes u, v. One can see that if H contains all forced edges and satisfies

---

[1] Recall that since the relation $\{(u, v) \in V \times V : \text{no 2-cut separates } u, v\}$ is transitive, this is equivalent to having a tree of 3-demands on R_C.

(i,ii) then the number of edge-disjoint uv-paths in H is larger by exactly 1 than their number in $H[C]$—the additional path (that exists in H but not in $H[C]$) goes along the cycle of the cactus that contains C, and there is exactly one such path. Thus, the demand $r(u, v) = 3$ is equivalent to requiring two edge-disjoint paths from u to v in R_C (in addition to including all forced edges).

We therefore obtain an equivalent 3-SND instance by replacing the single relative 3-demand st by a set D of 3-demands that form a clique (or, which is equivalent, a tree) on the set R_C of attachment nodes of every st-relevant 3-class C. These new demands can be computed in polynomial time, and they are ordinary 3-demands, since each R_C is a 3-edge-connected subset in G. This concludes the proof of Theorem 3.

Proof of Theorem 1. We can now describe a 2-approximation for 3-RSND. We treat each demand in the 3-RSND instance as its own instance of SD-3-RSND, solve each SD-3-RSND instance, and include the edges of each solution in our output.

4   SD-k-RSND

We give a recursive $2^{O(k^2)}$-approximation algorithm for SD-k-RSND for arbitrary constant k (Theorem 2). The algorithm is a generalization of the SD-3-RSND algorithm from [16]. At a high level, the main idea is to partition the input graph using a hierarchy of important separators, prove a structure theorem that characterizes the required connectivity guarantees within each component of the hierarchy, and then achieve these guarantees using a variety of subroutines.

4.1   Hierarchical Chain Decomposition

We first define important separators and describe how to use them to construct a hierarchical k-chain decomposition of G.

Definition 3. *Let X and Y be vertex sets of a graph G. An (X, Y)-separator of G is a set of edges S such that there is no path between any vertex $x \in X$ and any vertex $y \in Y$ in $G \setminus S$. An (X, Y)-separator S is minimal if no subset $S' \subset S$ is also an (X, Y)-separator. If $X = \{x\}$ and $Y = \{y\}$, we say that S is an (x, y)-separator.*

Definition 4 (Definition 20 of [16]). *Let S be an (X, Y)-separator of graph G, and let R be the vertices reachable from X in $G \setminus S$. Then S is an important (X, Y)-separator if S is minimal and there is no (X, Y)-separator S' such that $|S'| \leq |S|$ and $R' \subset R$, where R' is the set of vertices reachable from X in $G \setminus S'$.*

In Sect. 4.1 of [16], the authors describe how to construct the "$s - t$ 2-chain" of a graph G.[2] Here, we define the (X, Y) h-*chain* of G similarly, where X and Y are vertex sets and $h > 0$ is an integer.

---

[2] Note that all separator-based chain definitions given in this section are unrelated to the cactus-based chain definitions in Sect. 3.

First, if there are no important (X, Y)-separators of size h in G, then the (X, Y) h-chain of G is simply G and we're done (the chain is a single component, G, with no separators). If such an important separator exists, then we first find an important (X, Y)-separator S_0^h of size h in G, and we let R_0^h be the set of vertices reachable from any vertex $x \in X$ in $G \backslash S_0^h$. We let $V_{(0,r)}^h$ be the vertices in R_0^h incident on S_0^h, and let $V_{(1,\ell)}^h$ be the nodes in $V \setminus R_0^h$ incident on S_0^h. We then proceed inductively. Given $V_{(i,\ell)}^h$, if there is no important $(V_{(i,\ell)}^h, Y)$-separator of size h in $G \setminus (\cup_{j=0}^{i-1} R_j^h)$ then the chain is finished. Otherwise, let S_i^h be such a separator, let R_i^h be the nodes reachable from $V_{(i,\ell)}^h$ in $(G \setminus (\cup_{j=0}^{i-1} R_j^h)) \setminus S_i^h$, let $V_{(i,r)}^h$ be the nodes in R_i^h incident on S_i^h, and let $V_{(i+1,\ell)}^h$ be the nodes in $V \setminus (\cup_{j=0}^{i} R_j^h)$ incident on S_i^h. After this process completes we have our (X, Y) h-chain, consisting of components $R_0^h, \dots, R_p^h$ along with important separators $S_0^h, \dots, S_{p-1}^h$ between the components.

Next we note that by Lemma 21 of [16], we can find an important (X, Y)-separator of size h in polynomial time as long as h is a constant.

Lemma 3 Lemma 21 of [16]). *Let $d \geq 0$. An important (X, Y)-separator of size d can be found in time $4^d \cdot n^{O(1)}$ (if one exists), where $n = |V|$.*

Constructing the Hierarchical k-chain Decomposition. Now we describe how to construct the *hierarchical k-chain decomposition* of G. We start by creating the (s, t) 2-chain of G. We say that each component of the (s, t) 2-chain is a *2-component* of G in the hierarchical chain decomposition.

We then proceed inductively. Let R_i^h be an h-component of the hierarchical k-chain decomposition. If $h = k - 1$, then the decomposition is finished. Otherwise, build the $(V_{(i,\ell)}^h, V_{(i,r)}^h)$ $(h+1)$-chain of R_i^h. The $(h+1)$-chain consists of $(h+1)$-components. Component R_i^h is the parent of these $(h + 1)$-components. After this process completes we have our hierarchical k-chain decomposition of G.

The set of all h-components can be ordered as follows: The h-component that contains s is the first component while the h-component that contains t is last. All other h-components are adjacent via a left important separator and a right important separator to a left neighbor h-component and a right neighbor h-component, respectively.

4.2 Structure Theorem

Preliminaries. We say a subgraph H satisfies the RSND demand (X, Y, d) on input graph G if the following is true: If there is a path from at least one vertex in X to at least one vertex in Y in $G \setminus F$, where F is a set of at most $d - 1$ edges, then there is a path from at least one vertex in X to at least one vertex in Y in $H \setminus F$. Going forward, if $V_{(i,\ell)}^h = \{s\}$, then we consider S_{i-1}^h to be the empty set. Similarly, if $V_{(i,r)}^h = \{t\}$, then S_i^h is the empty set.

Fix an h-component R_i^h and let X be a vertex set such that $X \subseteq V_{(i,\ell)}^h$. We say that S_X is the set of edges in S_{i-1}^h incident on vertices in X. Similarly, if

Y is a vertex set such that $Y \subseteq V^h_{(i,r)}$, we say that S_Y is the set of edges in S^h_i incident on vertices in Y. We will also use S to denote the set of all edges in an important separator in the decomposition. Let H be a subgraph of G. We will also say that $G^h_i = G[R^h_i]$ is the subgraph of G induced by the vertex set R^h_i, and that $H^h_i = H[R^h_i]$ is the subgraph of H induced by R^h_i.

We can now use the hierarchical chain construction to give a structure lemma that characterizes feasible solutions. The lemma states that a subgraph H of G is a feasible solution to SD-k-RSND if and only if in the hierarchical k-chain decomposition of G, all edges in S are in H, and certain connectivity requirements between groups of vertices in $V^h_{(i,\ell)}$ and in $V^h_{(i,r)}$ are met in H^h_i for each component R^h_i in the decomposition.

Theorem 5 (Structure Theorem). *Subgraph H is a feasible solution to SD-k-RSND if and only if all edges in S are included in H, and for each h-component R^h_i in the hierarchical k-chain decomposition of G, subgraph H^h_i satisfies the following:*

1 H^h_i is a feasible solution to RSND on subgraph G^h_i with demands

$$\Big\{ (X, Y, d) \; : \; X \subseteq V^h_{(i,\ell)},\ Y \subseteq V^h_{(i,r)},\ (X, Y) \neq \big(V^h_{(i,\ell)}, V^h_{(i,r)}\big),$$
$$d = \max(0, k + |S_X| + |S_Y| - |S^h_{i-1}| - |S^h_i|) \Big\}.$$

2 H^h_i is a feasible solution to RSND on subgraph G^h_i with demand

$$\Big(V^h_{(i,\ell)}, V^h_{(i,r)}, h + 1 \Big).$$

3 H^h_i is a feasible solution to RSND on subgraph G^h_i with demand

$$\Big(V^h_{(i,\ell)}, V^h_{(i,r)}, k - 1 \Big).$$

The proof of this structure theorem is long and involved; due to space constraints it can be found in the full version [17].

4.3   Algorithm and Analysis

Algorithm. We can now use this Structure Theorem to give a $2^{O(k^2)}$-approximation for SD-k-RSND. We first create the hierarchical k-chain decomposition of G in polynomial time, as described in Sect. 4.1. Within each component we run a set of algorithms to satisfy the RSND demands stated in Theorem 5. Our solution, H, includes the outputs of each of these algorithms along with S, the set of all edges in the separators of the hierarchical k-chain decomposition. We now describe the set of algorithms run on each component in the hierarchical k-chain decomposition. Fix a component R^h_i of the decomposition and let $X \subseteq V^h_{(i,\ell)}$, $Y \subseteq V^h_{(i,r)}$, and $d = \max(0, k + |S_X| + |S_Y| - |S_{i-1}| - |S_i|)$:

- **Base Case (Shortest st Path).** For each X, Y pair such that $d = 1$, contract the vertices in X and contract the vertices in Y to create super nodes x and y, respectively. We first check in polynomial time if x and y are connected in $G_i^h = G[R_i^h]$. If they are connected, then we create an instance of the Weighted st Shortest-Path problem on G_i^h (in polynomial time), using x and y as our source and destination nodes. For each edge $e \in E(R_i^h)$, set the weight of e to $w(e)$. Run a polynomial-time Weighted st Shortest-Path algorithm on this instance (e.g. Dijkstra's algorithm), and add to H all edges in the output of the algorithm.
- **Recursive Step.** For each X, Y pair such that $1 < d < k$, we create an instance of SD-d-RSND on G_i^h. Contract the vertices in X, and in Y, to create super nodes x and y, respectively. For each $e \in E(R_i^h)$, set the cost of e to $w(e)$. The set of RSND demands is just $\{(x, y, d)\}$. Run the recursive polynomial-time SD-d-RSND algorithm on this instance, where $d < k$. Add to H all edges in the output of the algorithm.
- **Final Recursive Step.** We create an SD-$(k-1)$-RSND instance on G_i^h. Contract the vertices in $V_{(i,\ell)}^h$, and in $V_{(i,r)}^h$, to create super nodes v_ℓ and v_r, respectively. For each $e \in E(R_i^h)$, set the cost of e to $w(e)$. The set of RSND demands is just $\{(v_\ell, v_r, k-1)\}$. Run the recursive SD-$(k-1)$-RSND algorithm on this instance. Add to H all edges in the output of the algorithm.
- **Min-Cost Flow.** We create an instance of the Min-Cost Flow problem on G_i^h. Contract the vertices in $V_{(i,\ell)}^h$, and in $V_{(i,r)}^h$, to create super nodes v_ℓ and v_r, respectively. Let v_ℓ be the source and v_r the sink. For each $e \in E(R_i^h)$, set the capacity of e to 1 and the cost of e to $w(e)$. Require a minimum flow of $h + 1$, and run a poly-time Min-Cost Flow algorithm on this instance. Since all capacities are integer the algorithm will return an integral flow, so we add to H all edges with non-zero flow.

Analysis. All missing proofs from this section can be found in the full version [17]. The following lemma is essentially directly from Theorem 5 and the description of the algorithm.

Lemma 4. *Let H be the output of the algorithm given in Sect. 4.3. Subgraph H is a feasible solution to the SD-k-RSND problem.*

Let H^* denote the optimal solution, and for any set of edges $A \subseteq E$, let $w(A) = \sum_{e \in A} w(e)$. The next lemma follows from combining the approximation ratios of each of the subroutines used in the algorithm and solving the recurrence.

Lemma 5. $w(H) \leq 2^{O(k^2)} * w(H^*)$.

Theorem 2 is directly implied by Lemmas 4 and 5 together with the observation that the algorithm runs in polynomial time.

References

1. Adjiashvili, D., Hommelsheim, F., Mühlenthaler, M.: Flexible graph connectivity: approximating network design problems between 1- and 2-connectivity (2020)
2. Adjiashvili, D., Hommelsheim, F., Mühlenthaler, M., Schaudt, O.: Fault-tolerant edge-disjoint paths - beyond uniform faults (2020)
3. Bansal, I., Cheriyan, J., Grout, L., Ibrahimpur, S.: Improved approximation algorithms by generalizing the primal-dual method beyond uncrossable functions (2022)
4. Bilò, D., Gualà, L., Leucci, S., Proietti, G.: Multiple-edge-fault-tolerant approximate shortest-path trees (2016). https://doi.org/10.48550/ARXIV.1601.04169
5. Bodwin, G., Dinitz, M., Nazari, Y.: Vertex fault-tolerant emulators. In: Braverman, M. (ed.) 13th Innovations in Theoretical Computer Science Conference, ITCS 2022. LIPIcs, vol. 215, pp. 25:1–25:22. Schloss Dagstuhl - Leibniz-Zentrum für Informatik (2022). https://doi.org/10.4230/LIPIcs.ITCS.2022.25
6. Bodwin, G., Dinitz, M., Nazari, Y.: Epic fail: emulators can tolerate polynomially many edge faults for free. In: 14th Innovations in Theoretical Computer Science Conference, ITCS 2023 (2023)
7. Bodwin, G., Dinitz, M., Parter, M., Williams, V.V.: Optimal vertex fault tolerant spanners (for fixed stretch). In: Czumaj, A. (ed.) Proceedings of the Twenty-Ninth Annual ACM-SIAM Symposium on Discrete Algorithms, SODA 2018, New Orleans, LA, USA, 7–10 January 2018, pp. 1884–1900. SIAM (2018)
8. Bodwin, G., Dinitz, M., Robelle, C.: Optimal vertex fault-tolerant spanners in polynomial time. In: Naor, J.S., Buchbinder, N. (eds.) Proceedings of the 2022 ACM-SIAM Symposium on Discrete Algorithms, SODA 2022, pp. 2924–2938. SIAM (2022). https://doi.org/10.1137/1.9781611976465.174
9. Bodwin, G., Patel, S.: A trivial yet optimal solution to vertex fault tolerant spanners. In: Proceedings of the 2019 ACM Symposium on Principles of Distributed Computing, PODC 2019, pp. 541–543. Association for Computing Machinery, New York (2019). https://doi.org/10.1145/3293611.3331588
10. Boyd, S., Cheriyan, J., Haddadan, A., Ibrahimpur, S.: Approximation algorithms for flexible graph connectivity (2022)
11. Chechik, S., Langberg, M., Peleg, D., Roditty, L.: Fault tolerant spanners for general graphs. SIAM J. Comput. 39(7), 3403–3423 (2010)
12. Chekuri, C., Jain, R.: Approximating flexible graph connectivity via räcke tree based rounding (2022)
13. Chekuri, C., Jain, R.: Augmentation based approximation algorithms for flexible network design (2022)
14. Cheriyan, J., Laekhanukit, B., Naves, G., Vetta, A.: Approximating rooted Steiner networks. ACM Trans. Algorithms 11(2), 8:1–8:22 (2014)
15. Cheriyan, J., Thurimella, R.: Approximating minimum-size k-connected spanning subgraphs via matching. SIAM J. Comput. 30(2), 528–560 (2000). https://doi.org/10.1137/S009753979833920X
16. Dinitz, M., Koranteng, A., Kortsarz, G.: Relative survivable network design. In: APPROX-RANDOM, vol. 245, pp. 41:1–41:19 (2022)
17. Dinitz, M., Koranteng, A., Kortsarz, G., Nutov, Z.: Improved approximations for relative survivable network design (2023). https://doi.org/10.48550/arXiv.2304.06656
18. Dinitz, M., Krauthgamer, R.: Fault-tolerant spanners: better and simpler. In: Proceedings of the 30th Annual ACM Symposium on Principles of Distributed Computing, PODC 2011, San Jose, CA, USA, 6–8 June 2011, pp. 169–178 (2011)

19. Dinitz, M., Robelle, C.: Efficient and simple algorithms for fault-tolerant spanners. In: Emek, Y., Cachin, C. (eds.) ACM Symposium on Principles of Distributed Computing, PODC 2020, pp. 493–500. ACM (2020). https://doi.org/10.1145/3382734.3405735

20. Dinitz, Y., Westbrook, J.: Maintaining the classes of 4-edge-connectivity in a graph on-line. Algorithmica **20**, 242–276 (1998)

21. Dinitz, Y., Nutov, Z.: A 2-level cactus model for the system of minimum and minimum+ 1 edge-cuts in a graph and its incremental maintenance. In: Proceedings of the Twenty-Seventh Annual ACM Symposium on Theory of Computing, pp. 509–518 (1995)

22. Feldmann, A.E., Mukherjee, A., van Leeuwen, E.J.: The parameterized complexity of the survivable network design problem. In: SOSA, pp. 37–56 (2022)

23. Gabow, H.N., Goemans, M.X., Tardos, É., Williamson, D.P.: Approximating the smallest k-edge connected spanning subgraph by LP-rounding. Networks **53**(4), 345–357 (2009)

24. Henzinger, M.R.: A static 2-approximation algorithm for vertex connectivity and incremental approximation algorithms for edge and vertex connectivity. J. Algorithms **24**(1), 194–220 (1997)

25. Jain, K.: A factor 2 approximation algorithm for the generalized Steiner network problem. Combinatorica **21**(1), 39–60 (2001). https://doi.org/10.1007/s004930170004

26. Khandekar, R., Kortsarz, G., Nutov, Z.: Approximating fault-tolerant Group-Steiner problems. Theor. Comput. Sci. **416**, 55–64 (2012)

27. Lo, O.S., Schmidt, J.M., Thorup, M.: Compact cactus representations of all non-trivial min-cuts. Discret. Appl. Math. **303**, 296–304 (2021)

28. Marx, D.: Important separators and parameterized algorithms. In: Kolman, P., Kratochvíl, J. (eds.) WG 2011. LNCS, vol. 6986, pp. 5–10. Springer, Heidelberg (2011). https://doi.org/10.1007/978-3-642-25870-1_2

29. Poutre, J.A.L.: Maintenance of 2- and 3-edge-connected components of graphs II. SIAM J. Comput. **29**(5), 1521–1549 (2000)

30. Williamson, D.P., Goemans, M.X., Mihail, M., Vazirani, V.V.: A primal-dual approximation algorithm for generalized Steiner network problems. Combinatorica **15**(3), 435–454 (1995). https://doi.org/10.1007/BF01299747

Independent Set in k-Claw-Free Graphs: Conditional χ-Boundedness and the Power of LP/SDP Relaxations

Parinya Chalermsook[1], Ameet Gadekar[1(✉)], Kamyar Khodamoradi[2], and Joachim Spoerhase[3]

[1] Aalto University, Espoo, Finland
{parinya.chalermsook,ameet.gadekar}@aalto.fi
[2] University of Regina, Regina, Canada
kamyar.khodamoradi@uregina.ca
[3] University of Sheffield, Sheffield, UK
j.spoerhase@sheffield.ac.uk

Abstract. This paper studies k-claw-free graphs, exploring the connection between an extremal combinatorics question and the power of a convex program in approximating the maximum-weight independent set in this graph class. For the extremal question, we consider the notion, that we call *conditional χ-boundedness* of a graph: Given a graph G that is assumed to contain an independent set of a certain (constant) size, we are interested in upper bounding the chromatic number in terms of the clique number of G. This question, besides being interesting on its own, has algorithmic implications (which have been relatively neglected in the literature) on the performance of SDP relaxations in estimating the value of maximum-weight independent set.

For $k = 3$, Chudnovsky and Seymour (JCTB 2010) prove that any 3-claw-free graph G with an independent set of size three must satisfy $\chi(G) \leq 2\omega(G)$. Their result implies a factor 2-estimation algorithm for the maximum weight independent set via an SDP relaxation (providing the first non-trivial result for maximum-weight independent set in such graphs via a convex relaxation). An obvious open question is whether a similar conditional χ-boundedness phenomenon holds for any k-claw-free graph. Our main result answers this question negatively. We further present some evidence that our construction could be useful in studying more broadly the power of convex relaxations in the context of approximating maximum weight independent set in k-claw free graphs. In particular, we prove a lower bound on families of convex programs that are stronger than known convex relaxations used algorithmically in this context.

This research was partially done during the trimester on Discrete Optimization at Hausdorff Research Institute for Mathematics (HIM) in Bonn, Germany. The research has received funding from the European Research Council (ERC) under the European Union's Horizon 2020 research and innovation programme (grant agreement No 759557) and from Academy of Finland (grant number 310415). Kamyar Khodamoradi was supported by Deutsche Forschungsgemeinschaft (project number 399223600).

© The Author(s), under exclusive license to Springer Nature Switzerland AG 2023
J. Byrka and A. Wiese (Eds.): WAOA 2023, LNCS 14297, pp. 205–218, 2023.
https://doi.org/10.1007/978-3-031-49815-2_15

Keywords: χ-boundedness · Convex relaxation · Ramsey theory

1   Introduction

For $k \geq 3$, a graph is said to be k-claw-free if the neighborhood of each vertex does not contain an independent set of size k. This paper focuses on an extremal question in k-claw-free graphs and its connection to the power of convex programs in estimating the maximum weight independent set (MWIS) in such graphs. The study of such relation originated already around 50 years ago when Lovász defined the notion of perfect graphs based on graph extremal properties and showed connections to (exact) semi-definite programming formulations for optimization problems [16,24]. Such connections are known to be "approximation preserving" so they imply a connection between standard Ramsey-type theorems (and χ-boundedness) in approximating the cardinality (resp. the weight) of the maximum independent set. Several approximation algorithms in geometric intersection graphs have been successfully derived in this framework [6,9,10,23].

Most prior works that extend perfect graphs rely on the notion of the clique constrained stable set polytope (QSTAB)—a convex relaxation that can be optimized via semi-definite programs[1]. The power of QSTAB is captured precisely by standard extremal bounds. For example, the χ-boundedness $\chi(G) \leq \gamma\omega(G)$ in a "natural" graph class is (roughly) equivalent to QSTAB providing γ-estimate on the weight of maximum independent set in the same graph class [8]. Despite successful cases of the extremal approach, QSTAB fails unexpectedly in graph classes such as k-claw-free graphs: For any $k \geq 3$, a simple greedy algorithm immediately gives a factor $(k-1)$ approximation for MWIS, while QSTAB (and other known convex relaxations) is unable to give $f(k)$ approximation for any function f (see our full version [7]).

This work is an attempt to better understand the power of convex relaxations for approximating MWIS in k-claw-free graphs. MWIS on k-claw-free graphs contains many well-known (open) problems as special cases, such as set packing [12,14,17,18] and independent set in sparse graphs [1,2], for which QSTAB has been shown to perform relatively well in terms of approximating the problems. Somewhat surprisingly, the study of convex relaxations for MWIS on k-claw-free graphs has been absent from the literature.

This paper is inspired by the following theorem of Chudnovsky and Seymour.

Theorem 1 (Chudnovsky–Seymour [13]). *For every connected claw-free ($k = 3$) graph G with $\alpha(G) \geq 3$, it holds that $\chi(G) \leq 2\omega(G)$.*

Remark that, the condition $\alpha(G) \geq 3$ is necessary, for otherwise, we can have $\chi(G) = \widetilde{\Omega}(\omega(G)^2)$ for claw-free graphs, contradicting the above theorem[2]. Their

---

[1] Recall that the polytope is defined as $\mathsf{QSTAB}(G) = \{x \in [0,1]^{|V(G)|} : \sum_{i \in Q} x_i \leq 1 (\forall \text{clique } Q)\}$. Optimizing this itself is NP-hard, but we can optimize an SDP whose solution is feasible for QSTAB.

[2] The notation $\widetilde{\Omega}$ hides asymptotically smaller terms.

theorem, in particular, implies that the sum-of-squares (SoS) hierarchy—a family of increasingly tight convex relaxations—gives an efficient 2-estimation algorithm for MWIS in claw-free graphs. An obvious open question is whether the above theorem can be generalized to k-claw-free graphs.

1.1   Our Contributions

Our results are stated via our new notion of *conditional χ-boundedness*. We say that G is (t, γ)-conditionally χ-bounded if $\chi(G) \leq \gamma\omega(G)$ whenever $\alpha(G) \geq t$. Moreover, a graph class is (t, γ)-conditionally χ-bounded if every graph in that class is. The following theorem (which is a simple combination of known facts) connects SoS to an extremal question of Chudnovsky and Seymour.

Theorem 2. *Consider a graph class that is closed under clique replacement[3]. If the graph class is (t, γ)-conditionally χ-bounded, then t rounds of SoS gives a factor γ-estimation for MWIS.*

In particular, if $t = O(1)$, then SoS gives a γ-estimation algorithm in polynomial time. Our first main contribution is to rule out this approach, i.e., refuting the possibility to generalize Theorem 1. We show the following result (please refer Theorem 5 for a precise statement).

Theorem 3 (Simplified). *For every $k \geq 4$, there exists an infinite family of graphs $\{G_n\}$ such that G_n is a connected k-claw-free graph on n vertices with*

$$\alpha(G_n) = \Omega\left(\frac{n}{\log n}\right) \text{ and } \chi(G) \geq f(k)\left(\frac{\omega(G)}{\log \omega(G)}\right)^{k/2}, \text{ for some function } f.$$

This lower bound almost matches the upper bound provided by [32]. We remark that both upper and lower bounds are tight w.r.t. the state-of-the-art bounds on Ramsey number, that is, improving in either direction requires asymptotically improving the bound on the Ramsey number of a graph.

Given this theorem, the next obvious open question is whether there is a lower bound on the performance of SoS that "separates" SoS from the simple greedy factor-k approximation algorithm. While we do not manage to settle this question, we instead show in the theorem below that our construction for Theorem 3 can be used to construct a bad example for a Sherali–Adams strengthening of $\mathsf{QSTAB}(G) = \{x \in [0, 1]^{|V(G)|} : \sum_{i \in Q} x_i \leq 1(\forall \text{ clique} Q)\}$.

Theorem 4 (Integrality gap of Sherali–Adams on QSTAB). *Let $k \geq 4$. For $0 < \epsilon \leq 1/3$, there exists an infinite family of graphs $\{G_n\}$ such that G_n is a connected k-claw-free graph on n vertices with[4] $\alpha(G_n) = \Omega(n^\epsilon)$ and the integrality gap of Sherali–Adams hierarchy on $\mathsf{QSTAB}(G_n)$ is $\Omega_k(n^\epsilon)$, even after $\Omega_k(n^{1-2\epsilon})$ rounds.*

---

[3] A clique replacement operation on graph G replaces any vertex v with a clique K_v of arbitrary size and connects each vertex in a clique to every neighbor of v. It is easy to see that k-claw-free graphs are closed under clique replacement.

[4] The notations O_k, Θ_k, Ω_k hide multiplicative functions in k.

We remark that this theorem can be contrasted with special cases of k-claw-free graphs, e.g., in the bounded degree setting [2], where the (poly-logarithmic rounds of) Sherali–Adams strengthening of QSTAB provides an optimal approximation factor under the unique games conjecture (UGC) [1].

Discussion of Previous Work on Convex Programs. Let us compare our result of Theorem 4 with the bounds of Chan and Lau [11]. In particular, [11] considers k-SetPacking, a special case of maximum independent set in $(k+1)$-claw-free graph, where we are given a k-uniform hypergraph H on n vertices and we are asked to find a maximum matching in H. In their work, [11] show that Sherali–Adams on the standard LP for k-SetPacking has integrality gap of at least $k - 2$, even after $\Omega_k(n)$ rounds. Additionally, they show that, for constant k, QSTAB for k-SetPacking can be captured by a polynomial size LP and has integrality gap of at most $(k+1)/2$. In contrast, our result of Theorem 4 is for more general problem of maximum independent set in k-claw-free graphs, and yields integrality gap of $\Omega_k(n^\epsilon)$, which is a function of n, for Sherali–Adams on QSTAB with rounds $\Omega_k(n^{1-2\epsilon})$, which is a stronger program than that of [11]. However, we remark that, the parameters of Theorem 4 can be adjusted to yield an integrality gap of $g(k)$, for any function g, for Sherali–Adams on QSTAB with $\Omega_k(n)$ number of rounds, which is linear in n (refer our full version [7] for more details). Thus, our results yield a larger integrality gap for a stronger program for maximum independent set in k-claw-free graphs compared to that of [11].

Overview of Techniques: We give a high-level overview for the proof of Theorem 3. For simplicity, let us focus on the case of $k = 5$ and a slightly weaker bound. The first component of our construction is a Ramsey graph. It is known that there exists an n-vertex graph H such that $\alpha(H) = 2$ and there is no clique of size $c\sqrt{n \log n}$ for some constant c [19]. Therefore this graph is 3-claw-free (since there is no independent set of size 3) and $\chi(G) \geq n/\alpha(G) = n/2 \geq \widetilde{\Omega}(\omega(G)^2)$. So, this graph has almost all our desired properties except that $\alpha(H)$ is very small while we need large independent sets. We will "compose" several copies of H together to obtain a final graph G, ensuring that (i) $\alpha(G)$ can be made arbitrarily large, (ii) $\omega(G)$ is roughly the same as $\omega(H)$, and (iii) the graph remains 5-claw-free.

The key component of our composition step is a special graph operation that we call *bi-conflict composition.* In particular, given any two graphs H_1, H_2 having n vertices each, the graph $D = bcc(H_1, H_2)$ is obtained as follows. First, construct a graph D' by connecting H_1 and H_2 by an arbitrary matching $M: |M| = n$. Next, define D where $V(D) = M$ and for each $e, e' \in M$, we have $(e, e') \in E(D)$ if and only if they are not induced matching edges in D'. There are two simple properties of graph D that are crucial to our analysis.

- (P1) the maximum claw in D is at most $\min\{\alpha(H_1), \alpha(H_2)\}$.
- (P2) the maximum clique in D is $\omega(D) \leq \omega(H_1) + \omega(H_2) + 1$.

The bi-conflict composition will be used as an analytical tool in our construction. Take q copies of Ramsey graphs $H_1, H_2, \ldots, H_q$ on n vertices. For each $i =$

$1, \ldots, q-1$, connect H_i to H_{i+1} by a matching M_i: $|M_i| = n$. Call this graph G'. Our final graph G has vertices $V(G) = \bigcup_i V(M_i)$ and the edges are defined such that $(e, e') \in E(G)$ if e and e' form an induced matching. Notice that the structure of graph G for vertices that correspond to M_i-edges is roughly the same as $bcc(H_i, H_{i+1})$, which allows us to invoke properties (P1) and (P2).

Notice that $\alpha(G) = \Omega(q)$ (Pick one matching edge from each M_i when i is odd). Property (P1) guarantees that the maximum claw is at most 4 (therefore G is 5-claw-free) and property (P2) guarantees that the value of maximum clique is at most $O(\sqrt{n \log n})$. By choosing q to be sufficiently large, we would be done. We remark that this construction gives a non-trivial lower bound (albeit weaker than Theorem 3) for $k \geq 5$. To make the result work for $k = 4$, we need to compose the copies of the Ramsey graph more carefully.

1.2 Conclusion and Open Problems

Our main contribution is to initiate the study of convex relaxation aspects of independent set in k-claw-free graphs. The main open question is whether there is any convex relaxation approach that gives a reasonable approximation guarantee for MWIS in k-claw-free graphs (for $k \geq 4$). Conceptually, we made explicit the implication of Chudnovsky and Seymour's theorem (which can be seen as conditional χ-boundedness), that SoS gives a reasonable approximation guarantee for claw-free graphs. We refute the possibility of generalizing such a result to k-claw-free graphs for all $k \geq 4$ and present evidence that this graph family might be a bad instance for SoS.

1.3 Further Related Work

A graph class is said to be χ-bounded if $\chi(G)$ can be upper bounded by $f(\omega(G))$ where f is some function. The concept of χ-boundedness has been studied extensively in graph theory (see, e.g., the survey [29] and references therein). In algorithms and optimization, we are mostly concerned with χ-boundedness where f is a linear or close to linear function. Roughly, the ratio $\chi(G)/\omega(G)$ captures the integrality gap of a convex programming relaxation QSTAB in estimating the value of maximum independent set of a graph [8].

As for approximating the maximum weight of the independent set of k-claw-free graphs, a local search algorithm due to Berman [3] had remained the best known approximation algorithm for the past two decades. Berman's algorithm achieved a factor of $\frac{k}{2} + \epsilon$ in polynomial time. In 2021, Neuwohner [25] broke the barrier of $\frac{k}{2}$ by improving the factor to $\frac{k}{2} - \frac{1}{63,700,992}$. In a very recent work, Thiery and Ward [31] improved the factor to $\frac{k}{2} - \delta_k$ for a constant $\delta_k \geq 0.214$. On the other hand, the problem is NP-hard to approximate better than a factor of $\Omega(k/\log k)$ [17][5].

---

[5] In fact, the hardness even holds for a special case of the problem, namely the unweighted k-set packing problem.

With respect to convex programs, Chan and Lau [11] study the power of standard LP and Sherali–Adams on k-SetPacking, which is a special case of maximum independent set in $(k + 1)$-claw-free graph. In k-SetPacking, we are given a k-uniform hypergraph H on n vertices and we are asked to find a maximum matching in H. In their work, [11] show that Sherali–Adams on the standard LP for k-SetPacking has integrality gap of at least $k - 2$, even after $\Omega(n/k^3)$ rounds. Additionally, they show that, for constant k, QSTAB for k-SetPacking can be captured by a polynomial size LP and has integrality gap of at most $(k + 1)/2$.

Organization. Section 2 explains the basic graph-theoretic terminologies. Our graph theoretic result is proved in Sect. 3. All convex relaxation results, as well as the connection with the notion of conditional χ-boundedness, are proved in Sect. 4.

2   Preliminaries

We follow standard graph theoretic notation. Given a graph $G = (V(G), E(G))$, $M \subseteq E(G)$ is a *matching* if no pair of edges in M share a vertex. Further, a matching $M \subseteq E(G)$ is said to be an *induced matching* if it is an induced subgraph of G. Finally, for a matching M, $e_i \neq e_j \in M$ is said to be an *intersecting matching pair* if e_i and e_j do not form an induced matching; and M is an *intersecting matching* if every pair in M is an intersecting matching pair.

For $k \geq 3$, a *k-claw* is the graph $K_{1,k}$. For a k-claw T, the vertex with degree k is called the *central vertex* of T, and the remaining vertices of degree one are called *leaves* of T. A graph G is said to be *k-claw-free* if there exists no k-claw as an induced subgraph. For a graph G, $\alpha(G)$ is the size of maximum independent set in G, $\omega(G)$ is the size of maximum clique in G, and $\chi(G)$ is the chromatic number of G. For weighted case, $\alpha(G)$ and $\omega(G)$ represent the maximum weight independent set and maximum weight clique in G respectively.

The notations O_k, Θ_k, Ω_k hide multiplicative functions in k.

In the *maximum-weight independent set* problem (MWIS), we are given a graph G together with weights $\{w_v\}_{v \in V(G)}$ on the vertices, and our goal is to find an independent set $S \subseteq V(G)$ with maximum total weights.

3   Graph Theoretic Result

In this section we prove the following theorem.

Theorem 5. *For $k \geq 4$, there exists n_0 depending on k such that for infinitely many $n \geq n_0$, there exists a connected k-claw-free graph G on n vertices with*

$$\alpha(G) = \Omega\left(\frac{n}{\log n}\right) \text{ and } \chi(G) \geq f(k)\left(\frac{\omega(G)}{\log \omega(G)}\right)^{k/2}, \text{ for some } f(k).$$

To this end, we will use known results about Ramsey graphs, which we introduce in Sect. 3.1. We explain our construction and analysis in Sects. 3.2.

3.1 Ramsey Graphs

Let $R(s,t)$, for $s \geq 3$ denote the Ramsey number, i.e., $R(s,t)$ is the minimum number such that any graph on $R(s,t)$ vertices has either an independent set of size s or a clique of size t.

Theorem 6 ([4,5,19]). *For any $s \geq 3$ and for sufficiently large t, $R(s,t) \geq c'_s \cdot t^{\frac{s+1}{2}} (\log t)^{\frac{1}{s-2} - \frac{s+1}{2}}$, for some constant c'_s depending only on s.*

Thus, the above theorem implies that for $s \geq 3$ and sufficiently large t, there is a graph on $\lceil R(s,t) - 1 \rceil$ vertices that has neither an independent set of size s nor a clique of size t. We call such a graph an (s,t)-*Ramsey graph*.

Corollary 1. *For $s \geq 3$ and sufficiently large t, there is an (s,t)-Ramsey graph on $c_s \cdot t^{\frac{s+1}{2}} (\log t)^{\frac{1}{s-2} - \frac{s+1}{2}}$ vertices, for some positive constant c_s depending only on s.*

3.2 Graph Construction

Lemma 1. *Let $k \geq 4, p \geq 1, \tau \geq 3$, and let $H = (V(H), E(H))$ be a $(k-1,\tau)$-Ramsey graph. Then, there exists a connected k-claw-free graph G on $\Theta(p|V(H)|)$ vertices such that $p \leq \alpha(G) \leq 3pk$ and $\omega(G) \leq 3\tau$.*

Proof. We construct the graph G in two steps. In the first step, we construct an auxiliary graph $G' = (V(G'), E(G'))$ using $(k-1,\tau)$-Ramsey graph H. Finally, in the second step, using this auxiliary graph G', we construct our graph G.

Construction of Auxiliary Graph G'. Let $Q = K_{\tau-1}$ be the complete graph on $\tau - 1$ vertices. We construct the graph G' in two steps: In the first step, we use H and Q to create a graph B, which we call a *block*. In the second step, we construct the graph G' using p copies of B. For $i \in [p]$, we construct block B_i using Q and two copies of H as follows (see Fig. 1). Let H_i^1 and H_i^2 be copies of H, and let Q_i be a copy of Q. We connect each vertex in H_i^1 with its respective copy in H_i^2 by an edge. Then, we pick an arbitrary set of $\tau - 1$ vertices from H_i^2 and add an (arbitrary) matching between this set and the vertices of Q_i. Finally, we connect each B_i to $B_{(i+1) \bmod p}$ as follows (see Fig. 2). We pick an arbitrary set of $\tau - 1$ vertices from $H_{(i+1) \bmod p}^1$ and add a matching between this set and the vertices of Q_i. This completes the construction of G'. Notice that G' forms a ring structure consisting of B_is. We call the new edges that we added in our construction *matching edges*.

Construction of G. For every matching-edge e_i in G', we create a vertex v_i in G. Then, $(v_i, v_j) \in E(G)$ if e_i and e_j is an intersecting matching pair (i.e., they do not form an induced matching).

Analysis. Let $n = |V(G)|$, then note $n = p(q_k(\tau) + 2(\tau - 1))$. Hence, $3pq_k(\tau) > n \geq pq_k(\tau)$. Since the Ramsey graph H and $Q = K_{\tau-1}$ used in construction of B are connected, we have that G' is connected. This implies that G is connected.

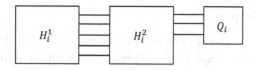

Fig. 1. One block B_i of the graph G'

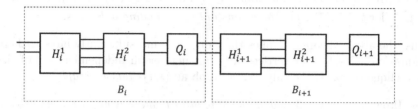

Fig. 2. Connecting block B_i and B_{i+1} in G'

Next we bound $\alpha(G)$. First note that $\alpha(G) \geq p\alpha(H)$ since from each block B_i we can pick matching edges between H_i^1 and H_i^2 that have endpoints on the vertices of $\alpha(H)$. The vertices in G corresponding to these matching edges form an independent set. For the upper bound, we will show that $\alpha(G) < 3p\alpha(H)$. Suppose for contradiction $\alpha(G) \geq 3p\alpha(H)$. Let I be an independent set in G with $|I| \geq 3p\alpha(H)$. Let M be the matching-edges of G' corresponding to the vertices in I. Let $M_i \subseteq M$ be the edges of M with both endpoints in B_i. Since B_i and $B_{(i+1) \bmod p}$ are connected by Q_i (which is a complete graph), the number of edges of M that are between B_is is at most p. Hence, the number of edges of M that lie completely within some B_i is at least $M - p$. Since $|M| \geq 3p\alpha(H)$, it must be that there is a block B_i such that $|M_i| \geq (3p\alpha(H) - p)/p \geq 2\alpha(H)$. Since the edges of M_i should have an endpoint in H_i^2, it implies that $\alpha(H_i^2) \geq |M_i| \geq 2\alpha(H)$, which is a contradiction since H_i^2 is a copy of H. Thus, we have that $p\alpha(H) \leq \alpha(G) < 3p\alpha(H)$.

For bounding $\omega(G)$, let $\mathcal{C}$ be a clique of maximum size in G. We claim that $|\mathcal{C}| \leq 3\tau$, which implies that $\omega(G) \leq 3\tau$. Let $E_\mathcal{C}$ be the matching-edges of G' corresponding to $\mathcal{C}$. Let $E_\mathcal{C}' \subseteq E_\mathcal{C}$ be the edges which have one endpoint in some copy of Q, and let $E_\mathcal{C}'' = E_\mathcal{C} \setminus E_\mathcal{C}'$. Suppose $E_\mathcal{C}' \neq \emptyset$, then consider an edge $e' \in E_\mathcal{C}'$ and suppose e' is incident on Q_i some for $i \in [p]$. Then, observe that every edge in $E_\mathcal{C}'$ must also be incident on Q_i since $E_\mathcal{C}'$ is an intersecting matching and edges incident on Q_i can not intersect with edges incident on $Q_j, j \neq i \in [p]$ by construction. Thus, $|E_\mathcal{C}'| \leq 2|Q| < 2\tau$. Hence, when $E_\mathcal{C}'' = \emptyset$, we have that $|E_\mathcal{C}| < 2\tau$ implying $|\mathcal{C}| < 2\tau$, as desired. For the other case when $E_\mathcal{C}'' \neq \emptyset$ then consider $e'' \in E_\mathcal{C}''$. We will bound the number of neighbors in G of e'' corresponding to the edges of $E_\mathcal{C}'$ and $E_\mathcal{C}''$. First, note that e'' has at most $|Q| < \tau$ neighbors in $E_\mathcal{C}'$ since e'' can intersect with at most $|Q| = \tau - 1$ many edges whose one end point is in Q_i. Next, we will show that e'' has at most $2(\tau - 1)$ neighbors in $E_\mathcal{C}''$, which implies that $|E_\mathcal{C}''| < 2\tau$, and hence $|\mathcal{C}| = |E_\mathcal{C}| = |E_\mathcal{C}'| + |E_\mathcal{C}''| < 3\tau$, as desired. To see this, note that e'' is incident on two adjacent copies of H, and hence the number of (matching) edges between these copies of H that form an intersecting matching together with e'' is at $2\omega(H) = 2(\tau - 1)$.

Finally, we show that G is k-claw-free. Suppose for contradiction, there is a k-claw T in G with central vertex $v \in T$ and leaves $v_1, \cdots, v_k \in T$. Let $M_T = \{e, e_1, \cdots, e_k\}$ be the matching-edges in G' corresponding to T with e corresponding to v and e_i corresponding to v_i, for $i \in [k]$. Also, let $L_T = \{e_1, \cdots, e_k\}$. First consider the case when e is incident between a copy H' of H and a copy Q' of Q. Let $I \subseteq V(H')$ be the endpoints of edges of L_T in H'. We claim that $|I| \geq k - 1$. To see this, note that for every edge $e_j \in L_T$ that is not incident on Q' must have one endpoint in H' since e_j and e form an intersecting matching pair. On the other hand, since Q' is a clique, there can be at most one edge in L_T with endpoint in Q' implying $|I| \geq k - 1$. As edges in L_T form an induced matching in G', we have that I is an independent set in H'. But then $|I| \geq k - 1$ which is a contradiction since $\alpha(H') < k - 1$. Now, consider the other case when e is between two copies H' and H'' of H. Let $L'_T \subseteq L_T$ be the edges that are between H' and H'', and let $L''_T = L_T \setminus L'_T$. Then, note that $|L'_T| < |L_T| = k$, since otherwise the endpoints of L'_T in H' (or H'') form an independent set of size k in H' (or H'' resp.), leading to the contradiction to the fact that $\alpha(H) < k - 1$. Suppose $(k - 2) \leq |L'_T| \leq (k - 1)$, and let $e_j \in L''_T$. Then, since e_j and e form an intersecting matching pair, assume, without loss of generality, e_j has an endpoint in H'. But then, since every edge of $L'_T \cup \{e_j\}$ has one endpoint in H', the endpoints of $L'_T \cup \{e_j\}$ in H' form an independent set of size $k - 1$ in H', leading to a contradiction. Finally, if $|L'_T| < k - 2$, then there must be at least two edges $e_i, e_j \in L''_T$ incident on one of the two copies of Q adjacent to H' and H''. But since Q is a clique, this means $(v_i, v_j) \in E(G)$ contradicting the fact that T is a k-claw. $\square$

3.3   Proof of Theorem 5

For $k \geq 4$, let H be a $(k - 1, t)$-Ramsey graph obtained from Corollary 1, for every sufficiently large t. Let $q_k(t) = |V(H)| = c_k t^{\frac{k}{2}} (\log t)^{\frac{1}{k-3} - \frac{k}{2}}$. Then, using $p = 2^{q_k(t)}$ and $\tau = t$ along with graph H, Lemma 1 produces a graph G on $n := \Theta(2^{q_k(t)} q_k(t))$ vertices such that $\alpha \geq 2^{q_k(t)} = \Omega(n/\log n)$, and $\alpha(G) \leq 3k2^{q_k(t)} = O(nk/\log n)$, and $\omega(G) \leq 3t$. Hence, we have

$$\chi(G) \geq \frac{n}{\alpha(G)} = \Omega\left(\frac{q_k(t)}{k}\right) = \Omega\left(\frac{c_k}{3k}\left(\frac{t}{\log t}\right)^{k/2}\right) = f(k)\left(\frac{\omega(G)}{\log \omega(G)}\right)^{k/2},$$

for some $f(k)$.

4   Convex Relaxation Results

4.1   Convex Relaxation Prelims

We explain only necessary terminologies to prove our results. For a complete exposition on sum-of-squares and related convex relaxations, we refer the readers to excellent survey papers [15,22,28]. Let K_G be the polytope $\{x \in [0,1]^{V(G)} : x_i + x_j \leq 1, \forall (i,j) \in E(G)\}$. The standard LP relaxation for MWIS $\max\{\sum_v w_v x_v : x \in K_G\}$ is known to have integrality gap of at least $\Omega(n)$ on n-vertex graphs[6].

Sum-of-Squares. The sum-of-square hierarchies (or Lasserre hierarchies) [20,21, 26,27] can be applied to (increasingly) tighten any linear program (captured by the level in the hierarchy). For any $t \geq 1$, the t-th level of SoS can be computed in time $n^{O(t)}$.

Now we formally define SoS, following the treatment of Rothvoss [28].

Definition 1. *Define the t-th level of SoS hierarchy* $\mathsf{SoS}_t(K_G)$ *as the set of vectors* $z \in \mathbb{R}^{2^{V(G)}}$ *that satisfy:*

$$M_t(z) := (z_{I \cup J})_{|I|,|J| \leq t} \succeq 0 \qquad and$$

$$M_t^{ij}(z) := (z_{I \cup J} - z_{I \cup J \cup \{i\}} - z_{I \cup J \cup \{j\}})_{|I|,|J| \leq t} \succeq 0 \quad (\forall (i,j) \in E(G))$$

Let $\mathsf{SoS}_t^{proj}(K_G) = \{(z_i)_{i \in V(G)} : z \in \mathsf{SoS}_t(K_G)\}$ *be the projection on the original variables.*

It is standard to view 1 as $z_\emptyset = 1$ (so that we have variables z_I for all subsets).

Proposition 1 Lemma 8 in [28]). *If any solution $x \in K_G$ contains at most t ones, then any $z \in \mathsf{SoS}_t^{proj}(K_G)$ is a convex combination of integer solutions in $K_G \cap \{0,1\}^{V(G)}$.*

Corollary 2. *Let* $z^* = \arg\max\{\sum_{i \in V(G)} w_i z_i : z \in \mathsf{SoS}_t(K_G)\}$. *If* $\alpha(G) \leq t$, *then the objective value of z^* is exactly the value of maximum weight independent set in G.*

Our next proposition states that any feasible solution of SoS (at level at least two) can be projected into a feasible solution for QSTAB. The proof is somewhat of a folklore nature. Since it has never been written anywhere in the form we need, we provide a proof in our full version [7] for completeness.

Proposition 2. *Let $z \in \mathsf{SoS}_t^{proj}(G)$ for $t \geq 2$. Then $z \in \mathsf{QSTAB}(G)$.*

Sherali–Adams. Another standard way to increasingly tighten a convex relaxation (such as K_G) is via Sherali–Adams hierarchies [30]. Let $G = (V, E)$ be a graph with $V = [n]$ of (unweighted) maximum independent set problem. The QSTAB LP for G is as follows.

$$(\mathsf{QSTAB}(G)) \max \sum_{i \in [n]} x_i$$

$$\text{s.t. } 1 - \sum_{i \in Q} x_i \geq 0 \quad \forall \text{ clique } Q$$

$$x_i \geq 0 \quad \forall i \in [n]$$

---

[6] Consider the clique K_n on n vertices and LP assignment $x_i = 1/2$ for vertex $i \in K_n$.

For $\ell \geq 1$, the Sherali Adams hierarchy applied on $\mathsf{QSTAB}(G)$ is as follows.

$(\mathsf{SA}_\ell^+(G))$

$$\max \quad \sum_{i \in [n]} y_{\{i\}}$$

s.t. $\forall S, T \subseteq [n], S \cap T = \emptyset, |S \cup T| \leq \ell$ following holds.

$$\sum_{T' \subseteq T} (-1)^{|T'|} y_{S \cup T'} \;\; - \sum_{i \in Q} \sum_{T' \subseteq T} (-1)^{|T'|} y_{S \cup T' \cup \{i\}} \geq 0 \;\; \forall \text{ clique } Q \quad (1)$$

$$\sum_{T' \subseteq T} (-1)^{|T'|} y_{S \cup T'} \;\; \geq 0 \tag{2}$$

$$\sum_{T' \subseteq T} (-1)^{|T'|} y_{S \cup T' \cup \{i\}} \geq 0 \quad \forall i \in [n] \tag{3}$$

$$y_\emptyset \quad = 1 \tag{4}$$

It can be shown that $\mathsf{SoS}_{\ell+2}$ is at least as strong as $\mathsf{SA}_\ell^+(G)$. The formal statement is encapsulated in the following proposition.

Proposition 3. *Let* $z \in \mathsf{SoS}_{\ell+2}(K_G)$*. Then the solution* $\{z_I\}_{|I| \leq \ell}$ *is feasible for* $\mathsf{SA}_\ell^+(G)$*.*

4.2   Conditional χ-Boundedness and SoS

In this section, we prove Theorem 2. Let H be a graph. A clique replacement on graph H replaces a vertex $v \in V(H)$ by a clique K_v of arbitrary size and connects any vertex $u \in K_v$ to all neighbors of v. It is an easy exercise to check that k-claw-free graphs are closed under clique replacements.

Now we proceed to prove Theorem 2. Let $\mathcal{G}$ be a graph class that is closed under clique replacement. Consider an instance $G \in \mathcal{G}$ and an optimal solution $z^* = \arg\max\{\sum_{i \in V(G)} w_i z_i : z \in \mathsf{SoS}_t(K_G)\}$. If $\alpha(G) \leq t$, we would be done, due to Corollary 2. Otherwise, we consider the projection z of z^* on $V(G)$, so $z \in \mathsf{QSTAB}(G)$ (due to Proposition 2). By Theorem 1 in [8], there is an independent set in G whose weight is at least $\frac{1}{\gamma} \cdot (\sum_{i \in V(G)} w_i z_i)$, which implies that the integrality gap of this convex relaxation is at most γ.

4.3   Integrality Gap of Sherali–Adams on QSTAB

In this section, we will show large integrality gap even for the unweighted version of the problem. We first show the following theorem.

Theorem 7. *For any graph G on n vertices and $\ell \geq 1$, the integrality gap of $\mathsf{SA}_\ell^+(G)$ is at least $\frac{n}{\alpha(G)(\omega(G)+\ell)}$.*

Proof. To this end, we show the following lemma.

Lemma 2. *Define $\hat{y}$ as follows. For $A \subseteq [n]$, define* $\hat{y}_A = \begin{cases} 1 & \text{if } A = \emptyset \\ \frac{1}{\omega(G)+\ell} & \text{if } |A| = 1 \\ 0 & \text{otherwise} \end{cases}$.

Then, $\hat{y}$ is a feasible solution to $\mathsf{SA}_\ell^+(G)$.

Proof. For $S, T \subseteq [n], |S \cup T| \leq \ell, S \cap T = \emptyset$, let $J_{S,T}(y) = \sum_{T' \subseteq T} (-1)^{|T'|} y_{S \cup T'}$. We will first show that $\hat{y}$ satisfies constraint (1)of $\mathsf{SA}_\ell^+(G)$. Fix some clique Q, then we will show that the left hand side of constraint (1): $\sum_{T' \subseteq T} (-1)^{|T'|} y_{S \cup T'} - \sum_{i \in Q} \sum_{T' \subseteq T} (-1)^{|T'|} y_{S \cup T' \cup \{i\}}$ is at least 0. For S, consider the two cases: when $S \neq \emptyset$ and when $\tilde{S} = \emptyset$. For the first case, we have $|S| \geq 1$ and hence $J_{S,T}(\hat{y}) = \hat{y}_S$. Since $\hat{y}_S = 0$ for $|S| \geq 2$, this means $J_{S,T}(\hat{y}) = 0$ for $|S| \geq 2$, as required. For $S = \{a\}, a \in [n]$, this means $J_{S,T}(\hat{y}) = \hat{y}_a$. Hence, we have

$$\sum_{T' \subseteq T} (-1)^{|T'|} y_{S \cup T'} - \sum_{i \in Q} \sum_{T' \subseteq T} (-1)^{|T'|} y_{S \cup T' \cup \{i\}} = \hat{y}_{\{a\}} - \sum_{i \in Q} \hat{y}_{\{a\} \cup \{i\}}$$

Now if $a \in Q$, this term is $\hat{y}_{\{a\}} - \hat{y}_{\{a\}} = 0$, and if $a \notin Q$, this term is $\hat{y}_{\{a\}} \geq 0$, due to our construction of $\hat{y}$.

For the second case when $S = \emptyset$, we have that $J_{S,T}(\hat{y}) = \hat{y}_{\{\emptyset\}} - \sum_{j \in T} \hat{y}_{\{j\}}$. Hence,

$$\sum_{T' \subseteq T} (-1)^{|T'|} y_{S \cup T'} - \sum_{i \in Q} \sum_{T' \subseteq T} (-1)^{|T'|} y_{S \cup T' \cup \{i\}}$$

$$= \hat{y}_{\{\emptyset\}} - \sum_{j \in T} \hat{y}_{\{j\}} - \sum_{i \in Q} \hat{y}_{\{i\}} + \sum_{i \in Q} \sum_{j \in T} \hat{y}_{\{i\} \cup \{j\}}$$

$$= \hat{y}_{\{\emptyset\}} - \sum_{j \in T} \hat{y}_{\{j\}} - \sum_{i \in Q} \hat{y}_{\{i\}} + \sum_{i \in Q \setminus T} \sum_{j \in T} \hat{y}_{\{i,j\}} + \sum_{i \in Q \cap T} \sum_{j \in T} \hat{y}_{\{i\} \cup \{j\}}$$

$$= \hat{y}_{\{\emptyset\}} - \left(\sum_{j \in T} \hat{y}_{\{j\}} + \sum_{i \in Q} \hat{y}_{\{i\}} - \sum_{i \in Q \cap T} \hat{y}_{\{i\}} \right)$$

$$= \hat{y}_{\{\emptyset\}} - \sum_{i \in Q \cup T} \hat{y}_{\{i\}} = 1 - |Q \cup T| \frac{1}{\omega(G) + \ell} \geq 0,$$

since $|Q \cup T| \leq \omega(G) + \ell$.

Next consider constraint (2). From the above observation, we have

$$J_{S,T}(\hat{y}) = \begin{cases} 0 & \text{if } |S| \geq 2 \\ \hat{y}_a & \text{if } S = \{a\} \\ 1 - \sum_{t \in T} \hat{y}_t & \text{if } S = \emptyset \end{cases}$$

Noting the fact that $|T| \leq \ell$ and $\hat{y}_t = 1/(\omega(G)+\ell)$, we have that $J_{S,T}(\hat{y}) \geq 0$. Finally consider constraint (3), and let $J_{S,T,i}(y) = \sum_{T' \subseteq T} (-1)^{|T'|} y_{S \cup T' \cup \{i\}}$, for $i \in [n]$. Now note that $J_{S,T,i}(\hat{y}) = 0$ for $|S| \geq 2$ as before. Hence, first consider the case when $|S| = 1$. When $S = \{i\}$ then $J_{S,T,i}(\hat{y}) = \hat{y}_{\{i\}} \geq 0$, otherwise for $S = \{j\}, j \neq i$, we have $J_{S,T,i}(\hat{y}) = 0$. Finally, when $S = \emptyset$ then $J_{S,T,i}(\hat{y}) = \sum_{T' \subseteq T} (-1)^{|T'|} \hat{y}_{T' \cup \{i\}} = \hat{y}_i$ if $i \notin T$ otherwise $J_{S,T,i}(\hat{y}) = \hat{y}_i - \hat{y}_i = 0$, if $i \in T$. $\square$

Proof of Theorem 4. For given constant $k \geq 4$, $0 < \epsilon \leq 1/3$, and sufficiently large n, we will show a connected k-claw-free graph G_n on $\Theta(n)$ vertices such that $\alpha(G_n) = \Omega(n^\epsilon)$ and the integrality gap of $\mathsf{SA}_\ell^+(G_n)$, for $\ell = \Theta_k(n^{1-2\epsilon})$, is at least $\Omega_k(n^\epsilon)$. To this end, let t be such that Corollary 1 yields a $(k-1, t)$-Ramsey graph H_n on $\Theta(n^{1-\epsilon})$ vertices. Note that $t = O_k(n^{\frac{1-\epsilon}{k/2}} \log n) = O_k(n^{\frac{1-\epsilon}{2}} \log n)$, since $k \geq 4$. Let G_n be the graph obtained from Lemma 1 with given value of k, $p = \Theta(n^\epsilon), \tau = t$, and H_n. Note that G_n has $\Theta(n)$ vertices. Then, we have that $\Omega(n^\epsilon) \leq \alpha(G_n) \leq O_k(n^\epsilon)$ and $\omega(G_n) = O_k(n^{(1-\epsilon)/2} \log n)$. Now using $\ell = \Theta_k(n^{1-2\epsilon})$ in Theorem 7, the integrality gap of $\mathsf{SA}_\ell^+(G_n) \geq \frac{n}{\alpha(G)(\omega(G)+\ell)} = \Omega_k(n^\epsilon)$, since $\omega(G) = O(\ell)$. $\square$

References

1. Austrin, P., Khot, S., Safra, M.: Inapproximability of vertex cover and independent set in bounded degree graphs. Theory Comput. **7**(1), 27–43 (2011)
2. Bansal, N., Gupta, A., Guruganesh, G.: On the Lovász theta function for independent sets in sparse graphs. In: Proceedings of the Forty-Seventh Annual ACM Symposium on Theory of Computing, pp. 193–200 (2015)
3. Berman, P.: A d/2 approximation for maximum weight independent set in d-claw free graphs. In: SWAT 2000. LNCS, vol. 1851, pp. 214–219. Springer, Heidelberg (2000). https://doi.org/10.1007/3-540-44985-X_19
4. Bohman, T.: The triangle-free process. Adv. Math. **221**(5), 1653–1677 (2009). https://doi.org/10.1016/j.aim.2009.02.018. https://www.sciencedirect.com/science/article/pii/S0001870809000620
5. Bohman, T., Keevash, P.: The early evolution of the h-free process. Invent. Math. **181**(2), 291–336 (2010)
6. Chalermsook, P., Chuzhoy, J.: Maximum independent set of rectangles. In: Proceedings of the Twentieth Annual ACM-SIAM Symposium on Discrete Algorithms, pp. 892–901. SIAM (2009)
7. Chalermsook, P., Gadekar, A., Khodamoradi, K., Spoerhase, J.: Independent set in k-claw-free graphs: conditional χ-boundedness and the power of LP/SDP relaxations (2023)
8. Chalermsook, P., Vaz, D.: A note on fractional coloring and the integrality gap of LP for maximum weight independent set. Electron. Notes Discret. Math. **55**, 113–116 (2016)
9. Chalermsook, P., Walczak, B.: Coloring and maximum weight independent set of rectangles. In: Proceedings of the 2021 ACM-SIAM Symposium on Discrete Algorithms (SODA), pp. 860–868. SIAM (2021)
10. Chan, T.M., Har-Peled, S.: Approximation algorithms for maximum independent set of pseudo-disks. In: Proceedings of the Twenty-Fifth Annual Symposium on Computational Geometry, pp. 333–340 (2009)
11. Chan, Y.H., Lau, L.C.: On Linear and Semidefinite Programming Relaxations for Hypergraph Matching, pp. 1500–1511. https://doi.org/10.1137/1.9781611973075.122. https://epubs.siam.org/doi/abs/10.1137/1.9781611973075.122
12. Chandra, B., Halldórsson, M.M.: Greedy local improvement and weighted set packing approximation. J. Algorithms **39**(2), 223–240 (2001)
13. Chudnovsky, M., Seymour, P.: Claw-free graphs VI. colouring. J. Comb. Theory Ser. B **100**(6), 560–572 (2010). https://doi.org/10.1016/j.jctb.2010.04.005. https://www.sciencedirect.com/science/article/pii/S009589561000064X
14. Cygan, M.: Improved approximation for 3-dimensional matching via bounded pathwidth local search. In: 2013 IEEE 54th Annual Symposium on Foundations of Computer Science, pp. 509–518. IEEE (2013)
15. Fleming, N., Kothari, P., Pitassi, T., et al.: Semialgebraic proofs and efficient algorithm design. Found. Trends® Theor. Comput. Sci. **14**(1–2), 1–221 (2019)
16. Grötschel, M., Lovász, L., Schrijver, A.: Polynomial algorithms for perfect graphs. In: North-Holland Mathematics Studies, vol. 88, pp. 325–356. Elsevier (1984)
17. Hazan, E., Safra, S., Schwartz, O.: On the complexity of approximating k-set packing. Comput. Complex. **15**(1), 20–39 (2006)
18. Hochbaum, D.S.: Efficient bounds for the stable set, vertex cover and set packing problems. Discret. Appl. Math. **6**(3), 243–254 (1983)

19. Kim, J.H.: The Ramsey number $R(3,t)$ has order of magnitude $t^2/\log t$. Random Struct. Algorithms **7**(3), 173–207 (1995)
20. Lasserre, J.B.: An explicit exact SDP relaxation for nonlinear 0-1 programs. In: Aardal, K., Gerards, B. (eds.) IPCO 2001. LNCS, vol. 2081, pp. 293–303. Springer, Heidelberg (2001). https://doi.org/10.1007/3-540-45535-3_23
21. Lasserre, J.B.: Global optimization with polynomials and the problem of moments. SIAM J. Optim. **11**(3), 796–817 (2001)
22. Laurent, M.: A comparison of the sherali-adams, lovász-schrijver, and lasserre relaxations for 0–1 programming. Math. Oper. Res. **28**(3), 470–496 (2003)
23. Lewin-Eytan, L., Naor, J.S., Orda, A.: Routing and admission control in networks with advance reservations. In: Jansen, K., Leonardi, S., Vazirani, V. (eds.) APPROX 2002. LNCS, vol. 2462, pp. 215–228. Springer, Heidelberg (2002). https://doi.org/10.1007/3-540-45753-4_19
24. Lovász, L.: A characterization of perfect graphs. J. Comb. Theory Ser. B **13**(2), 95–98 (1972)
25. Neuwohner, M.: An improved approximation algorithm for the maximum weight independent set problem in d-claw free graphs. In: Bläser, M., Monmege, B. (eds.) 38th International Symposium on Theoretical Aspects of Computer Science, STACS 2021, Saarbrücken, Germany, 16–19 March 2021 (Virtual Conference). LIPIcs, vol. 187, pp. 53:1–53:20. Schloss Dagstuhl - Leibniz-Zentrum für Informatik (2021). https://doi.org/10.4230/LIPIcs.STACS.2021.53
26. Parrilo, P.A.: Structured semidefinite programs and semialgebraic geometry methods in robustness and optimization. California Institute of Technology (2000)
27. Parrilo, P.A.: Semidefinite programming relaxations for semialgebraic problems. Math. Program. **96**, 293–320 (2003)
28. Rothvoß, T.: The Lasserre hierarchy in approximation algorithms. Lecture Notes for the MAPSP, pp. 1–25 (2013)
29. Scott, A., Seymour, P.: A survey of χ-boundedness. J. Graph Theory **95**(3), 473–504 (2020)
30. Sherali, H.D., Adams, W.P.: A hierarchy of relaxations between the continuous and convex hull representations for zero-one programming problems. SIAM J. Discret. Math. **3**(3), 411–430 (1990)
31. Thiery, T., Ward, J.: An improved approximation for maximum weighted k-set packing. In: Bansal, N., Nagarajan, V. (eds.) Proceedings of the 2023 ACM-SIAM Symposium on Discrete Algorithms, SODA 2023, Florence, Italy, 22–25 January 2023, pp. 1138–1162. SIAM (2023). https://doi.org/10.1137/1.9781611977554.ch42
32. Yang, J.: On coloring claw-free graphs (2007). https://www.mathcs.bethel.edu/yang/papers/clawfree.pdf

Fast Parallel Algorithms for Submodular p-Superseparable Maximization

Philip Cervenjak[✉] [iD], Junhao Gan[iD], and Anthony Wirth[iD]

School of Computing and Information Systems, The University of Melbourne,
Parkville, VIC, Australia
pcervenjak@student.unimelb.edu.au, {junhao.gan,awirth}@unimelb.edu.au

Abstract. Maximizing a non-negative, monontone, submodular function f over n elements under a cardinality constraint k (SMCC) is a well-studied NP-hard problem. It has important applications in, e.g., machine learning and influence maximization. Though the theoretical problem admits polynomial-time approximation algorithms, solving it in practice often involves frequently querying submodular functions that are expensive to compute. This has motivated significant research into designing parallel approximation algorithms in the *adaptive complexity model*; adaptive complexity (adaptivity) measures the number of sequential rounds of poly(n) function queries an algorithm requires. The state-of-the-art algorithms can achieve $(1 - \frac{1}{e} - \varepsilon)$-approximate solutions with $O(\frac{1}{\varepsilon^2} \log n)$ adaptivity, which approaches the known adaptivity lower-bounds. However, the $O(\frac{1}{\varepsilon^2} \log n)$ adaptivity only applies to maximizing worst-case functions that are unlikely to appear in practice. Thus, in this paper, we consider the special class of *p-superseparable* submodular functions, which places a reasonable constraint on f, based on the parameter p, and is more amenable to maximization, while also having real-world applicability. Our main contribution is the algorithm LS+GS, a finer-grained version of the existing LS+PGB algorithm, designed for instances of SMCC when f is p-superseparable; it achieves an expected $(1 - \frac{1}{e} - \varepsilon)$-approximate solution with $O(\frac{1}{\varepsilon^2} \log(pk))$ adaptivity *independent of n*. Additionally, unrelated to p-superseparability, our LS+GS algorithm uses only $O(\varepsilon^{-1} n + \varepsilon^{-2} \log n)$ oracle queries, which has an improved dependence on ε^{-1} over the state-of-the-art LS+PGB; this is achieved through the design of a novel thresholding subroutine.

Keywords: parallel algorithms · approximation algorithms · submodular maximization

© The Author(s), under exclusive license to Springer Nature Switzerland AG 2023
J. Byrka and A. Wiese (Eds.): WAOA 2023, LNCS 14297, pp. 219–233, 2023.
https://doi.org/10.1007/978-3-031-49815-2_16

1   Introduction

Submodular functions are an important class of set functions that capture a wide range of real-world applications that, informally, exhibit the property of "diminishing marginal gains" or "diminishing returns". In this paper, we consider maximizing *non-negative, monotone, submodular* functions $f : 2^{\mathcal{N}} \to \mathbb{R}_{\geq 0}$, defined on a ground set $\mathcal{N}$ of n elements, under a cardinality constraint (SMCC). The goal of SMCC is to select a subset $S \subseteq \mathcal{N}$ of size $|S| \leq k$ that maximizes $f(S)$. As a convention in the literature, we assume that, for any $S \subseteq \mathcal{N}$, the value of $f(S)$ can *only* be accessed via *queries* to a *value oracle*.

Solving SMCC is important for a wide range of applications, including machine learning (e.g., active learning [24], clustering [8], data summarization [20], and feature selection [14]), information gathering [15], network monitoring [18], sensor placement [16], and influence maximization [13].

The Greedy Algorithm. As is true for most interesting variants of submodular maximization, SMCC is unfortunately an NP-hard problem. Even worse, the best approximation that can be achieved with a polynomial number of oracle queries is $1 - 1/e$, unless P = NP [21]. Interestingly, the "best" such approximation ratio can be achieved by a simple *greedy* algorithm [22]. Specifically, the greedy algorithm starts with a solution $S = \varnothing$ and performs k iterations, in each of which the element with the largest *marginal gain* with respect to S is added to S. In its raw form, the greedy algorithm queries f $O(kn)$ times, and it is strongly *sequential*: it has to perform k iterations one by one.

The Adaptive Complexity Model. In practice, querying the oracle for a set's value, i.e., evaluating $f(S)$, can be time consuming and it is often the main bottleneck of the overall running time of an algorithm. This has motivated significant research into designing *parallelizable* algorithms for SMCC under the *adaptive complexity model* [2–7,9,10,12], where the *efficiency* of an algorithm is measured by *the number of queries* and in each round, an algorithm is allowed to perform a polynomial number, poly(n), of independent oracle queries in parallel. Each such round is called an *adaptive round* and the total number of adaptive rounds required is called the *adaptive complexity* (or *adaptivity*) of the algorithm. The smaller an algorithm's adaptivity is, the more parallelizable the algorithm is. Clearly, the adaptivity of the greedy algorithm is $O(k)$.

The State-of-the-Art Adaptive Algorithm. The goal of all existing adaptive algorithms [2–7,9,10,12] for SMCC is to beat the $O(k)$ adaptivity bound of the standard greedy algorithm, and ideally, to beat the query complexity $O(kn)$ at the same time. The state-of-the-art algorithm, due to Chen et al. [7], called LS+PGB, achieves a $(1 - \frac{1}{e} - \varepsilon)$-approximation with an adaptive complexity of $O(\frac{1}{\varepsilon^2} \log(\frac{n}{\varepsilon}))$ and a query complexity of $O(\frac{n}{\varepsilon^2})$. Assuming $k \in \omega(\frac{1}{\varepsilon^2} \log n)$, LS+PGB achieves $o(k)$ adaptivity and $o(kn)$ query complexity simultaneously, improving the naive greedy algorithm.

Known Lower Bounds. For SMCC, Balkanski and Singer [3] initially proved that $\Omega(\frac{\log n}{\log \log n})$ adaptive rounds are required to achieve a $\frac{1}{\log n}$-approximation. Li et al. [19] later proved lower bounds for achieving a $(1 - \frac{1}{e} - \varepsilon)$-approximation in two cases of $\varepsilon > 0$: when $\varepsilon > \frac{c'}{\log n}$, $\Omega(\frac{1}{\varepsilon})$ rounds are required; when $\varepsilon < \frac{c'}{\log n}$, $\Omega(\frac{\log^{2/3} n}{\varepsilon^{1/3}})$ rounds are required, where c' is an absolute constant. Kuhnle [17] proved that $\Omega(\frac{n}{k})$ queries are required to achieve a constant-factor approximation, even when queries can be made to infeasible sets.

Our Research Question. In spite of the aforementioned progress, SMCC is a general problem formulation and, thus, captures difficult problem instances that are not likely to appear in practice. Analogously, although the greedy algorithm for Set Cover achieves only an $O(\log n)$ approximation factor on n elements, the well known tight example is bespoke, and in practice greedy performs well [11]. It would be of theoretical and practical interest if there were a useful class of submodular functions that can be maximised in fewer adaptive rounds than what is needed for the worst-case functions, especially since real-world submodular functions can be computationally expensive to query. This motivates our main research question:

> *Is there an interesting class of SMCC instances that admits*
> *algorithms with $o(\log n)$ adaptive complexity,*
> *while achieving reasonable approximation?*

We address our research question by considering the class of *p-superseparable* submodular functions ($p \in [1, n]$ is a class parameter); in particular, we design highly parallel approximation algorithms for SMCC when f is assumed to be p-superseparable. This class of functions belongs in the super-class of *p-separable* submodular functions[1], introduced by Skowron [23] for the purpose of showing fixed-parameter tractable (FPT) algorithms for SMCC. As Skowron shows, p-superseparable submodular functions capture several useful real-world functions, such as those found in election and recommendation systems. They also capture the Max-k (Weighted) Coverage problem with element frequencies upper-bounded by p. We now outline our contributions one by one.

1.1 Our Contributions

Parallel Algorithm for p-Superseparable SMCC. Our first contribution is the algorithm LINEARSEQ+GREEDYSAMPLING (LS+GS) and its subroutine GREEDYSAMPLING (GS) for p-superseparable SMCC, with theoretical guarantees stated in Theorems 1 and 2 respectively.

LS+GS is essentially a finer-grained version, parameterised by p, of LS+PGB [7] that exploits the p-superseparability of f to achieve an adaptive complexity of $O(\frac{1}{\varepsilon^2} \log(\frac{pk}{\varepsilon}))$. For example, assuming $p, k \in O(\text{polylog}\, n)$ as well as

---

[1] Our work does not focus on the other two classes of p-separable functions, which are p-subseparable and rev-p-subseparable functions.

$k \in \omega(\frac{1}{\varepsilon^2} \log n)$ (since otherwise the greedy algorithm would have adaptivity as good as the existing algorithms), the adaptivity of our algorithm is bounded by $O(\frac{1}{\varepsilon^2} \log \log n)$. Under this setting, our algorithm's adaptive complexity beats the $O(\frac{1}{\varepsilon^2} \log n)$ adaptive complexity of the existing algorithms for general SMCC, as well as the $O(k) = O(\text{polylog}\, n)$ adaptive complexity of the greedy algorithm. We summarize the performance guarantees of LS+GS, compared with LS+PGB, in the Table 1 below. For simplicity and fair comparisons, we assume $k \in \omega(\frac{1}{\varepsilon^2} \log n)$ in Table 1.

Table 1. Key performance indicators for SMCC algorithms.

Algorithm	Approx.	Adaptivity	Expected queries
LS+PGB [7]	$1 - \frac{1}{e} - \varepsilon$	$O\left(\frac{1}{\varepsilon^2} \log n\right)$	$O\left(\frac{n}{\varepsilon^2}\right)$
general LS+GS (full paper, ours)	$1 - \frac{1}{e} - \varepsilon$	$O\left(\frac{1}{\varepsilon^2} \log n\right)$	$O\left(\frac{n}{\varepsilon}\right)$
p-supersep. LS+GS (Theorem 1, ours)	expected $1 - \frac{1}{e} - \varepsilon$	$O\left(\frac{1}{\varepsilon^2} \log (pk)\right)$	$O\left(n + \frac{pk}{\varepsilon^2}\right)$

Additionally, for general SMCC, LS+GS has an expected query complexity of $O(\frac{n}{\varepsilon} + \frac{\log n}{\varepsilon^2})$. This improves the dependence on ε^{-1} in the $O(\frac{n}{\varepsilon^2})$ expected query complexity of LS+PGB [7]. The improved query complexity is due to GS using a novel thresholding procedure, which we outline next. We formally state the theoretical guarantees of LS+GS for general SMCC in the full paper.

Parallel Thresholding Procedure for SMCC. Our second contribution is the procedure THRESHOLDBLOCKSEQ (TBS), with theoretical guarantees stated in Theorem 3. TBS is used by GS for the task of selecting a set of elements whose average marginal gain approximately satisfies a given threshold τ.

The significance of TBS is that, given n' input elements and an error term ε', it achieves an expected query complexity of $O(n' + \frac{\log n'}{\varepsilon'})$. This improves the dependence on ε'^{-1} in the query complexity: existing procedures for the same task [1,7,10,12] perform $O(\frac{n'}{\varepsilon'})$ queries. Note that TBS does not rely on p-superseparability for its improved query complexity; indeed, the improved query complexity of TBS is what leads to the $O(\frac{n}{\varepsilon} + \frac{\log n}{\varepsilon^2})$ expected query complexity of LS+GS for general SMCC. This also means that TBS can replace the existing procedures that are used as subroutines for solving general SMCC. We summarize the guarantees of TBS in Table 2 below (m is an input parameter satisfying $n' \le m \le n$, and δ is a failure probability term).

Simple Parallel Algorithm for p-Superseparable SMCC. Finally, we introduce LOWADAPTOP (LAT) for p-superseparable SMCC, with theoretical guarantees stated in the full paper. LAT works by simply running the existing procedure LOWADAPLINEARSEQ [7] on the set of top-$\lceil \frac{pk}{1-\alpha} + k \rceil$ elements by value. LAT achieves an $\alpha(5 + O(\varepsilon))^{-1}$-approximation, an adaptive complexity of $O(\frac{1}{\varepsilon^3} \log(\frac{p}{1-\alpha}))$, and a query complexity of $O(\frac{1}{1-\alpha}(\frac{pk}{\varepsilon^3} + \frac{p}{\varepsilon^4}))$.

Table 2. Key performance indicators for parallel thresholding procedures.

Procedure	Adaptivity	Expected queries
THRESHOLDSEQ (Theorem 2 of [7])	$O\left(\frac{1}{\varepsilon'} \log\left(\frac{m}{\delta}\right)\right)$	$O\left(\frac{n'}{\varepsilon'}\right)$
THRESHOLDBLOCKSEQ (Theorem 3, ours)	$O\left(\frac{1}{\varepsilon'} \log\left(\frac{m}{\delta}\right)\right)$	$O\left(n' + \frac{\log n'}{\varepsilon'}\right)$

1.2 Our Techniques

Our algorithms are based on the state-of-the art framework of Chen et al. [7], particularly the LS+PGB algorithm. Importantly, our two key techniques exploit p-superseparability to achieve adaptive complexities *independent of* n, rather depending on parameters p and k.

The first key technique is to run an existing algorithm on a limited number of "top-valued" elements so as to bound the adaptive complexity of the algorithm. It follows from the p-superseparability of f that the set of top-$\lceil \frac{pk}{1-\alpha} + k \rceil$ valued elements contains a k-size α-approximation of the optimal solution, leading to a good approximation overall. This result is stated in Lemma 1 and follows from Theorem 1 of Skowron [23]. We use this technique in our main algorithm LS+GS and our algorithm LAT.

The second key technique is to uniformly-at-random sample elements from a sufficiently large set $\mathcal{G}$ of valuable elements. The p-superseparability of f ensures that newly sampled elements, on average, remain valuable even after previously sampled elements are added to a solution, bypassing the need to make sequential oracle queries; this is formally stated in Lemma 4. We use this technique in GS, a subroutine of LS+GS.

Furthermore, independent of p-superseparability, we develop a new "element filtering" technique, used by TBS to achieve its improved expected query complexity of $O(n' + \frac{\log n'}{\varepsilon'})$. The key insight in TBS is to avoid repeated filtering queries on *all* remaining elements, and instead mainly perform the filtering queries on small random samples or "blocks" of the remaining elements.

1.3 Paper Structure

We give additional related work in the full paper. We present preliminaries in Sect. 2, including the definition and intuition of p-superseparability, and an overview of the state-of-the-art algorithm LS+PGB. We then give the descriptions of LS+GS and GS in Sect. 3, TBS in Sect. 4, and LAT in the full paper, while also deferring all formal analyses to the full paper. Finally, we give some conclusions in Sect. 5.

2 Preliminaries

Denote by $\mathcal{N}$ the *ground set* of elements. For every set $S \subseteq \mathcal{N}$ and a *set function* f, $f(S)$ is called the *value* of S. Given two sets $S, T \subseteq \mathcal{N}$, we define $f(S \mid T)$ to be the *marginal gain* of S to T, i.e., $f(S \mid T) := f(S \cup T) - f(T)$. When expressing the value, or marginal gain, of a singleton set, $\{x\}$, we abuse notation and use $f(x)$ and $f(x \mid S)$, rather than $f(\{x\})$ and $f(\{x\} \mid S)$. We denote by $\mathcal{O}$ the optimal solution to the SMCC problem, and $\mathrm{OPT} := f(\mathcal{O})$.

Definition 1 (Submodular; monotone; non-negative). *Set function f is submodular iff $\forall S, T \subseteq \mathcal{N}$ such that $S \subseteq T$, and $\forall x \in \mathcal{N} \backslash T$: $f(x \mid S) \geq f(x \mid T)$; monotone iff $\forall S, T \subseteq \mathcal{N}$: $f(S \mid T) \geq 0$; non-negative iff $\forall S \subseteq \mathcal{N}$: $f(S) \geq 0$.*

2.1 p-Superseparable Functions

Our interest is in the class of p-superseparable functions, Definition 2, introduced by Skowron [23]. A larger p represents a more general class, so a smaller p yields stronger results. When f is non-negative and submodular, the smallest sensible value for p is 1; on the other hand, every monotone f is n-superseparable. We give some background and applications for p-separable functions in the full version of the paper.

Definition 2 (p-superseparable set function [23]). *A set function, f, is p-superseparable iff $\forall S \subseteq \mathcal{N}$:*

$$\sum_{x \in \mathcal{N}} f(x \mid S) \geq \sum_{x \in \mathcal{N}} f(x) - p f(S). \tag{2.1}$$

Intuition of p-Superseparable Functions. Intuitively, dividing both sides of Eq. (2.1) by n, the left-hand side becomes the average marginal gain of an element x to S, while the right-hand side becomes the average individual value of an element minus the average "loss" or "overlap" due to S. Hence, p-superseparability ensures that, *on average*, a single element x loses at most $\frac{p}{n} f(S)$ from its individual value to give $f(x \mid S)$.

2.2 An Overview of the State-of-the-Art Algorithm

We now give an overview of LS+PGB by Chen et al. [7], which is the starting point for our algorithm LS+GS as proposed in Sect. 3 (see Table 1). LS+PGB comprises two procedures performed in sequence, where PGB invokes THRESH-OLDSEQ as a subroutine. Our outline includes our notation rather than that of Chen et al. [7].

LINEARSEQ **(LS).** This is a pre-processing procedure whose purpose, given an approximation error $\hat{\varepsilon}$, is to obtain a value Γ satisfying $\Gamma \leq \mathrm{OPT} \leq \frac{\Gamma}{\beta}$ for $\beta = (4 + O(\hat{\varepsilon}))^{-1}$. It uses $O(\frac{1}{\hat{\varepsilon}^3} \log n)$ adaptive rounds and $O((\frac{1}{\hat{\varepsilon} k} + 1)\frac{n}{\hat{\varepsilon}^3})$ expected queries. The quantity $\hat{\varepsilon}$ can be set constant without affecting the main approximation error, ε.

PARALLELGREEDYBOOST (PGB). With the previously obtained Γ and β, PGB initializes a threshold τ to upper-bound the average value of an optimal solution element, i.e., $\tau = \frac{\Gamma}{\beta k} \geq \frac{\text{OPT}}{k}$. It then uses a diminishing-threshold strategy to achieve the final $(1 - \frac{1}{e} - \varepsilon)$-approximation, while using $O(\frac{1}{\varepsilon^2} \log(\frac{n}{\varepsilon}))$ adaptive rounds and $O(\frac{n}{\varepsilon^2})$ expected queries. It is crucial that β is a constant, as this helps to bound the number of threshold diminutions over a while-loop.

THRESHOLDSEQ (TS). For each threshold τ considered and a current solution S, PGB calls TS to select a set of elements T whose marginal gain to S approximately satisfies $\tau|T|$; PGB then appends T to S. Given a failure probability term δ, TS uses $O(\frac{1}{\varepsilon'} \log(\frac{n}{\delta}))$ adaptive rounds and $O(\frac{n}{\varepsilon'})$ expected queries. TS performs a loop, in which each iteration appends elements to T using an improved version of the *adaptive sequencing* technique. Adaptive sequencing was introduced by Balkanski et al. for monotone submodular maximization under a matroid constraint [1], and refined in the FAST algorithm by Breuer et al. for SMCC [5]. We describe TS in more detail in Sect. 4 so as to directly compare it with our improved procedure TBS.

3   Parallel Algorithm for p-Superseparable SMCC

In this section, we introduce our main parallel approximation algorithm and its components, and formalize the performance guarantees.

3.1   LINEARSEQ+GREEDYSAMPLING

We begin with LS+GS, pseudocode in Algorithm 1, for p-superseparable SMCC, with theoretical guarantees in Theorem 1 below. The performance guarantees of LS+GS for the *general* SMCC problem are given in the full paper.

Theorem 1. *Let (f, k) be an instance of SMCC where f is p-superseparable. Suppose LS+GS (Algorithm 1) is run such that $0 < \varepsilon < 1 - \frac{1}{e}$, $0 < \hat{\varepsilon} < \frac{1}{2}$, and $0 < \alpha < 1$, where $\hat{\varepsilon}$ and α are constants. Then, with probability $1 - O(\frac{\varepsilon}{pk})$, LS+GS achieves: (i) a solution S satisfying $|S| \leq k$ and $\mathbb{E}[f(S)] \geq (1 - \frac{1}{e} - \varepsilon)\text{OPT}$, (ii) an adaptive complexity of $O(\frac{1}{\varepsilon^2} \log(\frac{pk}{\varepsilon}))$, and (iii) an expected query complexity of $O(n + \frac{pk}{\varepsilon^2} + \frac{1}{\varepsilon^2} \log(\frac{pk}{\varepsilon}))$.*

Description of LS+GS. Based on LS+PGB [7], our two key modifications achieve an adaptive complexity dependent on p and k rather than n.

1. To obtain an initial value, Γ, satisfying $\Gamma \leq \text{OPT} \leq \frac{\Gamma}{\beta}$ for a constant β, instead of running LINEARSEQ on $\mathcal{N}$, LINEARSEQ is run only on the set $\mathcal{A}$ of top-$\lceil \frac{pk}{1-\alpha} + k \rceil$ elements by individual value, thus using only $O(\frac{1}{\hat{\varepsilon}^3} \log(\frac{pk}{1-\alpha}))$ adaptive rounds; $\hat{\varepsilon}$ and α can be constant.
2. Instead of running PGB, our procedure GS is run on $\mathcal{N}$ (taking Γ and β in its input). GS uses only $O(\frac{1}{\varepsilon^2} \log(\frac{pk}{\varepsilon}))$ adaptive rounds at the expense of returning an *expected* $(1 - \frac{1}{e} - \varepsilon)$-approximation.

Algorithm 1

1: **procedure** LINEARSEQ+GREEDYSAMPLING$(f, \mathcal{N}, p, k, \alpha, \hat{\varepsilon}, \varepsilon)$
2:     **Input:** value oracle $f \colon 2^{\mathcal{N}} \to \mathbb{R}_{\geq 0}$, ground set $\mathcal{N}$, parameter p such that f is p-superseparable, cardinality constraint k, initial approximation error $\hat{\varepsilon}$, initial approximation term α, approximation error ε
3:     **Output:** set S satisfying $\mathbb{E}[f(S)] \geq (1 - \frac{1}{e} - \varepsilon) \text{OPT}$
4:     $\mathcal{A} \leftarrow$ set of top-$\left\lceil \frac{pk}{1-\alpha} + k \right\rceil$ elements $x \in \mathcal{N}$ by value $f(x)$
5:     $Q \leftarrow$ LINEARSEQ$(f, \mathcal{A}, k, \hat{\varepsilon})$
6:     $\Gamma \leftarrow f(Q)$
7:     **if** $|\mathcal{A}| < |\mathcal{N}|$ **then**
8:         $\beta \leftarrow \alpha \left(4 + \frac{4(2-\hat{\varepsilon})\hat{\varepsilon}}{(1-\hat{\varepsilon})(1-2\hat{\varepsilon})} \right)^{-1}$
9:     **else**
10:         $\beta \leftarrow \left(4 + \frac{4(2-\hat{\varepsilon})\hat{\varepsilon}}{(1-\hat{\varepsilon})(1-2\hat{\varepsilon})} \right)^{-1}$
11:     $S \leftarrow$ GREEDYSAMPLING$(f, \mathcal{N}, p, k, \beta, \Gamma, \varepsilon)$
12:     **return** S

Deriving Theorem 1. At a high-level, we obtain the guarantees in Theorem 1 by combining the guarantees of running LINEARSEQ on the set $\mathcal{A}$ of top-valued elements and the guarantees of GREEDYSAMPLING when β is constant (see Theorem 2 below), as they are run one at a time.

We can ensure β is constant by setting $\hat{\varepsilon}$ and α constant and by Lemma 1 below, which follows from Theorem 1 of Skowron [23]. This lemma guarantees that, for p-superseparable SMCC, running an approximation algorithm such as LINEARSEQ on a sufficiently large set of "top-valued" elements $\mathcal{A}$ only worsens its approximation by a factor of α, since the optimal solution within the top-valued elements is an α-approximation of the optimal solution in $\mathcal{N}$. Note also that if $\mathcal{A} = \mathcal{N}$, then β defaults to $(4 + O(\hat{\varepsilon}))^{-1}$.

Lemma 1 (Best k-size subset in top-valued elements [23]). *Let (f, k) be an instance of SMCC where f is p-superseparable. Further, let α be a parameter such that $0 < \alpha < 1$, let $\mathcal{A}$ be the set of top-$\lceil \frac{pk}{1-\alpha} + k \rceil$ elements $x \in \mathcal{N}$ by value $f(x)$, and let $A^* \subseteq \mathcal{A}$ be a k-size subset that maximizes f. Then $f(A^*) \geq \alpha \text{OPT}$.*

3.2 GREEDYSAMPLING

GREEDYSAMPLING (GS), pseudocode in Algorithm 2, is the greedy-thresholding procedure called by LS+GS to find an expected $(1 - \frac{1}{e} - \varepsilon)$-approximate solution in $O(\frac{1}{\varepsilon^2} \log(\frac{pk}{\varepsilon}))$ adaptive rounds and $O(n + \frac{pk}{\varepsilon^2} + \frac{1}{\varepsilon^2} \log(\frac{pk}{\varepsilon}))$ expected queries for constant β. Theoretical guarantees are given in Theorem 2.

Theorem 2. *Let (f, k) be an instance of SMCC where f is p-superseparable. Suppose GS (Algorithm 2) is run such that $\Gamma \leq \mathrm{OPT} \leq \frac{\Gamma}{\beta}$ and $0 < \varepsilon < 1 - \frac{1}{e}$. Then, with probability $1 - \frac{\beta\varepsilon}{6pk}$, GS achieves: (i) a solution S satisfying $|S| \leq k$ and $\mathbb{E}[f(S)] \geq (1 - \frac{1}{e} - \varepsilon)\mathrm{OPT}$, (ii) an adaptive complexity of $O(\frac{\log(\beta^{-1})}{\varepsilon^2} \log(\frac{pk \log(\beta^{-1})}{\beta\varepsilon}))$, and (iii) a query complexity of $O(n + \frac{(1-\beta)pk}{\varepsilon^2} + \frac{\log(\beta^{-1})}{\varepsilon^2} \log(\frac{pk}{\varepsilon}))$.*

We give the description of GS and then explain how it achieves its adaptive complexity bounds with reference to Lemma 4. This is the key lemma showing that, for p-superseparable SMCC, sampling uniformly-at-random from sufficiently many high-value elements does not decrease the expected marginal gain of the remaining elements too much, giving a final sample with good expected marginal gain.

Description of GREEDYSAMPLING. GS takes in its input the values Γ and β such that $\Gamma \leq \mathrm{OPT} \leq \frac{\Gamma}{\beta}$. In Line 4, GS initializes the solution $S_0 \leftarrow \varnothing$, the threshold $\tau_0 \leftarrow \frac{\Gamma}{\beta k} \geq \frac{\mathrm{OPT}}{k}$, and $m_{\max} \leftarrow \frac{3pk}{\beta\varepsilon/2} + k - 1$, which is the maximum number of elements that can be passed to TBS over the entire run of GS. For each element $x \in \mathcal{N}$, Line 6 assigns $w_x \leftarrow f(x)$; these are used to build $\mathcal{G}_i$ in Line 10.

After the initialisation steps, GS performs the steps below in each iteration, i, of the Line 7 loop. GS differs from PGB [7] in steps 2 and 3.

1. Assigns the threshold τ_i by geometrically diminishing the previous threshold τ_{i-1} (Line 9).
2. Constructs the set $\mathcal{G}_i$ of elements $x \in \mathcal{N} \setminus S_{i-1}$ with $f(x) = w_x \geq \tau_i$ (Line 10).
3. If $|\mathcal{G}_i| \geq m_i$, GS uniformly-at-random samples a set T_i of size $k - |S_{i-1}|$ from $\mathcal{G}_i$ (Line 13). Otherwise, if $|\mathcal{G}_i| < m_i$, GS runs TBS on $\mathcal{G}_i$ to obtain T_i (Line 15). Either way, the expected marginal gain $f(T_i \mid S_{i-1})$ is approximately $|T_i|\tau_i$. This step is crucial to bounding the adaptive complexity as explained below.
4. Produces S_i by adding the set of new elements T_i to S_{i-1} (Line 16).

The Line 7 loop breaks if S_i satisfies the cardinality constraint or if the threshold is too small to add elements with significant marginal gain.

Bounding the Adaptive Complexity via p-Superseparability. The key idea behind the $O(\frac{1}{\varepsilon^2} \log(\frac{pk}{\varepsilon}))$ adaptive complexity of GS is to bound the adaptive complexity of each iteration i in two cases for $|\mathcal{G}_i|$.

When $|\mathcal{G}_i| \geq m_i$, *no adaptive rounds* are needed to sample $T_i \subseteq \mathcal{G}_i$ in Line 13. This is because Lemma 4 ensures that, when f is p-superseparable and $|\mathcal{G}_i| \geq m_i$, T_i has good expected marginal gain $f(T_i \mid S_{i-1})$.

Otherwise, when $|\mathcal{G}_i| < m_i$ (where $m_i \leq m_{\max} = O(\frac{pk}{\varepsilon})$), running TBS on $\mathcal{G}_i$ has a bounded adaptive complexity of $O(\frac{1}{\varepsilon} \log(\frac{pk}{\varepsilon}))$.

The overall adaptive complexity of GS follows since the case $|\mathcal{G}_i| < m_i$ (in which TBS is called) may occur in every Line 7 iteration, and the number of such iterations is bounded by $O(\log_{1-\varepsilon}(\beta))$, which is $O(\frac{1}{\varepsilon})$ for constant β.

Lemma 4. *Suppose GS (Algorithm 2) is run such that $\Gamma \leq \mathrm{OPT} \leq \frac{\Gamma}{\beta}$. Further, suppose that in some iteration u, Line 13 is executed so that T_u is assigned an ordered, uniform-at-random sample of $k - |S_{u-1}|$ elements from $\mathcal{G}_u$ (without replacement). Then T_u satisfies*

$$\mathbb{E}[f(T_u \mid S_{u-1})] \geq |T_u| \left(1 - \frac{\varepsilon}{2}\right) \tau_u.$$

Algorithm 2

1: **procedure** GREEDYSAMPLING($f, \mathcal{N}, p, k, \beta, \Gamma, \varepsilon$)

2:   **Input:** value oracle $f : 2^{\mathcal{N}} \to \mathbb{R}_{\geq 0}$, ground set $\mathcal{N}$, value p such that f is p-superseparable, cardinality constraint k, initial approximation factor β, value Γ such that $\Gamma \leq \mathrm{OPT} \leq \frac{\Gamma}{\beta}$, approximation error ε

3:   **Output:** set $S \subseteq \mathcal{N}$ satisfying $\mathbb{E}[f(S)] \geq (1 - \frac{1}{e} - \varepsilon)\,\mathrm{OPT}$

4:   $i \leftarrow 0, S_0 \leftarrow \varnothing, \tau_0 \leftarrow \frac{\Gamma}{\beta k}, m_{\max} \leftarrow \frac{3pk}{\beta \varepsilon/2} + k - 1,$

$$\delta \leftarrow \left(\log_{1-\varepsilon}\left(\frac{\beta}{3}\right)\right)^{-1}$$

5:   **for** $x \in \mathcal{N}$ **do**

6:     $w_x \leftarrow f(x)$

7:   **while** $|S_i| < k$ and $\tau_i \geq \frac{\Gamma}{(1-\varepsilon)3k}$ **do**

8:     $i \leftarrow i + 1$

9:     $\tau_i \leftarrow (1 - \varepsilon)^i \frac{\Gamma}{\beta k}$

10:     $\mathcal{G}_i \leftarrow \{x \in \mathcal{N} \setminus S_{i-1} : w_x \geq \tau_i\}$

11:     $m_i \leftarrow \frac{pk}{(1-\varepsilon)^i \varepsilon/2} + k - |S_{i-1}| - 1$

12:     **if** $|\mathcal{G}_i| \geq m_i$ **then**

13:       $T_i \leftarrow$ unif.-at-rand. sample of $k - |S_{i-1}|$ elements from $\mathcal{G}_i$

14:     **else**

15:       $T_i \leftarrow \mathrm{TBS}(f(S_{i-1} \cup \cdot), \mathcal{G}_i, \min\{m_{\max}, n\}, k - |S_{i-1}|, \frac{\varepsilon}{3}, \delta, \tau_i)$

16:     $S_i \leftarrow S_{i-1} \cup T_i$

17:   **return** S_i

---

4   Parallel Thresholding Procedure for SMCC

In this section, we propose THRESHOLDBLOCKSEQ (TBS), with pseudocode given in Algorithm 3. This is the subroutine used by GREEDYSAMPLING in Line 15. We formally state the performance guarantees of TBS in Theorem 3 below.

Theorem 3. *Suppose TBS (Algorithm 3) is run such that g is monotone submodular, $n' = |\mathcal{G}| \leq m$, $0 < \varepsilon' < 1$, and $0 < \delta < 1$. Then, with probability $1 - \frac{\delta}{m}$, TBS achieves: (i) an adaptive complexity of $O(\frac{1}{\varepsilon'} \log(\frac{m}{\delta}))$, (ii) an expected query complexity of $O(n' + \frac{\log n'}{\varepsilon'})$, (iii) an output set T satisfying $|T| \leq k'$ and $g(T \mid \varnothing) \geq \frac{1-\varepsilon'}{1+\varepsilon'} \tau |T|$, and (iv) in case $|T| < k'$, for all $x \in \mathcal{G}: g(x \mid T) < \tau$.*

Given a value oracle g, an input set $\mathcal{G}$, a value $m \geq |\mathcal{G}|$, an error term ε', and a probability term δ, the purpose of TBS is to return a set $T \subseteq \mathcal{G}$ satisfying $g(T \mid \varnothing) \geq \frac{1-\varepsilon'}{1+\varepsilon'} \tau |T|$ in $O(\frac{1}{\varepsilon'} \log(\frac{m}{\delta}))$ adaptive rounds. The task of finding a set whose average marginal gain is above some threshold is common in many algorithms for submodular maximisation; in fact, TBS is an improved version of the THRESHOLDSEQ (TS) procedure by Chen et al. [7] for performing this task and can serve to replace it. The main feature of TBS is its query complexity of $O(n' + \frac{\log n'}{\varepsilon'})$, which has an improved dependence on ε'^{-1} over that of TS.

For the purpose of comparison, we briefly describe TS and point out its main inefficiency, which leads to its $O(\frac{n'}{\varepsilon'})$ query complexity. After this, we describe the steps in TS, and finally explain how these steps work to achieve its improved query complexity of $O(n' + \frac{\log n'}{\varepsilon'})$.

Description of THRESHOLDSEQ. TS works by updating a solution T over a loop. Each loop iteration uses an improved *adaptive sequencing* technique to update T. Specifically, each loop iteration (1) queries g over all previously remaining elements x to filter out those x with $g(x \mid T) < \tau$, (2) uniformly-at-random permutes the remaining elements, and then (3) adds an appropriate prefix of the remaining elements to T. By adding this prefix, at least $\frac{\varepsilon'}{2}$ proportion of the remaining elements now have $g(x \mid T) < \tau$ (with probability $\geq \frac{1}{2}$) and are, thus, filtered out in the next iteration.

Query Complexity of THRESHOLDSEQ. The main inefficiency in TS is due to performing filtering queries over *all* remaining elements when only $\frac{\varepsilon'}{2}$ of these elements are likely to be filtered out; in other words, $1 - \frac{\varepsilon'}{2}$ proportion of elements that TS queries will *not* be filtered out and, thus, will appear in the next iteration to be queried again. So over all iterations, the expected query complexity due to filtering steps is essentially $O\left(\sum_{j=0}^{\infty} n' \left(1 - \frac{\varepsilon'}{2}\right)^j\right) = O\left(\frac{n'}{\varepsilon'}\right)$.

Description of THRESHOLDBLOCKSEQ. TBS works by updating a solution $T_{i,j}$ over an *outer* loop (Line 5) that contains a nested *inner* loop (Line 12). Each outer iteration i updates the set of remaining elements E_i by filtering out those $x \in E_{i-1}$ with $g(x \mid T_{i-1,j}) < \tau$ (Line 7). Each inner iteration j uniformly-at-random samples a "block" B of size $O(\lceil \varepsilon' |E_i| \rceil)$ from E_i (Line 14), and filters out those $x \in B$ with $g(x \mid T_{i,j-1}) < \tau$ to give B^* (Line 15); that is, B^* is obtained by rejection sampling. Then the inner iteration adds an appropriate prefix $P_{\lambda_{\text{best}}} \subseteq B^*$ to $T_{i,j-1}$, giving $T_{i,j}$ (Lines 18 to 28).

Achieving the Query Complexity of THRESHOLDBLOCKSEQ. At a high level, TBS uses the same adaptive sequencing technique as TS, but improves the query complexity's dependence on ε'^{-1} essentially because each *outer* iteration (which performs "filtering" queries over all remaining x) is only executed when a *constant* ϕ proportion of x are likely to be filtered out, i.e., satisfy $g(x \mid T_{i-1,j}) < \tau$. The fact that ϕ proportion of x are likely to satisfy this is achieved by the *inner* loop.

Below, we give a simplified explanation of why TBS has an expected query complexity of only $O(n' + \frac{\log n'}{\varepsilon'})$, with details in the full paper. Note that TBS performs $numOut \in O(\log(\frac{m}{\delta}))$ outer iterations and $numIn \in O(\frac{1}{\varepsilon'})$ inner iterations (Line 4), the latter being important to our explanation.

- In each *inner* iteration j, adding the prefix $P_{\lambda_{\text{best}}}$ causes $\geq \frac{\varepsilon'}{4}$ proportion of $x \in E_i$ with $g(x \mid T_{i,j-1}) \geq \tau$ to have $g(x \mid T_{i,j}) < \tau$ (with probability $\geq \frac{1}{2}$). Further, each *inner* iteration uses $O(|B|) = O(\lceil \varepsilon' |E_i| \rceil)$ queries since queries are only made on $x \in B$ or on prefixes of B^*, the filtered subset of B.
- In each *outer* iteration i, with probability $\geq \frac{1}{2}$, $O(\frac{1}{\varepsilon'})$ inner iterations will successfully cause $\geq \frac{\varepsilon'}{4}$ proportion of $x \in E_i$ with $g(x \mid T_{i,j-1}) \geq \tau$ to have $g(x \mid T_{i,j}) < \tau$. Thus, by the start of the next *outer* iteration $i+1$, at least $1 - (1 - \frac{\varepsilon'}{4})^{O(\frac{1}{\varepsilon'})}$ proportion of $x \in E_i$ has $g(x \mid T_{i,j}) < \tau$ (with probability $\geq \frac{1}{2}$). This proportion is at least a constant ϕ for all $\varepsilon' > 0$, so this next *outer* iteration will filter out a constant ϕ proportion of $x \in E_i$.
- Furthermore, over a single *outer* iteration i, only $O(numIn |B|)$ queries are made, which evaluates to $O(\frac{1}{\varepsilon'} \lceil \varepsilon' |E_i| \rceil) = O(|E_i| + \frac{1}{\varepsilon'})$ queries.
- Initially $|E_0| = n'$. So due to the reduction of constant ϕ proportion of $|E_i|$ in each outer iteration, the overall expected query complexity of TBS is essentially $O\left(\sum_{j=0}^{\infty} n'(1-\phi)^j + \frac{1}{\varepsilon'} \log_{\frac{1}{1-\phi}}(n')\right) = O\left(n' + \frac{\log n'}{\varepsilon'}\right)$.

5   Conclusions

In this paper, we propose highly parallel algorithms for p-superseparable SMCC that achieve adaptive complexities independent of n, but dependent on parameters p and k, with the main algorithm being LS+GS. We also propose a new procedure THRESHOLDBLOCKSEQ, a subroutine of GREEDYSAMPLING, which is the key to improving the existing state-of-the-art query complexity of our LS+GS, not only for the p-superseparable SMCC, but also for the general case. An interesting research direction is to design an algorithm whose adaptivity depends on p and k without the need of prior knowledge on the value of p, as our LS+GS needs to know this value to set parameters appropriately. Also, our simple algorithm for p-superseparable SMCC hints at the possibility of a $(5+O(\varepsilon))^{-1}$-approximation algorithm that only requires $O(\frac{1}{\varepsilon^3} \log p)$ rounds, removing the α term in the

Algorithm 3

1: **procedure** THRESHOLDBLOCKSEQ$(g, \mathcal{G}, m, k', \varepsilon', \delta, \tau)$

2:     **Input:** value oracle $g \colon 2^{\mathcal{G}} \to \mathbb{R}_{\geq 0}$, set of elements $\mathcal{G}$, quantity m such that $|\mathcal{G}| \leq m$, cardinality constraint k', error term ε', failure probability term δ, marginal gain threshold τ

3:     **Output:** set $T \subseteq \mathcal{G}$ satisfying $g(T \mid \varnothing) \geq \frac{1-\varepsilon'}{1+\varepsilon'} \tau |T|$

4:     $i \leftarrow 0$, $j \leftarrow 0$, $T_{0,0} \leftarrow \varnothing$, $E_0 \leftarrow \mathcal{G}$, $\phi \leftarrow 0.11879$, $numOut \leftarrow \left\lceil 4 \left(1 + \frac{1}{\phi}\right) \log\left(\frac{m}{\delta}\right) \right\rceil$, $numIn \leftarrow \left\lceil 4 \left(1 + \frac{4}{\varepsilon'}\right) \log\left(\frac{2}{1-\phi}\right) \right\rceil$

5:     **for** $numOut$ iterations **do**

6:         $i \leftarrow i + 1$

7:         $E_i \leftarrow \{x \in E_{i-1} : g(x \mid T_{i-1,j}) \geq \tau\}$

8:         **if** $|E_i| = 0$ **then**

9:             **return** $T_{i-1,j}$

10:         $T_{i,0} = T_{i-1,j}$

11:         $j \leftarrow 0$

12:         **for** $numIn$ iterations **do**

13:             $j \leftarrow j + 1$

14:             $B \leftarrow$ unif.-at-rand. sample of $\left\lceil \frac{\varepsilon'}{4(1-4\phi)} |E_i| \right\rceil$ elements from E_i

15:             $B^* \leftarrow \{x \in B : g(x \mid T_{i,j-1}) \geq \tau\}$

16:             **if** $|B^*| = 0$ **then**

17:                 **continue to next iteration**

18:             $\{b_1, \ldots, b_{|B^*|}\} \leftarrow$ unif.-at-rand. permutation of B^*

19:             $maxSize \leftarrow \min\{k' - |T_{i,j-1}|, |B^*|\}$

20:             $\Lambda \leftarrow \{\lfloor (1+\varepsilon')^h \rfloor : 1 \leq \lfloor (1+\varepsilon')^h \rfloor \leq maxSize, h \in \mathbb{N}\} \cup \{maxSize\}$

21:             **for** λ in Λ **do**

22:                 $P_\lambda \leftarrow \{b_1, \ldots, b_\lambda\}$

23:             $\Lambda^* \leftarrow \{\lambda \in \Lambda : g(P_\lambda \mid T_{i,j-1}) \geq (1-\varepsilon')\tau |P_\lambda|\}$

24:             **if** $\max \Lambda^* < maxSize$ **then**

25:                 $\lambda_{\text{best}} \leftarrow \min\{\lambda \in \Lambda : \forall \lambda^* \in \Lambda^*, \lambda > \lambda^*\}$

26:             **else**

27:                 $\lambda_{\text{best}} \leftarrow maxSize$

28:             $T_{i,j} \leftarrow T_{i,j-1} \cup P_{\lambda_{\text{best}}}$

29:             **if** $|T_{i,j}| = k'$ **then**

30:                 **return** $T_{i,j}$

31:     **return** failure

current approximation factor. Finally, it is also worth conducting experiments to compare our algorithms against the existing parallel algorithms for general SMCC, especially on those submodular p-superseparable functions with small values of p.

Acknowledgements. This work was in part supported by ARC Discovery Early Career Researcher Award (DECRA) DE190101118 and the University of Melbourne Faculty of Engineering and Information Technology, and School of Computing and Information Systems.

References

1. Balkanski, E., Rubinstein, A., Singer, Y.: An optimal approximation for submodular maximization under a matroid constraint in the adaptive complexity model. In: 51st STOC, pp. 66–77. ACM (2019)
2. Balkanski, E., Rubinstein, A., Singer, Y.: An exponential speedup in parallel running time for submodular maximization without loss in approximation. In: 30th SODA, pp. 283–302. ACM-SIAM (2019)
3. Balkanski, E., Singer, Y.: The adaptive complexity of maximizing a submodular function. In: 50th STOC, pp. 1138–1151. ACM (2018)
4. Balkanski, E., Singer, Y.: Approximation guarantees for adaptive sampling. In: 35th ICML, pp. 384–393. PMLR (2018)
5. Breuer, A., Balkanski, E., Singer, Y.: The FAST algorithm for submodular maximization. In: 37th ICML, pp. 1134–1143. PMLR (2020)
6. Chekuri, C., Quanrud, K.: Submodular function maximization in parallel via the multilinear relaxation. In: 30th SODA, pp. 303–322. ACM-SIAM (2019)
7. Chen, Y., Dey, T., Kuhnle, A.: Best of both worlds: practical and theoretically optimal submodular maximization in parallel. In: Advances in Neural Information Processing Systems, vol. 34, pp. 25528–25539 (2021)
8. Dueck, D., Frey, B.J.: Non-metric affinity propagation for unsupervised image categorization. In: 11th CCV, pp. 1–8. IEEE (2007)
9. Ene, A., Nguyen, H.L.: Submodular maximization with nearly-optimal approximation and adaptivity in nearly-linear time. In: 30th SODA, pp. 274–282. ACM-SIAM (2019)
10. Fahrbach, M., Mirrokni, V., Zadimoghaddam, M.: Submodular maximization with nearly optimal approximation, adaptivity and query complexity. In: 30th SODA, pp. 255–273. ACM-SIAM (2019)
11. Grossman, T., Wool, A.: Computational experience with approximation algorithms for the set covering problem. Eur. J. Oper. Res. **101**(1), 81–92 (1997)
12. Kazemi, E., Mitrovic, M., Zadimoghaddam, M., Lattanzi, S., Karbasi, A.: Submodular streaming in all its glory: tight approximation, minimum memory and low adaptive complexity. In: 36th ICML, pp. 3311–3320. PMLR (2019)
13. Kempe, D., Kleinberg, J., Tardos, É.: Maximizing the spread of influence through a social network. In: 9th KDD, pp. 137–146. ACM (2003)
14. Khanna, R., Elenberg, E., Dimakis, A., Negahban, S., Ghosh, J.: Scalable greedy feature selection via weak submodularity. In: 20th AISTATS, pp. 1560–1568. PMLR (2017)
15. Krause, A., Guestrin, C.: Submodularity and its applications in optimized information gathering. ACM Trans. Intell. Syst. Technol. (TIST) **2**(4), 1–20 (2011)
16. Krause, A., Singh, A., Guestrin, C.: Near-optimal sensor placements in Gaussian processes: theory, efficient algorithms and empirical studies. J. Mach. Learn. Res. **9**(2) (2008)
17. Kuhnle, A.: Quick streaming algorithms for maximization of monotone submodular functions in linear time. In: 24th AISTATS, pp. 1360–1368. PMLR (2021)

18. Leskovec, J., Krause, A., Guestrin, C., Faloutsos, C., VanBriesen, J., Glance, N.: Cost-effective outbreak detection in networks. In: 13th KDD, pp. 420–429. ACM (2007)
19. Li, W., Liu, P., Vondrák, J.: A polynomial lower bound on adaptive complexity of submodular maximization. In: 52nd STOC, pp. 140–152. ACM (2020)
20. Mirzasoleiman, B., Badanidiyuru, A., Karbasi, A.: Fast constrained submodular maximization: personalized data summarization. In: 33rd ICML, pp. 1358–1367. PMLR (2016)
21. Nemhauser, G.L., Wolsey, L.A.: Best algorithms for approximating the maximum of a submodular set function. Math. Oper. Res. **3**(3), 177–188 (1978)
22. Nemhauser, G.L., Wolsey, L.A., Fisher, M.L.: An analysis of approximations for maximizing submodular set functions-I. Math. Program. **14**(1), 265–294 (1978)
23. Skowron, P.: FPT approximation schemes for maximizing submodular functions. Inf. Comput. **257**, 65–78 (2017)
24. Wei, K., Iyer, R., Bilmes, J.: Submodularity in data subset selection and active learning. In: 32nd ICML, pp. 1954–1963. PMLR (2015)

Author Index

© The Editor(s) (if applicable) and The Author(s), under exclusive license
to Springer Nature Switzerland AG 2023
J. Byrka and A. Wiese (Eds.): WAOA 2023, LNCS 14297, pp. 235–236, 2023.
https://doi.org/10.1007/978-3-031-49815-2

Printed in the United States
by Baker & Taylor Publisher Services

Printed in the United States
by Baker & Taylor Publisher Services